Materials Horizons: From Nature to Nanomaterials

Series Editor

Vijay Kumar Thakur, School of Aerospace, Transport and Manufacturing, Cranfield University, Cranfield, UK

Materials are an indispensable part of human civilization since the inception of life on earth. With the passage of time, innumerable new materials have been explored as well as developed and the search for new innovative materials continues briskly. Keeping in mind the immense perspectives of various classes of materials, this series aims at providing a comprehensive collection of works across the breadth of materials research at cutting-edge interface of materials science with physics, chemistry, biology and engineering.

This series covers a galaxy of materials ranging from natural materials to nanomaterials. Some of the topics include but not limited to: biological materials, biomimetic materials, ceramics, composites, coatings, functional materials, glasses, inorganic materials, inorganic-organic hybrids, metals, membranes, magnetic materials, manufacturing of materials, nanomaterials, organic materials and pigments to name a few. The series provides most timely and comprehensive information on advanced synthesis, processing, characterization, manufacturing and applications in a broad range of interdisciplinary fields in science, engineering and technology.

This series accepts both authored and edited works, including textbooks, monographs, reference works, and professional books. The books in this series will provide a deep insight into the state-of-art of Materials Horizons and serve students, academic, government and industrial scientists involved in all aspects of materials research.

Review Process

The proposal for each volume is reviewed by the following:

1. Responsible (in-house) editor
2. One external subject expert
3. One of the editorial board members.

The chapters in each volume are individually reviewed single blind by expert reviewers and the volume editor.

Ajay Chourasia · Biswajit Pal · Ashish Kapoor ·
R. Pradeep Kumar

Advances and Challenges in 3D Printed Concrete (3DPC)

A Holistic Approach

Springer

Ajay Chourasia
Structural Engineering
CSIR-Central Building Research Institute
Roorkee, Uttarakhand, India

Biswajit Pal
Structural Engineering
CSIR-Central Building Research Institute
Roorkee, Uttarakhand, India

Ashish Kapoor
Structural Engineering
CSIR-Central Building Research Institute
Roorkee, Uttarakhand, India

R. Pradeep Kumar
Structural Engineering
CSIR-Central Building Research Institute
Roorkee, Uttarakhand, India

ISSN 2524-5384 ISSN 2524-5392 (electronic)
Materials Horizons: From Nature to Nanomaterials
ISBN 978-981-95-8556-4 ISBN 978-981-95-8557-1 (eBook)
https://doi.org/10.1007/978-981-95-8557-1

This Springer imprint is published by the registered company Springer Nature Singapore Pte Ltd.
The registered company address is: 152 Beach Road, #21-01/04 Gateway East, Singapore 189721, Singapore

Preface

Advances and Challenges in 3D Printed Concrete (3DPC): A Holistic Approach offers a pioneering and authoritative account of 3D concrete printing (3DPC)—a disruptive innovation poised to redefine the future of construction. Uniquely comprehensive in scope, this book bridges the gap between foundational knowledge and cutting-edge advancements, making it the first consolidated academic resource that caters to a diverse readership, from students and early researchers to seasoned industry professionals and academics.

Structured into 13 meticulously curated chapters, the book commences with the conceptual framework and evolution of 3DPC, charting its trajectory from a novel construction technique to a viable mainstream technology. It then provides in-depth technical coverage of key domains, including material science, mix design, rheology, mechanical performance, buildability, durability under aggressive environments, fire resistance, and reinforcement strategies tailored to the layer-wise construction paradigm of 3DPC.

In addressing the broader innovation ecosystem, the book delves into sustainability assessments, performance benchmarking, advanced numerical modeling techniques, and the transformative potential of Artificial Intelligence (AI) and Machine Learning (ML) in optimizing design, process control, and predictive maintenance. These discussions are firmly grounded in contemporary research and real-world case studies, offering readers both theoretical insights and practical guidance.

While 3DPC promises significant benefits—including accelerated construction timelines, reduced labor dependency, enhanced design flexibility, and sustainability—its widespread adoption remains challenged by fragmented knowledge and a lack of integrative resources. Existing literature is often limited to fragmented proceedings or narrowly focused studies. This book distinguishes itself by systematically consolidating current developments, challenges, and future trajectories in one cohesive volume.

By offering a holistic view of 3D-printed concrete technology, this book fills a critical void in the literature. It serves as an indispensable reference for researchers, practitioners, and policymakers, and as a core textbook for advanced courses in construction technology, digital fabrication, and sustainable infrastructure.

Roorkee, India
September 2025

Ajay Chourasia

Acknowledgements

The ideas and perspectives presented in this book have been shaped by years of curiosity, learning, and continuous exploration in the field of sustainable construction and emerging technologies. Much of what we share here is grounded in the research we have conducted and the meaningful discussions we have had with colleagues over the past decade.

We are deeply grateful to our organizations—CSIR-CBRI, Roorkee, Ministry of Science and Technology, Government of India, and the Ministry of Housing and Urban Affairs, Government of India—for their financial support and encouragement, which enabled us to carry out the research activities that form the foundation of this book.

Our sincere thanks go to our colleagues, collaborators, and research partners, whose insights, hard work, and cooperation have significantly contributed to our academic journey and to the ideas we present here. We also extend our gratitude to Dr. Deepak Suryavanshi, whose guidance, support, and trust provided us with the opportunity to bring this book to publication. His patience during the inevitable delays has been greatly appreciated.

Lastly, and most importantly, we acknowledge the unwavering support of our families. Their encouragement, understanding, and sacrifices have sustained us—not only through the writing of this book but also throughout the many challenges that accompany academic life.

Contents

About the Authors

Dr. Ajay Chourasia is a Chief Scientist and Head of Structural Engineering and 3D Concrete Printing Group at CSIR—Central Building Research Institute, Roorkee. He did Ph.D. in Structural Engineering from Indian Institute of Technology Roorkee. His area of interest encompasses precast buildings, structural health monitoring, earthquake-resistant design, confined masonry, post-earthquake damage assessment, 3D concrete printing, etc. He has published more than 175 papers in reputed journals/conferences and 12 books on Earthquake Resistant Design and Construction, Guidelines for Confined Masonry, Structural Design and Detailing of Confined Masonry EWS Houses, and Good Construction Practices. He has filed five Indian patents and transferred five technologies to the industry.

Dr. Biswajit Pal is a Scientist at CSIR—Central Building Research Institute, Roorkee. He holds a Ph.D. in Structural Engineering from IISc Bangalore and an M.Tech. from IIT Guwahati. With over a decade of experience in civil and materials engineering, his research focuses on mathematical modeling of material and structural behavior under various conditions, fire structural engineering, and 3D concrete printing using agro-industrial waste. Dr. Pal has authored over 25+ publications in reputed journals and conferences. He has also contributed to technology transfer in sustainable and low-cost building materials and previously served as an Assistant Professor at NIT Silchar. He received the Most Interactive Participant award at an international conference in Christchurch, New Zealand.

Mr. Ashish Kapoor is a Scientist in the Department of Structural Engineering at CSIR-CBRI, Roorkee. He earned his B.E. in Civil Engineering from Maharishi Markandeshwar University and M.E. from NITTTR, Chandigarh. His research interests include 3D concrete printing and the rehabilitation and retrofitting of existing structures.

Prof. R. Pradeep Kumar is the Director of CSIR—Central Building Research Institute (CBRI), Roorkee. He holds an M.Tech. from IIT Kanpur, and Ph.D. from the University of Tokyo. With 22 years of experience in teaching and research, his expertise lies in nonlinear structural behavior and earthquake safety assessment. He has guided 16 Ph.D. scholars and published over 120 journal papers. His key contributions include developing national earthquake risk indexing, BIS standards for post-earthquake safety, and promoting Universal Human Values. He serves on several BIS and national code committees.

List of Figures

List of Tables

Chapter 1
Foundations and Frontiers of 3D-Printed Concrete in Construction

1.1 Overview

The field of construction is undergoing a profound transformation, driven by advances in automation, robotics, and digital fabrication technologies. One of the most promising developments at the forefront of this transformation is 3D-printed concrete (3DPC), also referred to as additive manufacturing of cementitious materials [1]. This technique represents a significant departure from traditional methods of concrete construction, which rely heavily on manual labor, rigid formwork, and prolonged curing periods. Instead, 3D-printed concrete embraces the principles of digital construction, using a layer-by-layer deposition method controlled by computer-generated models and executed by robotic systems.

At its core, the process involves translating a digital three-dimensional model into machine-readable instructions that guide a printer, typically configured as a gantry system or a robotic arm, across defined paths to deposit concrete precisely where needed. This model-driven approach enables the construction of highly complex geometries without the constraints imposed by formwork. As a result, structures that once required elaborate molds and extensive labor can now be fabricated with greater speed, accuracy, and design freedom. Architects and engineers are empowered to explore new frontiers in building design, incorporating intricate forms, integrated service conduits, and structurally optimized geometries that would be difficult or economically infeasible with conventional construction techniques.

The benefits of this approach are multifaceted [2]. Most notably, the elimination of formwork not only reduces material usage and waste but also shortens construction timelines and lowers labor costs. The precision afforded by digital control ensures that only the required quantity of material is deposited, contributing to sustainability by reducing the carbon footprint of the construction process. Furthermore, the ability to automate construction tasks paves the way for increased safety, particularly in hazardous or remote environments, by minimizing human involvement on-site. These

A. Chourasia et al., *Advances and Challenges in 3D Printed Concrete (3DPC)*, Materials Horizons: From Nature to Nanomaterials, https://doi.org/10.1007/978-981-95-8557-1_1

advantages collectively signal a shift toward more efficient, sustainable, and scalable methods of building construction.

Beyond these immediate efficiencies, 3D-printed concrete holds transformative potential in the long term. It facilitates the industrialization of construction, allowing for modular fabrication, mass customization, and on-demand production. It also enables the use of alternative binders and recycled materials, promoting environmentally responsible practices. As an intersection of materials science, structural engineering, robotics, and digital design, 3DPC is poised to redefine how civil infrastructure is conceptualized, designed, and delivered in the decades to come.

1.2 Detailed Process of 3D Concrete Printing

The process of 3D concrete printing begins with the generation of a detailed digital model using specialized design software. Tools such as parametric modeling platforms or building information modeling (BIM) systems are employed to create accurate representations of the desired structure. These digital models encapsulate the geometry, internal features, and construction sequencing needed for the printing process. Once finalized, the model undergoes a slicing operation, in which it is divided into a sequence of thin, horizontal cross sections. Each layer represents a discrete elevation of the structure and serves as a guide for material deposition.

This sliced model is then converted into executable printing instructions, commonly encoded as G-code. These instructions dictate the movement of the printer's nozzle, the rate of extrusion, and other critical parameters needed to ensure dimensional accuracy and material continuity. The printing hardware, which may be a fixed-frame gantry system or a multi-axis robotic arm, receives these instructions and begins the layer-by-layer construction of the object. As the printer moves, it deposits a continuous bead of cementitious material, carefully placing each layer on top of the previous one. The resolution and fidelity of the final structure depend on several factors, including the precision of the toolpath, the uniformity of extrusion, and the consistency of the concrete mix. Figure 1.1 presents the operational summary of 3D concrete printer

The concrete used in 3D printing differs significantly from traditional ready-mix concrete. It must be formulated to satisfy a set of stringent requirements, including adequate extrudability, shape retention, and early-age strength. To achieve these characteristics, the mix typically incorporates a carefully calibrated blend of fine aggregates, binders, water-reducing agents, and rheology-modifying admixtures. The water-to-binder ratio is kept low to enhance buildability, while additives such as hydroxypropyl methylcellulose or silica fume may be included to improve viscosity and cohesion. In certain cases, supplementary cementitious materials such as fly ash, ground granulated blast-furnace slag (GGBS), or even geopolymer binders are used to enhance sustainability and performance.

During the printing process, time is of the essence. Each layer must set quickly enough to support the weight of subsequent layers without deformation, yet remain

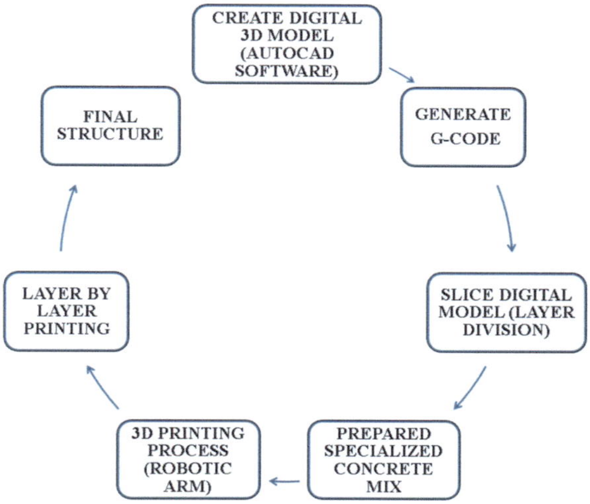

Fig. 1.1 Operational summary of 3D concrete printer

sufficiently moist to ensure good bonding with the next deposition. This delicate balance is maintained through continuous monitoring and, where applicable, through the use of embedded sensors or control systems. Advanced printing platforms are often equipped with real-time feedback loops, employing laser scanners, thermal cameras, or pressure sensors to monitor parameters such as layer height, extrusion rate, temperature, and setting behavior. These systems can make automatic adjustments to the printing parameters, compensating for deviations, and ensuring consistency across the entire structure.

Following the completion of printing, the structure enters the curing phase. Depending on the mix design and environmental conditions, curing may be achieved through conventional water curing, steam curing, or controlled atmosphere methods. Some printing systems integrate curing accelerators or inline infrared heating to initiate rapid strength gain during construction. In addition to curing, post-processing may also include surface finishing, the integration of mechanical reinforcements, or the embedding of functional elements such as conduits, sensors, or insulation materials.

In applications where structural performance is critical, reinforcement strategies must be employed to overcome the current limitations of unreinforced 3D-printed concrete. These strategies may include the pre-installation of reinforcement cages, post-tensioning cables, or the co-extrusion of reinforcing filaments during printing. Research in this area continues to evolve, with hybrid approaches combining printed elements and cast-in-place reinforcement systems showing considerable promise.

The evolution of 3D-printed concrete from laboratory-scale research to full-scale construction demonstrates the viability of this technology as a cornerstone of future construction practices. Early pilot projects across the globe have showcased the feasibility of using 3DPC for a variety of applications, including low-cost housing, pedestrian bridges, urban furniture, and emergency shelters. These deployments underscore the potential of additive manufacturing to deliver high-performance, customizable, and rapidly deployable infrastructure solutions.

As this technology matures, it is increasingly being integrated with other innovations under the umbrella of Industry 4.0, including the internet of things (IoT), artificial intelligence (AI), and digital twins. The convergence of these technologies promises not only greater automation but also predictive maintenance, lifecycle monitoring, and adaptive design capabilities. From an academic and engineering standpoint, 3D-printed concrete opens up new avenues for research in material development, structural optimization, thermal performance, and resilience under extreme loading conditions.

In conclusion, 3D-printed concrete offers a bold reimagining of how structures are designed and constructed. It merges the rigor of engineering with the creativity of digital design, supported by the precision of automation and the adaptability of advanced materials. For students, researchers, and professionals in civil engineering, it presents an opportunity to engage with an evolving field that challenges conventional paradigms and invites innovation at every level—from conceptual design to construction execution.

1.3 Advantages of 3D-Printed Concrete

1.3.1 Design Flexibility

Among the most profound advantages of 3D-printed concrete (3DPC) is its unprecedented design flexibility, which significantly surpasses the limitations of conventional construction techniques. Traditional methods are heavily dependent on fixed molds and labor-intensive formwork, which restrict the ability to realize geometrically complex or highly customized architectural elements. These constraints often make the execution of freeform or non-standard designs economically unfeasible. In contrast, 3D concrete printing is fundamentally moldless, enabling the direct digital fabrication of intricate, curvilinear, and topologically optimized geometries without additional tooling or manual rework. The integration of advanced parametric design software with automated extrusion technology empowers architects and engineers to explore and materialize innovative architectural visions with a high degree of freedom. This capability allows for the construction of components with complex contours, lattice structures, overhangs, and biomorphic shapes that are otherwise difficult to produce.

Figure 1.2 illustrates the range of forms that can be achieved through 3D concrete printing, underscoring its potential for reshaping architectural and structural design paradigms.

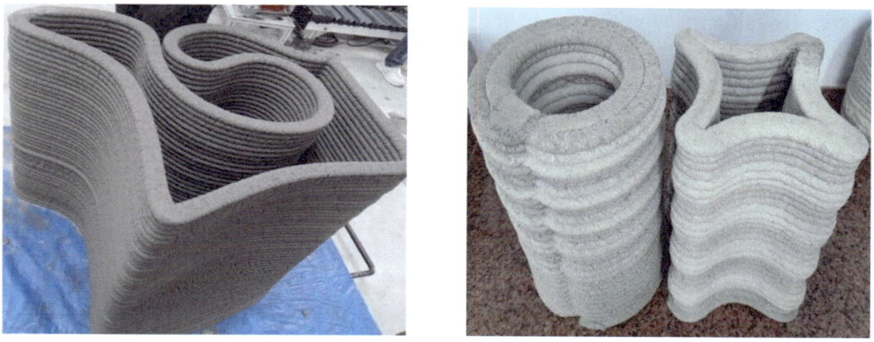

Fig. 1.2 Flexibility in terms of different shapes of 3D-printed concrete elements

1.3.2 Reduced Waste

A significant benefit of 3D-printed concrete lies in its material efficiency and waste reduction, which are direct outcomes of its additive manufacturing process. Traditional construction workflows often involve subtractive operations—cutting, trimming, and adjusting prefabricated components—which generate substantial quantities of offcuts, unused materials, and disposable formwork. These practices not only escalate material costs but also contribute to environmental degradation through increased construction waste. In contrast, 3D printing employs a layer-by-layer deposition strategy that ensures material is placed exclusively where it is structurally or functionally required, based on the digital model. This precision-driven approach virtually eliminates overuse and offcuts, thereby optimizing resource utilization. Moreover, the absence of single-use molds further reduces the volume of discarded material. By streamlining material consumption and minimizing site-generated debris, 3DPC aligns with resource-conscious construction practices and significantly improves the sustainability of building operations.

1.3.3 Speed of Construction

3D-printed concrete offers dramatic improvements in construction speed by automating and integrating multiple stages of the building process into a continuous and uninterrupted workflow. Conventional construction methods are characterized by sequential dependencies—formwork setup, reinforcement placement, concrete pouring, curing, and demolding—each of which introduces delays and requires coordinated manual intervention. In contrast, 3D printing bypasses many of these phases by directly translating digital building models into physical structures via continuous material extrusion. Once the printer is calibrated and the digital slicing of the model is complete, the printing process can proceed with minimal human oversight, drastically reducing construction time. This acceleration is especially beneficial in large-scale housing developments, modular infrastructure projects, and post-disaster reconstruction scenarios, where time-sensitive deployment is critical. Furthermore, the inherent repeatability and programmability of 3DPC allow for simultaneous or parallelized construction tasks, offering potential time savings that can far exceed those achievable through conventional methods.

1.3.4 Cost Efficiency

The economic viability of 3D-printed concrete is underpinned by its potential to reduce both direct construction expenditures and ancillary project costs. Traditional concrete construction is labor-intensive, requiring skilled workers for diverse tasks such as carpentry, reinforcement tying, form setting, and surface finishing. The high dependency on labor not only raises costs but also introduces variability in productivity and quality. In contrast, the automation embedded in 3D printing minimizes the reliance on manual labor by executing construction tasks robotically, thereby lowering labor-related costs and reducing the risk of human error. Additionally, the reduction in material waste, elimination of formwork, and streamlined logistics collectively contribute to significant cost savings. While the initial investment in 3D printing equipment and formulation of specialized concrete mixes may be substantial, the long-term operational savings—particularly in projects involving customization or mass repetition—render the technology economically competitive. In settings where speed, design uniqueness, or labor constraints are primary concerns, 3DPC emerges as a highly cost-effective solution.

1.3.5 Sustainability

Sustainability represents a cornerstone of 3D-printed concrete's value proposition, driven by its inherent material efficiency, adaptability to alternative binders, and potential for reducing the environmental footprint of construction. The precision of additive manufacturing ensures that resources are used judiciously, minimizing excess and promoting efficient building practices. Furthermore, the concrete mixtures developed for 3D printing can be engineered to include supplementary cementitious materials (SCMs) such as fly ash, silica fume, or ground granulated blast furnace slag (GGBS), thereby reducing dependence on ordinary Portland cement—a major contributor to greenhouse gas emissions. Recycled aggregates and locally sourced materials can also be integrated into the printing mix, decreasing transportation-related emissions and supporting regional resource cycles. In addition to these material-based advantages, the shorter construction timelines and reduced on-site activity help lower energy consumption and associated emissions. The ability to construct high-performance, thermally efficient envelopes using advanced design

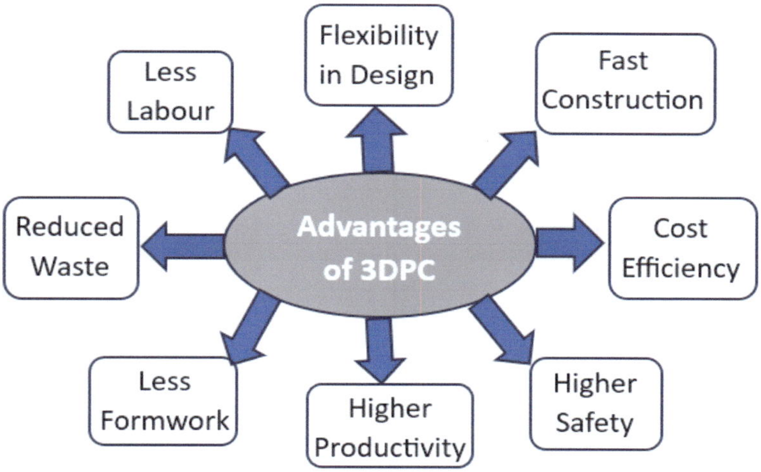

Fig. 1.3 Advantages of 3DPC in comparison to conventional concreting

strategies further enhances the environmental performance of 3D-printed buildings. Thus, the technology offers a multi-dimensional approach to sustainable construction, addressing both environmental and operational concerns.

As illustrated in Fig. 1.3, the cumulative advantages of 3D-printed concrete—ranging from geometric freedom and accelerated project delivery to cost-effectiveness and environmental responsibility—position it as a transformative construction technology. By integrating digital design, automation, and sustainable material science, 3DPC is poised to redefine the future of civil infrastructure and architectural design, enabling faster, cleaner, and smarter ways to build.

1.4 Historical Background

The origins of 3D printing—also referred to as *additive manufacturing*—can be traced to the early 1980s, when Charles Hull introduced the groundbreaking technology of stereolithography. This innovation enabled the creation of physical objects by sequentially layering materials according to digital models, laying the foundation for a broad spectrum of 3D printing techniques. Initially embraced in sectors such as aerospace, automotive, and biomedical industries, additive manufacturing gradually attracted interest in the construction domain. However, its adoption in the building sector remained limited until the early twenty-first century, when rapid advancements in automation, robotics, and material science catalyzed the development of large-scale 3D printing for construction. The journey from concept to application has been marked by critical technological breakthroughs, visionary experiments, and landmark projects that have shaped the trajectory of 3D-printed construction globally and in India.

1.4.1 Early Developments

1990s: Theoretical Foundations and Experimental Initiatives

The 1990s witnessed the conceptual emergence of 3D-printed construction. During this formative period, academic researchers and technologists began exploring the feasibility of applying additive manufacturing principles to building-scale fabrication. These early efforts were primarily theoretical and experimental, focusing on automating the construction process to address long-standing industry challenges, such as labor inefficiencies, material wastage, and extended project timelines. Investigations into robotic extrusion and layering techniques laid the groundwork for developing automated systems capable of constructing entire structures with minimal human intervention. The research emphasis was on establishing feasibility and envisioning a future wherein digital fabrication could enable faster, cheaper, and more precise construction methods.

2004: Contour Crafting by Behrokh Khoshnevis

A significant leap in the practical application of 3D printing in construction occurred in 2004 with the development of *contour crafting* by Professor Behrokh Khoshnevis at the University of Southern California. This process introduced one of the earliest large-scale 3D printing technologies tailored for building applications. Utilizing a computer-controlled gantry system, contour crafting involves the extrusion of cementitious material in successive layers to construct structural components directly from a digital blueprint. This innovation not only demonstrated the possibility of constructing full-scale walls and enclosures autonomously but also introduced new paradigms for improving speed, precision, and sustainability in construction. It marked a pivotal shift from conceptual feasibility to functional prototyping in additive manufacturing for the built environment.

1.4.2 Global Milestones in 3D-Printed Construction

2013: Winsun's Pioneering Efforts in China

In 2013, the Shanghai-based company *Yingchuang Building Technique Co. Ltd.* (widely known as *Winsun*) achieved a groundbreaking milestone by fabricating the world's first 3D-printed house (Fig. 1.4). The structure was printed in a controlled factory environment using a proprietary cement-based mixture and subsequently assembled on-site. This project demonstrated the practical advantages of off-site prefabrication, including speed, cost-effectiveness, and the potential for scalable

Fig. 1.4 Yingchuang building technique (Winsun) in China [3]

deployment of affordable housing. Winsun's innovation showcased how additive manufacturing could address housing shortages while promoting material efficiency and waste reduction. The company has since expanded its portfolio to include multi-unit housing and commercial buildings, receiving global recognition for advancing sustainable and rapid construction technologies.

2014: First Fully Functional 3D-Printed Office Building in Dubai

The year 2014 marked another historic advancement with the construction of the world's first fully functional 3D-printed office building in Dubai (Fig. 1.5). The 250-square-meter structure was printed using a specially engineered cementitious mix and was completed within 17 days, followed by just two days of on-site assembly. Executed by Winsun in collaboration with Dubai's municipal authorities, this project illustrated the commercial viability of additive construction methods for mid-scale projects. It emphasized the potential of 3D printing to dramatically reduce construction duration, minimize labor demands, and improve cost efficiency. Serving as a global demonstration project, the Dubai office building highlighted the adaptability of 3D-printed technology to varied climatic and urban conditions.

2016: MX3D Bridge in Amsterdam—Steel 3D Printing in Infrastructure

A significant technological diversification occurred in 2016 with the completion of the *MX3D Bridge* in Amsterdam (Fig. 1.6). Unlike previous efforts that focused on concrete-based materials, this project employed robotic arms to 3D print a 12-m pedestrian bridge using molten stainless steel. Developed by the Dutch company *MX3D* in collaboration with engineers and researchers, the bridge featured complex, biomimetic geometries that would have been difficult to realize using traditional

Fig. 1.5 Fully functional 3D printed office building in Dubai [4]

construction methods. The structure was also equipped with embedded sensors for real-time structural health monitoring, enabling continuous data collection on stress, strain, and environmental impact. This project not only validated the feasibility of metal additive manufacturing in infrastructure but also demonstrated its potential for intelligent design, customization, and adaptive performance in urban environments.

Fig. 1.6 MX3D bridge in Amsterdam [5]

1.4.3 *Emergence of 3D-Printed Construction in India*

2020: India's First 3D-Printed House at IIT Madras

India entered the realm of additive construction in 2020 with the successful fabrication of its first 3D-printed house, developed by *Tvasta Manufacturing Solutions*, a startup incubated at the Indian Institute of Technology (IIT), Madras. Located on the IIT Madras campus, the 600-square-foot structure (Fig. 1.7) was completed within five days using a regionally optimized concrete mix. The project underscored the applicability of 3D printing for rapid, low-cost housing in the Indian context, addressing local climatic conditions, material availability, and social housing demands. It served as a practical demonstration of how digital construction technologies can be adapted to suit the socioeconomic and environmental constraints of developing nations.

2021: Military Applications by the Indian Army

In 2021, India expanded the application of 3D printing to strategic infrastructure through collaborations between the *military engineering services (MES)* and Tvasta. The Indian Army implemented the construction of 3D-printed concrete bunkers and shelters in remote and difficult terrains (Fig. 1.8), demonstrating the feasibility of deploying rapid, robust infrastructure under adverse conditions. These structures were designed for high durability, quick deployment, and operational resilience—features critical to military and disaster-response scenarios. This initiative highlighted the broader utility of 3D printing in defense and emergency infrastructure, marking a significant evolution in India's additive construction capabilities.

Fig. 1.7 3D printed house at IIT Madras [6]

Fig. 1.8 Military 3D printed structures [7]

The historical development of 3D-printed construction reflects a dynamic interplay of innovation in materials, digital design, and automated fabrication. From its conceptual roots in the 1990s to cutting-edge implementations in residential, commercial, and infrastructure projects worldwide, 3D printing is transforming the way we design and build the built environment. As demonstrated by key projects across China, the UAE, the Netherlands, and India, the technology holds significant promise in addressing global challenges such as affordable housing, disaster resilience, and infrastructure modernization. Continued advancements in robotic systems, material science, and real-time monitoring are poised to further expand the scope and impact of additive manufacturing in civil engineering.

1.5 Other Notable Projects Around the World

The global adoption of 3D-printed construction has accelerated over the past decade, with a growing number of innovative projects that demonstrate its adaptability, sustainability, and cost-effectiveness across diverse geographies. These pioneering efforts showcase how additive manufacturing is redefining architectural design, material use, and housing delivery. The following case studies illustrate some of the most influential and technically significant 3D-printed construction projects worldwide.

Fig. 1.9 Project Milestone in Eindhoven, Netherlands [8]

1.5.1 Project Milestone, Eindhoven, Netherlands

Project Milestone (Fig. 1.9), situated in Eindhoven, Netherlands, is widely recognized as the world's first commercial housing development realized entirely through 3D printing. A collaborative initiative involving Eindhoven University of Technology, the municipality of Eindhoven, and various industrial partners, the project comprises a series of five architecturally unique dwellings. Each unit leverages the design freedom offered by additive manufacturing to create organic and non-standard geometries that are difficult to achieve with conventional construction techniques.

The first home in the series was completed and occupied in 2021, representing a significant advancement in the mainstream adoption of 3D-printed housing. The structures were fabricated using a concrete-based material specifically developed for strength, durability, and printability. Project Milestone also serves as a testbed for exploring energy-efficient design principles, lifecycle sustainability, and digital fabrication workflows. The project exemplifies how 3D-printed housing can contribute to sustainable urban development through innovative design, rapid construction, and material optimization.

1.5.2 TECLA Habitat, Massa Lombarda, Italy

The **TECLA Habitat** (Technology and Clay) project (Fig. 1.10), located in Massa Lombarda, Italy, represents a groundbreaking exploration of sustainable construction through locally sourced, biodegradable materials. Designed by Mario Cucinella Architects in partnership with WASP (World's Advanced Saving Project), TECLA was completed in 2021 as a prototype for eco-friendly, low-cost housing that can be constructed rapidly using raw earth.

Fig. 1.10 TECLA habitat in Massa Lombarda, Italy [9]

Comprising two interconnected dome-like modules, the structure was 3D-printed simultaneously using dual large-scale printers. The primary construction material—stabilized clay—was selected for its thermal inertia, recyclability, and minimal embodied energy. TECLA addresses key sustainability goals by eliminating the need for long-distance transportation of building materials and by promoting circular construction practices. The project demonstrates the potential of additive manufacturing to respond effectively to global housing crises, particularly in areas facing economic or environmental constraints.

1.5.3 Luyten 3D, Australia

Luyten 3D, an Australian construction technology company, has been at the forefront of developing resilient and sustainable housing solutions tailored to the region's challenging climate. One of its most notable achievements is the construction of the **Stargate House** (Fig. 1.11), designed to withstand extreme weather conditions such as intense heat, storms, and flooding that are prevalent in parts of Australia.

Using its proprietary **BUILDPRINTER™** technology and a high-performance geopolymer concrete mix, Luyten 3D has demonstrated the feasibility of rapidly deploying robust housing in both urban and remote locations. The emphasis on locally relevant solutions, energy-efficient construction, and affordable design positions the company as a leader in adapting 3D-printed construction to meet regional infrastructure and housing demands. Their work exemplifies how 3D printing can enhance climate resilience and contribute to sustainable development goals in harsh environments.

Fig. 1.11 Luyten 3D's Stargate House in Australia [10]

1.5.4 The Apis Cor House, Stupino, Russia

In 2017, the Russian company **Apis Cor** achieved global recognition by completing a fully functional 3D-printed house in just 24 hours (Fig. 1.12). Located in Stupino, Moscow region, the 37-square-meter structure was printed on-site using a mobile robotic 3D printer. This monolithic printing process involved the direct extrusion of a specially formulated concrete mix to construct the load-bearing walls and internal partitions in a single continuous operation.

The Apis Cor project underscored the rapid construction capabilities of additive manufacturing, especially in regions with extreme climatic conditions. The house was engineered to withstand Russia's harsh winters, with appropriate insulation and weatherproofing features. By minimizing labor and material waste while significantly reducing construction time, Apis Cor demonstrated the potential of 3D printing to deliver fast, cost-effective housing in remote or disaster-prone areas.

Fig. 1.12 The Apis Cor House in Stupino, Russia [11]

1.5.5 Habitat for Humanity, Tempe, Arizona, USA

In 2021, **Habitat for Humanity** unveiled its first 3D-printed house in the United States, located in Tempe, Arizona (Fig. 1.13). This 1700-square-foot home was constructed using the **COBOD BOD2** printer, which extruded a high-strength concrete mix layer-by-layer to form the structural walls. The project was undertaken as part of the organization's mission to create affordable, sustainable housing for families in need.

This initiative marked a significant step forward in integrating 3D printing into the affordable housing sector. By reducing construction time and labor costs, the technology offers a scalable solution to America's growing housing deficit. The Tempe home incorporates energy-efficient design principles and modern architectural features, highlighting how additive manufacturing can be both socially impactful and technically sophisticated. Habitat for Humanity's adoption of 3D printing sets a precedent for non-profit and government-led housing programs seeking innovative delivery models.

The projects highlighted in this section exemplify the transformative potential of 3D-printed construction across a range of climates, contexts, and building typologies. From energy-efficient urban housing in the Netherlands to climate-resilient shelters in Australia and low-cost, earthen dwellings in Italy, these initiatives demonstrate how additive manufacturing can address global challenges in housing affordability, sustainability, and resilience. As the technology matures and regulatory frameworks evolve, 3D-printed concrete is poised to play an increasingly pivotal role in redefining the construction landscape worldwide.

Fig. 1.13 Habitat for humanity 3D printed house in Tempe, Arizona, USA

1.6 Basic Differences with Mold Cast Construction

3D-printed concrete construction fundamentally diverges from traditional mold cast techniques in several key aspects—most notably in the areas of formwork, labor, and design customization. These differences underscore the disruptive potential of additive manufacturing in transforming conventional construction workflows into more efficient, adaptable, and sustainable processes.

1.6.1 Formwork

In conventional mold cast construction, formwork is indispensable for shaping concrete during curing. Typically constructed from materials such as timber, steel, or plastic, formwork must be custom-fabricated for each specific structural element. This process demands extensive time and skilled labor for tasks including assembly, alignment, bracing, and eventual removal. The removal phase itself can be particularly labor-intensive, and in many cases, the formwork cannot be reused—especially for complex or non-repetitive geometries—thereby increasing both material costs and construction waste.

Conversely, 3D-printed concrete construction eliminates the requirement for traditional formwork altogether. Instead, concrete is extruded layer-by-layer in predefined geometries based on digital building information models (BIM). This direct digital fabrication method not only streamlines construction time but also eliminates the cost, labor, and environmental burdens associated with formwork fabrication and dismantling. Moreover, the absence of mold constraints allows for increased geometric freedom, enabling architects and engineers to explore intricate and nonlinear designs that are economically and technically impractical using conventional methods.

1.6.2 Labor

Traditional mold cast construction is labor-intensive by nature, involving numerous manual operations requiring skilled craftsmanship. These include preparing and installing formwork, batching and pouring concrete, surface finishing, and ensuring dimensional accuracy and alignment. This heavy reliance on skilled human resources increases project costs and introduces variability and delays due to the dependency on workforce availability and coordination.

In contrast, 3D-printed concrete construction significantly reduces manual labor inputs. The process is highly automated, with computer-controlled extrusion systems executing the build plan with precision. Once the digital model is prepared and the printing equipment is calibrated, construction progresses with minimal human oversight. This automation drastically lowers labor costs, shortens project timelines, and reduces construction errors, all while enhancing the reproducibility and quality of output. As a result, 3D printing offers a compelling solution for regions facing skilled labor shortages or requiring rapid construction in challenging contexts.

1.6.3 Customization

Customization in traditional construction is a complex and costly undertaking. Achieving non-standard architectural features or bespoke geometries requires extensive alterations to formwork, leading to increased fabrication costs, longer lead times, and higher material wastage. For this reason, architectural innovation in mold-cast construction is often constrained by economic and logistical limitations.

3D-printed construction, however, enables seamless customization through digital design workflows. Adjustments to form, size, or architectural detail can be implemented effortlessly within the digital model, without necessitating any physical changes to the production setup. This adaptability supports mass customization and allows for individualized designs at minimal additional cost or time. Architects and engineers can thus incorporate complex surface textures, organic forms, or functionally optimized geometries, expanding the horizons of architectural and structural creativity. The integration of parametric design with 3D printing further enhances the ability to produce tailored structures that respond to site-specific, cultural, or environmental requirements.

In summary, 3D-printed concrete construction offers transformative advantages over traditional mold cast methods by eliminating formwork, reducing manual labor, and enabling high-level customization. These capabilities not only enhance construction efficiency and design freedom but also align with the goals of sustainable and adaptive urban development. As the technology matures, its adoption is poised to redefine the paradigms of architectural expression, structural engineering, and construction project delivery.

1.7 Continued Evolution and Future Prospects

Since its early demonstrations, the field of 3D-printed concrete has undergone rapid and remarkable development. Driven by advancements in digital design, robotics, materials engineering, and sustainability imperatives, 3D concrete printing (3DCP) is increasingly being adopted for diverse construction needs. Among the most notable developments is *Project Milestone* in Eindhoven, Netherlands—a pioneering initiative led by Eindhoven University of Technology. It represents the world's first commercially viable 3D-printed housing project, featuring five architecturally distinct homes. The first of these was completed and occupied in 2021, serving as a proof of concept for structurally sound, habitable buildings constructed using 3DCP techniques. The success of this project illustrates not only the feasibility of additive manufacturing in real-world housing but also its readiness for broader market adoption.

Beyond residential construction, 3D-printed concrete is being applied in infrastructure, emergency housing, military logistics, and public amenities. These diverse applications underscore the technology's potential to revolutionize conventional construction paradigms by enabling rapid deployment, mass customization, and more sustainable building practices.

As 3DCP technology matures, it is expected to take on a more central role in future construction ecosystems, supported by ongoing research, policy frameworks, and increased industrial collaboration.

1.7.1 Material Advancements

A cornerstone of 3D-printed concrete's progress lies in the evolution of tailor-made cementitious materials. Unlike conventional concrete, which is designed for pouring into molds, 3D printable concrete requires a unique rheological profile—exhibiting both *extrudability* (ease of flow through the nozzle) and *buildability* (ability to hold shape under successive layers). To achieve these properties, concrete mixes are increasingly being engineered with specialized admixtures.

Superplasticizers improve workability and reduce water content, maintaining high strength while enhancing pumpability. Accelerators are added to expedite setting, enabling vertical layering without collapse. Additionally, viscosity-modifying agents (VMAs) help maintain shape fidelity during and after extrusion.

To improve structural integrity and crack resistance, fiber-reinforced formulations—particularly those incorporating polypropylene, basalt, or glass fibers—are gaining traction. These not only enhance tensile strength but also mitigate shrinkage-induced cracking. Research is also exploring the use of *geopolymer concrete* and *cementitious composites* with reduced clinker content to lower carbon emissions and increase sustainability.

Ongoing innovations continue to optimize fresh-state and hardened-state properties, enabling the construction of increasingly complex, robust, and durable structures using 3D printing.

1.7.2 Robotics and Automation

The integration of advanced robotics and control systems has dramatically transformed the construction workflow in 3DCP. Modern printers utilize robotic arms, Cartesian gantries, or delta configurations equipped with multi-axis control, allowing for intricate geometries and high build precision. These robotic systems translate digital building information models (BIM) into toolpaths that guide the nozzle's movement in real time.

Artificial intelligence (AI) and machine learning are also playing pivotal roles. Embedded sensors and vision systems enable real-time monitoring of parameters such as layer adhesion, material extrusion rate, and print head positioning. AI algorithms dynamically adjust these variables, ensuring print quality, compensating for environmental conditions, and reducing error margins. This level of automation minimizes human intervention, enhances safety, and significantly reduces construction time and cost.

Moreover, the advent of *mobile 3D printers* capable of being deployed on-site—such as truck-mounted robotic arms or gantry frames—has expanded the feasibility of 3DCP in remote and challenging terrains, enabling construction where traditional methods are impractical.

1.7.3 Sustainability and Environmental Impact

Sustainability is at the forefront of the transition toward 3D-printed construction. Traditional building methods are resource-intensive and generate substantial waste and emissions. In contrast, 3DCP inherently promotes material efficiency and waste minimization. The additive layering technique ensures that material is placed precisely where needed, reducing overuse and formwork waste.

Efforts are underway to enhance the ecological footprint of 3D printable mixes by incorporating recycled and industrial by-products. Researchers have successfully integrated supplementary cementitious materials (SCMs) such as *fly ash*, *ground granulated blast-furnace slag (GGBS)*, and *silica fume* to replace portions of Portland cement. Recycled aggregates, waste plastic fillers, and even reclaimed glass are being explored to promote circular economy principles within construction.

Furthermore, 3DCP allows for the design of structurally optimized forms—such as cellular walls and ribbed components—that reduce the volume of materials required while maintaining load-bearing capacity. These innovations contribute not only to resource conservation but also to lowering embodied carbon and achieving climate-resilient infrastructure.

1.7.4 Advancements in Applications

The scope of 3D-printed concrete has broadened from experimental structures to mainstream applications. In the field of infrastructure, printed pedestrian bridges and retaining walls have showcased the technology's load-bearing capabilities and aesthetic potential. A landmark example is the *MX3D Bridge* in Amsterdam, which, although made from steel, exemplifies the design freedom possible with robotic 3D printing systems in construction.

In humanitarian contexts, 3DCP has emerged as a game-changer. Organizations like *New Story* and *ICON* are creating entire villages using 3D-printed homes to provide affordable, durable housing to underserved communities in Latin America and Africa. These structures are not only rapidly deployable but also adaptable to local climate and cultural preferences.

Other innovative uses include military bunkers, printed urban furniture, public toilets, and educational facilities. The technology's versatility enables construction in dense urban environments, post-disaster zones, and resource-scarce regions, reinforcing its broad applicability.

1.7.5 Global Adoption

Globally, countries are embracing 3D-printed concrete as a strategic solution to meet construction demands. In Europe, the *Horizon 2020* research framework has funded numerous projects aiming to refine materials and digital design tools for additive construction. Germany, the Netherlands, and Denmark have hosted multiple full-scale pilot projects to validate long-term performance.

In the United States, companies such as *ICON*, *Mighty Buildings*, and *Branch Technology* are leading innovations in off-site and on-site 3D-printed housing and components. They are exploring hybrid systems that combine printed and prefabricated elements to accelerate the construction lifecycle.

India, recognizing the potential of 3DCP in addressing housing shortages and disaster-resilient infrastructure, has taken pioneering steps. Noteworthy examples include the following:

- The **3D-printed house at IIT Madras**, designed by *Tvasta Manufacturing Solutions*, which demonstrates low-cost, rapid housing tailored to Indian climatic conditions.
- The **Indian Army's deployment of 3D-printed bunkers** in forward areas, enabling secure and rapid construction in difficult terrains.
- The **CSIR–Central Building Research Institute (CBRI)**, which is at the forefront of developing indigenous 3D printing materials and techniques suitable for Indian geographies, ensuring both affordability and sustainability.

Such initiatives highlight the increasing confidence in the viability and scalability of 3D-printed concrete technology in diverse national contexts.

1.7.6 Challenges and Future Scope

Despite its promise, several challenges must be overcome for 3D-printed concrete to achieve widespread commercial adoption. These include the following:

- **Regulatory Gaps**: Most building codes and standards do not yet accommodate printed structural elements, creating uncertainty in project approval.
- **Integration with Conventional Systems**: Electrical conduits, plumbing, insulation, and finishes often need to be manually added, increasing the need for hybridized approaches.
- **Material Variability**: Maintaining consistent mix performance in varied environmental conditions remains a technical hurdle.
- **Scaling and Mobility**: While mobile printers exist, logistical constraints limit their deployment at scale.

To address these issues, continued interdisciplinary research is essential. The development of universal testing standards, certification protocols, and modular plug-and-play systems for utilities will be critical.

Looking ahead, convergence with other technologies—such as **digital twins**, **augmented reality-assisted site operations**, and **autonomous robotics**—could revolutionize construction workflows. The combination of *performance-based design*, *real-time monitoring*, and *environmental responsiveness* will redefine how we conceptualize and build structures.

1.8 Summary

This opening chapter provided a comprehensive overview of the emergence, evolution, and expansive potential of 3D-printed concrete (3DPC) as a transformative construction technology. It began by tracing the historical development of additive manufacturing in construction, highlighting key milestones and early experimental efforts that laid the groundwork for today's advanced systems. From the conceptual beginnings of contour crafting to the first habitable 3D-printed structures, the chapter mapped the trajectory of this innovation within the broader context of digital fabrication and construction automation.

A detailed examination of the fundamental principles of 3DCP—such as layer-wise extrusion, robotic path control, and material behavior—laid the foundation for understanding the core technical aspects of the technology. Various printing systems, including gantry-based and robotic-arm configurations, were described, alongside the specialized requirements for printable concrete mixtures in terms of rheology, workability, and structural buildability.

The chapter further explored the wide-ranging applications of 3D-printed concrete across sectors—from affordable housing and emergency shelters to bridges, urban furniture, and military infrastructure—demonstrating its growing relevance to real-world construction challenges. Several pioneering global projects, such as Project Milestone (Netherlands), the MX3D bridge (Netherlands), and India's indigenously developed 3D-printed housing prototypes, showcased the technology's adaptability, scalability, and socio-economic utility.

Key advantages of 3DPC were articulated, including reduced construction time, minimized material waste, architectural freedom, and improved sustainability through optimized design and eco-friendly material alternatives. Simultaneously, the chapter acknowledged persistent challenges, such as regulatory uncertainty, integration with conventional construction systems, limitations in reinforcement techniques, and environmental variability in print quality.

In the final sections, the chapter projected the future trajectory of 3D-printed concrete, emphasizing ongoing research in smart materials, AI-driven automation, integration with digital twins, and the development of performance-based codes. It underlined the need a cohesive ecosystem involving academia, industry, policymakers, and practitioners to ensure that 3DCP matures into a reliable, standardized, and mainstream construction methodology.

In summary, Chap. 1 sets the stage for the subsequent technical and application-focused chapters by establishing a firm understanding of the foundational principles, transformative potential, and future directions of 3D-printed concrete in the built environment. As construction demands grow globally, this technology stands poised to redefine how we build—faster, smarter, and more sustainably.

References

1. Rangel B, Guimarães AS, Lino J, Santana L (2023) 3D printing for construction with alternative materials. Springer
2. Sanjayan JG, Nazari A, Nematollahi B (2019) 3D concrete printing technology: construction and building applications. Butterworth-Heinemann
3. WinSun (Yingchuang Building Technique Co.)–Housing Innovation Collaborative, (n.d.). https://housinginnovation.co/factory/winsun-yingchuang-building-technique-co/. Accessed 3 Dec 2025
4. World's first 3D-Printed Office Building Completed in Dubai (n.d.). https://newatlas.com/3d-printed-office-dubai-completed/43522/. Accessed 3 Dec 2025
5. MX3D: Amsterdam's 3D Printed Steel Bridge–Arup (n.d.). https://www.arup.com/projects/mx3d-bridge/. Accessed 3 Dec 2025
6. IIT Madras Startup Builds 'India's First' 3D Printed House|Technology News (n.d.). https://www.gadgets360.com/science/news/3d-printed-house-india-iit-madras-2424579. Accessed 3 Dec 2025
7. India Builds World's Highest 3D-Printed Bunker for Army at 11,000 Feet in Ladakh (n.d.). https://asiaone.co.in/india-builds-worlds-highest-3d-printed-bunker-for-army-at-11000-feet-in-ladakh/. Accessed 3 Dec 2025
8. Gallery of Project Milestone|Houben/Van Mierlo Architecten|Media–4 (n.d.). https://archello.com/story/95857/attachments/photos-videos/4. Accessed 3 Dec 2025
9. 3D Printed House TECLA–Eco-Housing|3D Printers|WASP (n.d.). https://www.3dwasp.com/en/3d-printed-house-tecla/. Accessed 3 Dec 2025
10. Luyten 3D to Supply 3D Print Concrete Mix to the US–Australian Manufacturing Forum (n.d.). https://www.aumanufacturing.com.au/luyten-3d-to-supply-3d-print-concrete-mix-to-the-us. Accessed 3 Dec 2025
11. Apis Cor Invented a $10,000 House That Can Be Built in One Day–Business Insider (n.d.). https://www.businessinsider.com/house-built-one-day-apis-cor-2017-3. Accessed 3 Dec 2025

Chapter 2
Technologies and Systems in 3D-Printed Concrete

2.1 Introduction

Additive manufacturing, particularly 3D concrete printing (3DCP), represents a paradigm shift in the construction industry through its integration of automation, digital design, and sustainable material usage. As a form of digital fabrication, 3D printing enables the layer-by-layer deposition of concrete, transforming conventional construction practices by reducing labor-intensive processes, material wastage, and project timelines.

This chapter presents an overview of various 3D printing technologies that are commonly applied to construction, with a focus on the interplay between printer typologies, mechanical components, and software systems. It begins with a classification of 3D concrete printers—gantry, robotic arm, delta, cable-driven, and hybrid systems—highlighting their capabilities and use-case advantages. The next section investigates the primary components of a 3D printer, e.g., the print head, material feeding mechanisms, gantry systems, and control architectures, which highlights the contribution of these elements to precision and scalability.

After that, the chapter explains the role of CAD-based software, tools for slicing the object to be printed, and the control systems that connect the digital model with its physical realization after printing. Thus, by describing the detailed workflow, from modeling to extrusion and post-processing, the current chapter provides the readers with an all-inclusive understanding of digital fabrication and how it is reforming the future of the construction sector in terms of sustainability, efficiency, and freedom in architectural design.

A. Chourasia et al., *Advances and Challenges in 3D Printed Concrete (3DPC)*, Materials Horizons: From Nature to Nanomaterials, https://doi.org/10.1007/978-981-95-8557-1_2

2.2 Types of 3D Printers for Concrete

3D concrete printing (3DCP) has various types of printers with respect to architectural characteristics and each of them suitable for a specific scale of construction, geometric complexities, and on-site environments. The main types of printers that are found to be commonly used in the field of 3D printing in the construction sector are gantry systems based printer, robotic arms, cable-driven mechanisms, delta configurations, and hybrid designs. Each of the systems has different types of operational principles, thus bringing variability in terms of advantages, and limitations.

2.2.1 *Gantry System Printers*

Gantry-based printers (Fig. 2.1) are among the most extensively accepted technologies in 3DCP owing to their higher positional accuracy and structural stability. Such type of printer works on a Cartesian coordinate basis; the fixed frame of it supports the rails along the three perpendicular directions, X, Y, and Z axes. The print head, which is supported on a horizontal gantry, typically moves in the vertical direction. Whereas, the base of it enables motion in the horizontal direction. Such type of configuration enables precise and uniform deposition of layer, thus making the gantry-based printers an ideal choice in the case of large-scale constructions like walls, slabs, etc. By offering higher repeatability, the gantry-type printer is found to be effective in controlling the prefabrication settings. However, the large framework for the mobility of it may pose challenges for an on-site application.

2.2.2 *Robotic Arm Printers*

In robotic arm printers (Fig. 2.2), multi-jointed and articulated arms are utilized to attain higher degrees of freedom and flexibility with respect to spatial context. In contrast to fixed-path-based gantry printers, robotic arm printers can be used for complicated geometries and at challenging angles. Thus, such types of printers can be useful for fabricating suitable intricate architectural geometries such as curved facades, decorative motifs, and lattice structures. In general, such types of printers are mounted on transportable podiums and thus are highly adjustable to dynamic construction sites and uneven type of terrains. The fast and precise ability of the printer also allows integrating the reinforcement or other components during the process of printing.

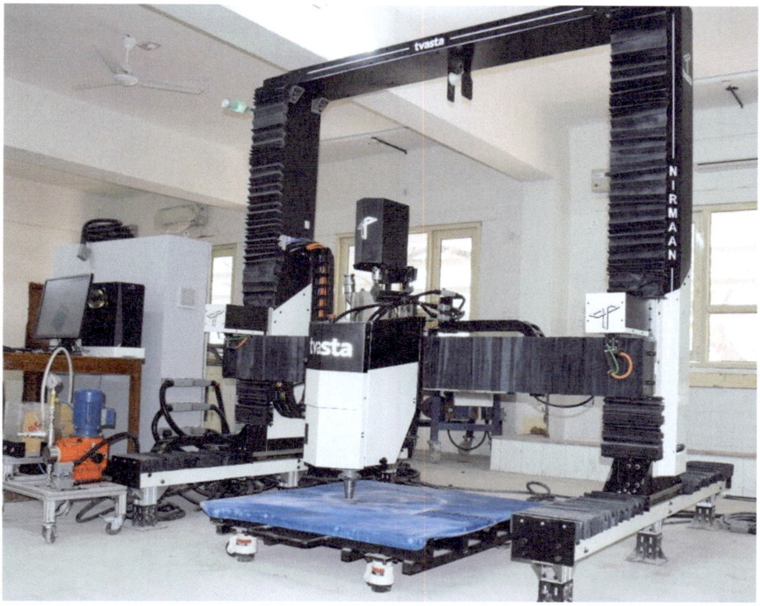

Fig. 2.1 Gantry system printer

2.2.3 Delta Printers

Delta printers (Fig. 2.3) utilize a distinctive triangular arm configuration, where three vertical columns control the print head via equidistant arms connected to a central effector. The parallel kinematics of this design result in fast and fluid print head movements, making delta printers efficient for repetitive and symmetrical geometries. Although less common in full-scale construction, their rapid response time and smooth motion are advantageous for fabricating smaller architectural components, molds, and prefabricated modules with consistent quality. The solid base and efficient mechanisms of the delta printer give various functional advantages in an environment where space is constrained.

2.2.4 Cable-Driven Parallel Robotic Printers

Cable-driven parallel robotic (CDPR)-based 3D printers (Fig. 2.4) represent a leading-edge innovative printer to print very large-scale concrete structures. Such systems are made of a mobile platform, known as a print head, which is suspended through multiple cables and are controlled by motorized winches. The cables which are fastened to a configurable frame, known as on-site poles, let the print head

Fig. 2.2 Robotic arm printer

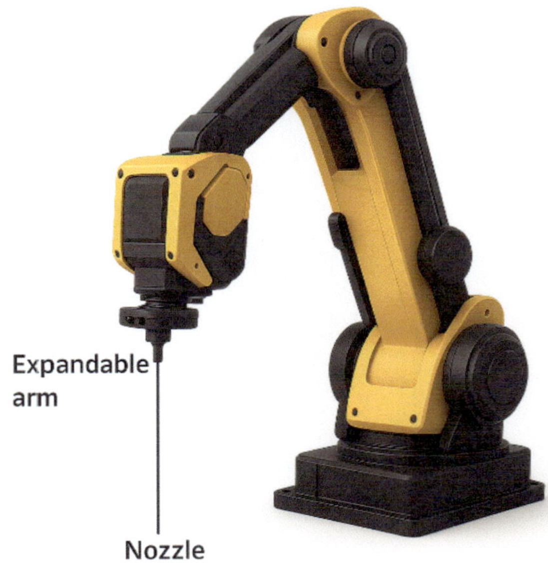

navigate through a large workspace with a higher speed and notable adaptability. The modular characteristic of the printer enables easy transportation and placement across diverse types of terrains, especially in the case of large-scale construction. Current research on CDPRs focuses on improving the positional accuracy and material handling capability of the printer by employing advanced control algorithms along with optimized rheological characteristics of concrete. Such type of technological advancement would improve the practical adaptability of the printer for large and monolithic infrastructural printing which is crucial for housing and disaster-relief shelters.

2.2.5 Hybrid Printers

A hybrid type of 3D printer integrates the technology of different printing systems and offers a wide variety of construction requirements under one umbrella. For

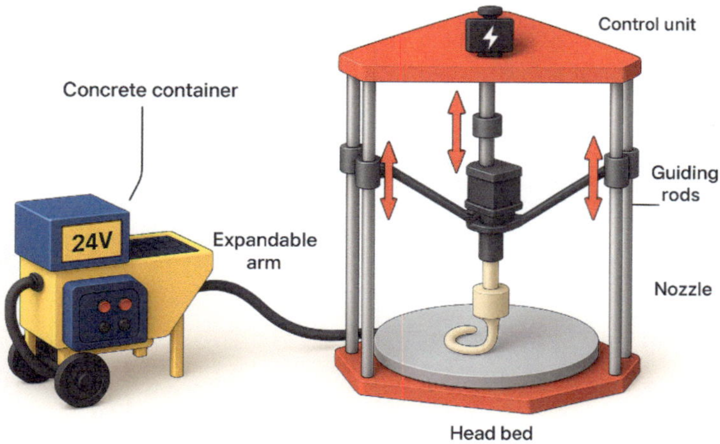

Fig. 2.3 Delta printer

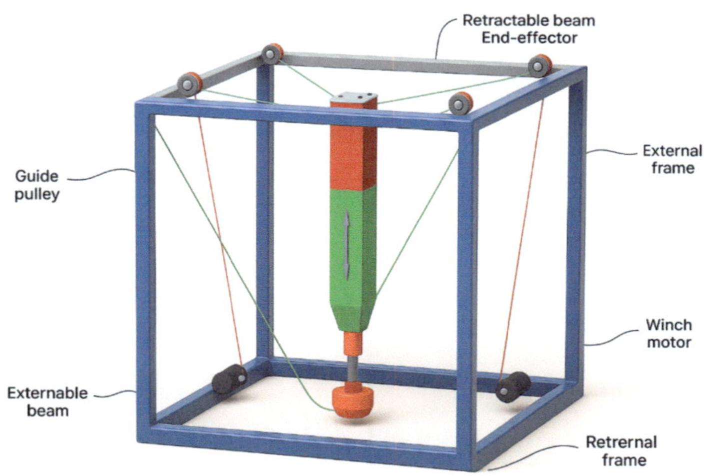

Fig. 2.4 Cable-driven parallel robotic 3D printer

instance, one type of hybrid printer combines the precision and rigidity of a gantry-based system along with the pronounced scope of a robotic arm. Such a type of configuration allows a continuous transition among the printing of large-scale structural components along with producing detailed features without the requirement to change equipment. Thus, hybrid systems are primarily advantageous for a customized type of construction works which offers both flexibility and scalability along with

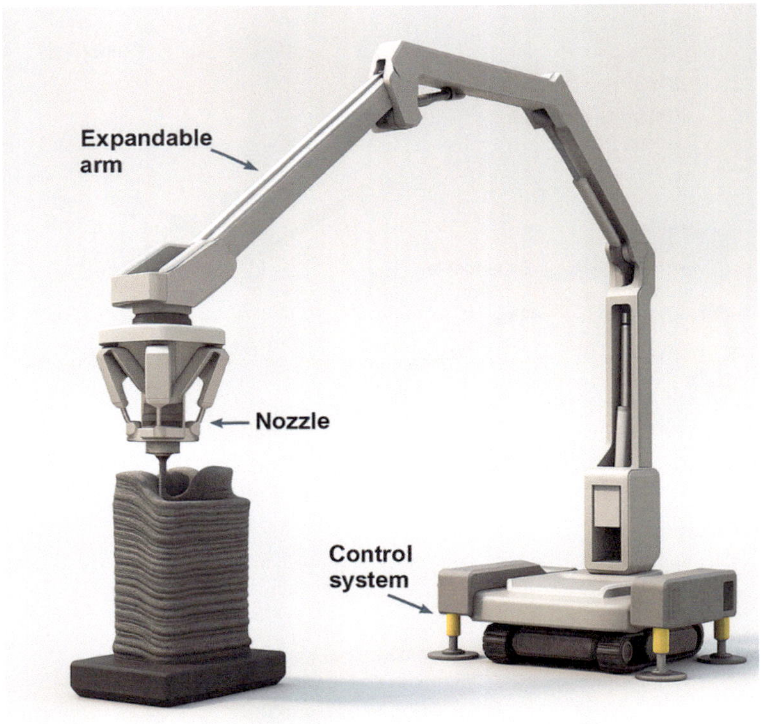

Fig. 2.5 Hybrid printer

sufficient accuracy. The versatility of such type of printer makes it a good candidate for construction that involves multi-phase works, or for on-site modifications, and complex experimental designs (Fig. 2.5).

2.3 Components of a Printer for 3D-Printed Concrete (3DPC)

The 3D printing of concrete—commonly referred to as 3D-printed concrete (3DPC)—is a transformative technology in construction that enables the creation of structural elements and entire buildings layer by layer, without the need for traditional formwork. For this process to function effectively, the printer must be composed of several interlinked subsystems that work together seamlessly. These systems include the preparation of concrete, its transportation to the print head, its precise extrusion onto the printing platform, and the automated control and monitoring of every stage of printing. Each component has a specific function, and the performance of the entire system relies on the efficient coordination among all parts. The following sections describe these components in detail (Fig. 2.6), offering an in-depth understanding of how they contribute to the overall functioning of a 3D concrete printer.

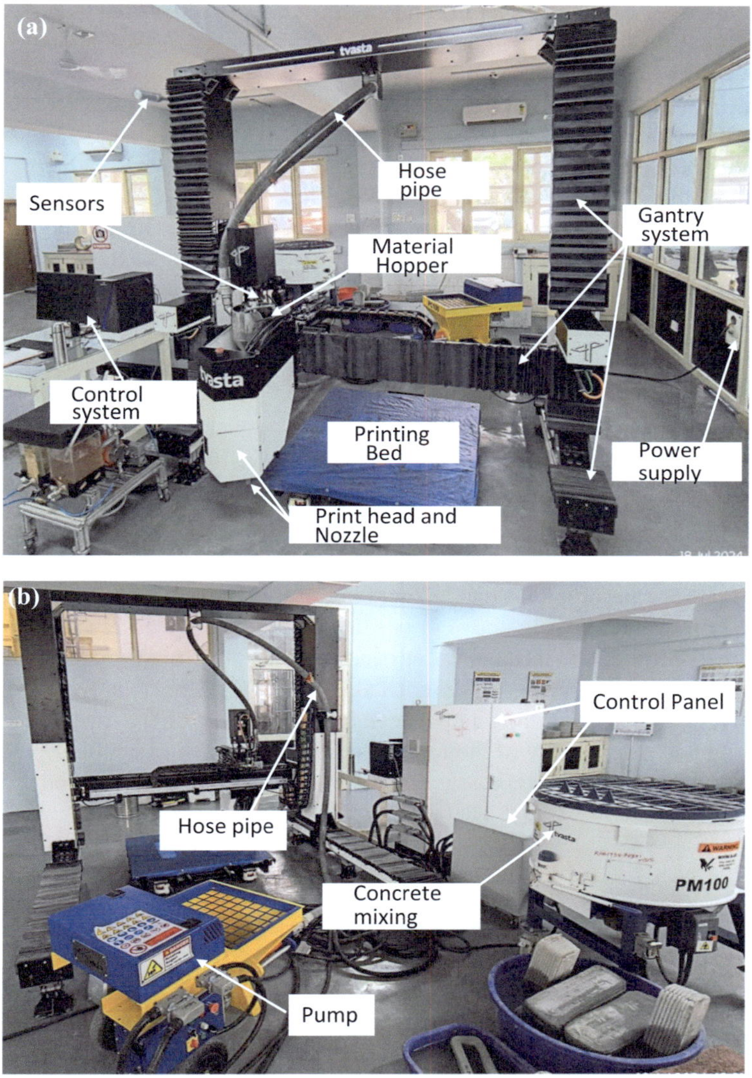

Fig. 2.6 Components of 3D printer

2.3.1 Concrete Mixing and Feeding System

The very first stage in the 3D concrete printing process begins with the preparation of a specialized concrete mixture that is suitable for extrusion. Unlike conventional concrete used for casting, printable concrete must have tailored rheological properties—it should be flowable enough to pass through the system, but stiff enough to hold its shape once extruded. This mixture is prepared using a concrete mixing unit, which typically includes a high-shear or pan-type mixer designed to uniformly combine all ingredients including cement, sand, water, additives, and fibers (if used). Once mixed, the fresh concrete is transferred into a feeding system—usually involving a hopper and pump—that delivers it to the print head. Continuous mixing and feeding are essential during the entire operation to prevent the material from settling or hardening before deposition. Thus, this system ensures that a consistent, homogeneous, and fresh mix reaches the extrusion point at the right time, maintaining the structural integrity and surface finish of the printed object.

2.3.2 Material Hopper

The hopper acts as an intermediate storage vessel that holds the freshly prepared concrete mix before it is pumped toward the nozzle. It plays a critical role in maintaining a steady supply of material and avoiding interruptions during printing, especially for large-scale structures that require long, continuous print cycles. The hopper is typically mounted directly above or beside the pump and is designed to feed the material using gravity or mechanical stirring mechanisms. In advanced setups, the hopper may also be equipped with internal agitators or vibrators to prevent the material from segregating or forming air pockets. Its size depends on the expected volume of the print and the capacity of the mixing system. By enabling uninterrupted flow, the hopper helps maintain a consistent build speed and layer quality throughout the printing process.

2.3.3 Hose Pipe (Delivery Tube)

Once the concrete mix leaves the hopper, it travels through a flexible yet durable pipe known as the hose or delivery tube. This component acts as the conduit that carries the concrete from the pump to the print head, sometimes over several meters of distance. Given the abrasive nature and high viscosity of concrete, the hose must be made from materials that are both wear-resistant and pressure-tolerant. Common materials include reinforced rubber or specially lined composites that can withstand the mechanical stress and friction. The hose also needs to maintain its flexibility to accommodate the motion of the print head without restricting flow. Any blockage or kink in the pipe could result in material inconsistency, extrusion delays, or even print failure, making the hose a deceptively critical component of the system.

2.3.4 Pump (Material Delivery System)

The pump is the driving force that pushes the concrete through the delivery system. Its function is to maintain a constant and controlled pressure that ensures the material reaches the print nozzle smoothly and continuously. Concrete, being a dense and coarse material, cannot be pumped like ordinary fluids. Therefore, special types of pumps are used—such as progressive cavity pumps (screw-type), piston pumps, or peristaltic pumps—that are designed to handle the high viscosity and abrasive properties of cementitious mixes. The choice of pump depends on the printer size, required flow rate, and specific application. A well-designed pump not only ensures a consistent delivery rate but also minimizes pulsation, which can cause inconsistencies in the printed layers. In essence, the pump must work in harmony with the mixer, hopper, and nozzle to provide a smooth and uninterrupted printing experience.

2.3.5 Print Head

The print head is the part of the printer that receives the pumped material and controls its final extrusion onto the build platform. It must be carefully engineered to accommodate the flow behavior, setting time, and particle size of the concrete. Inside the print head, the concrete may undergo additional conditioning, such as de-airing or shear alignment, before being pushed out through the nozzle. The print head ensures that the material is deposited in a controlled and accurate manner as dictated by the print design. Some advanced systems may include variable control mechanisms to adjust flow rate dynamically, based on the geometry of the structure being printed. The print head must also be robust, as it is exposed to significant mechanical stress and must move with high precision along all three spatial axes.

2.3.6 Printing Nozzle or Extruder

At the very end of the print head lies the nozzle or extruder, which is responsible for the final shaping and placement of the concrete material. The nozzle determines the width, thickness, and surface finish of each printed layer. A key innovation in 3DPC technology is the screw-type extruder, where a rotating screw continuously moves the material forward, providing more controlled and stable extrusion compared to traditional pressure-based nozzles. This type of extruder is particularly helpful when printing structures with varying cross sections or intricate geometries, as it allows

Fig. 2.7 Different shapes and sizes of nozzles

finer control over the extrusion speed and volume. The nozzle may also come in interchangeable shapes and sizes depending on the project requirements and must be cleaned and maintained regularly to prevent blockages (Fig. 2.7).

2.3.7 *Printing Bed or Platform*

The printing bed serves as the base on which the concrete is deposited layer by layer. It must be stable, level, and strong enough to support the growing structure during printing. The quality of the first few layers greatly depends on the condition of the bed, as poor adhesion or misalignment can lead to defects in the entire build. In some printer designs, the bed may move in one or more directions, aiding in large-scale or multi-component printing. In others, it remains fixed while the gantry system moves the nozzle. The bed surface may be prepared with bonding agents or textured finishes to improve layer adhesion. For field-scale printers, the ground itself may serve as the printing base, in which case surface leveling and compaction are critical.

2.3.8 *Frame and Movement System*

The frame of a 3D concrete printer is the physical skeleton that supports all mechanical and electronic components. It holds the movement system in place, which allows the print head to travel precisely along the designated print path. Movement may be achieved using rails, guideways, robotic arms, or gantry systems depending on the type of printer. These components must be rigid, stable, and precisely aligned to handle the weight of the printer and resist vibrations during operation. Accuracy in movement is essential for ensuring that each layer is printed in the correct position and thickness, especially when constructing tall or geometrically complex structures.

2.3.9 Gantry System

The gantry system is a specific type of movement framework that allows the print head to move across three-dimensional space—typically in the X, Y, and Z directions. It usually consists of motorized rails, pulleys, and bearings that guide the nozzle along its designated path. The gantry system must be sturdy enough to support the print head and flexible enough to move it with high precision. The speed and accuracy of the gantry system directly affect the quality of the printed layers and the overall efficiency of the printing process.

2.3.10 Control System

The control system is the computational core that governs the entire printing process. It reads the digital model of the structure—usually in the form of a G-code or similar instruction set—and translates it into precise commands for the pump, print head, gantry system, and other components. This system is composed of both hardware (controllers, processors, sensors) and software (printing algorithms, slicing software, feedback logic). It ensures synchronization among different subsystems, adjusts parameters such as flow rate or layer height on the fly, and monitors the real-time status of the print. More sophisticated control systems also include feedback loops from sensors that help in auto-correction and error handling during the printing process.

2.3.11 Support Structures

In some cases, the geometry of the printed object may include overhangs, bridges, or complex cantilevered elements that cannot support themselves during printing. For such features, temporary support structures are added, either manually or through software-generated print paths. These supports prevent deformation, sagging, or collapse during layer deposition and are either removed after printing or integrated into the structure if they are functional. In 3DPC, support structures are generally made from the same concrete mix and are printed simultaneously with the main object.

2.3.12 Power Supply

The entire printing system is powered by a stable and reliable electrical supply, which energizes motors, sensors, heaters (if used), and the control unit. Depending on the scale of the printer, the power system may include backup units, voltage regulators, and distributed power lines to ensure continuous operation without interruptions. Any failure in the power system can halt the printing process and potentially compromise the structural integrity of the build; hence, redundancy and quality in the power supply are essential.

2.3.13 Sensors and Feedback Systems

Modern 3D concrete printers are equipped with a variety of sensors to monitor and control key parameters during printing. These may include sensors for temperature, humidity, material flow rate, nozzle pressure, and print head position. The data from these sensors is fed into the control system, allowing real-time adjustments to maintain print quality. For example, if a reduction in flow rate is detected due to a blockage, the system may slow down the print head or initiate a cleaning cycle. Feedback systems improve automation, reduce material waste, and enhance the repeatability and reliability of the printed product.

2.4 Software and Design Tools for 3D Concrete Printing

The success of 3D concrete printing is rooted not just in hardware and materials but in the intelligent orchestration of software tools that bridge digital design and physical construction [1, 2]. These tools ensure that each component is precisely modeled, structurally sound, and ready for seamless layer-by-layer fabrication. This section elaborates on the key categories of software critical to the 3DCP process.

2.4.1 Computer-Aided Design (CAD) Software

CAD software is the starting point of any 3D-printed construction project. It enables the creation of accurate digital models that encapsulate the structure's geometry, dimensions, and functional details. These models serve as the digital blueprint for construction.

AutoCAD remains a widely used platform in the architectural and civil engineering world due to its versatility in both 2D and 3D drafting. Rhino (Rhinoceros) offers sophisticated capabilities for complex surface modeling, ideal for organic or free-form architectural forms. SolidWorks, on the other hand, is preferred for its parametric modeling and strong integration with simulation tools, making it ideal for components requiring mechanical performance validation.

Each of these software tools allows professionals to visualize and refine design iterations long before the concrete mix is ever pumped through a nozzle, ensuring that both aesthetics and structural feasibility are addressed at the earliest stage.

2.4.2 *Slicing Software*

Once a 3D model is finalized, slicing software translates this digital model into a language that 3D printers understand—G-code. This translation divides the model into hundreds or thousands of thin horizontal slices, each representing a layer to be printed.

Slicing software determines key printing parameters such as layer thickness, print speed, extrusion rates, and path planning. Cura is widely adopted for its intuitive interface and high degree of customization, while Simplify3D is appreciated for advanced features like manual support generation and detailed preview modes. Printer manufacturers also offer proprietary slicing programs that optimize performance specific to their hardware systems.

This software acts as a bridge between creative design and mechanical execution, dictating the precision and integrity of the physical build.

2.4.3 *Structural Analysis Tools*

Prior to printing, it is critical to evaluate how the structure will perform under expected loads and environmental conditions. Structural analysis software like ANSYS and ABAQUS simulates real-world forces on digital models, identifying potential weak points and allowing design adjustments to improve safety and performance.

ANSYS excels in multiphysics simulations, including stress distribution, thermal effects, and material nonlinearity, making it ideal for comprehensive safety validation. ABAQUS, with its superior nonlinear modeling capabilities, is particularly effective in predicting failure modes in complex geometries or under seismic loading.

By integrating these simulations early in the design workflow, engineers ensure that the printed structure will perform reliably once completed.

2.5 Workflow from Digital Design to Physical Construction

The process of constructing a building through 3D concrete printing is a sequential and collaborative workflow that integrates digital precision with construction site realities. This section explains each step in the journey from initial design to final structure.

2.5.1 Design Phase

The workflow begins with the design phase, where architects, engineers, and designers collaborate to create a digitally defined structure using CAD software. This digital model incorporates not just aesthetic and spatial considerations but also the functional and load-bearing requirements of the structure. Every aspect—curvature, wall thickness, thermal performance, and space utilization—is carefully calibrated. The goal is to ensure that the structure meets user needs, complies with codes, and is optimized for digital fabrication.

This phase may involve several iterations, using visual rendering, performance analysis, and structural simulation tools to validate the design from every angle before proceeding to production.

2.5.2 Preparation Phase

Once the design is finalized, the model is processed using slicing software to convert it into G-code instructions, defining the movement of the print head, the extrusion rate, and the layer height. This digital output is sent to the printer's control system.

Simultaneously, on-site preparations are undertaken—leveling the printing area, installing formwork (if necessary), and assembling and calibrating the 3D printer. The concrete mixing and feeding systems are tested, and safety checks are conducted to ensure a controlled and hazard-free environment.

This dual preparation—digital and physical—sets the stage for a smooth printing operation.

2.5.3 Printing Phase

The printing phase marks the transition from design to reality. The print head begins depositing the concrete layer by layer, precisely following the G-code path derived from the digital model. The success of this phase depends on the continuous and uniform flow of concrete, the printer's positional accuracy, and real-time monitoring.

Operators supervise the process closely to resolve any flow interruptions, nozzle blockages, or environmental influences (like temperature or wind) that could impact the print quality. Advanced sensors and feedback systems feed real-time data to the control software, allowing for dynamic adjustments that maintain accuracy and consistency throughout the build.

2.5.4 Post-processing Phase

After printing, the structure enters the post-processing stage, essential for achieving long-term strength and finish. The first step is curing, where the printed concrete gains strength and durability over time, often under controlled humidity and temperature conditions.

Surface treatments may follow, such as polishing, sealing, or aesthetic enhancements. Reinforcements—like steel rods or anchors—may be installed to augment structural capacity. Openings for doors, windows, and mechanical systems are refined, ensuring full integration with building services.

This phase ensures the printed object evolves from a raw build into a fully functional and code-compliant structure.

2.5.5 Continuous Layer Printing

Continuous layer printing is a core technique in 3D concrete printing wherein a seamless and uninterrupted stream of concrete is extruded to form successive layers of the structure. This method ensures a high degree of structural continuity by maintaining a steady and uniform material flow, which effectively reduces the chances of interlayer discontinuities such as delamination or cold joints.

The print quality in continuous layer printing hinges on the careful synchronization of key parameters: the extrusion flow rate, print head speed, and layer height. These are meticulously calibrated to produce layers of consistent thickness and mechanical integrity. Since there are no intermediate stops or segmentation between passes, the resulting build is smoother, faster, and structurally more monolithic compared to other printing techniques.

This approach is particularly advantageous when constructing large-scale structures with repetitive geometries, such as walls or shells, where speed and strength are of paramount importance. Furthermore, the reduction in printing interruptions minimizes thermal gradients and hydration differentials, leading to improved bond strength between layers and overall structural durability (Fig. 2.8).

2.5.6 Segmented Layer Printing

Segmented layer printing, as the name suggests, involves dividing each structural layer into smaller sections or segments that are printed individually. Unlike continuous printing, this technique allows for greater design flexibility and adaptability, especially when constructing geometrically complex components or multi-material elements within the same print layer.

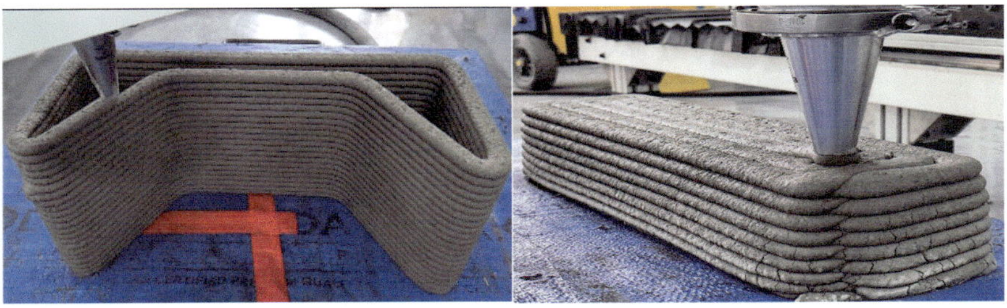

Fig. 2.8 Continuous layer printing

This approach is particularly beneficial when the structure demands differentiated mechanical properties—such as regions requiring higher strength, embedded insulation, or integration of service channels. Each segment can be independently customized with respect to material composition, reinforcement detailing, and printing strategy.

Segmented layer printing also enables more precise placement of embedded components like conduits, sensors, or reinforcement bars. The printing process requires a high degree of spatial coordination, ensuring each segment aligns and bonds accurately with its neighboring units. The controlled segmentation also facilitates the printing of overhangs and non-orthogonal geometries by allowing phased support integration (Fig. 2.9).

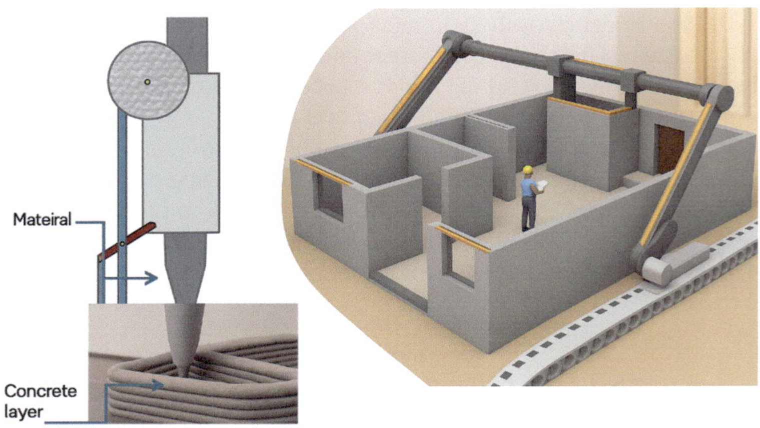

Fig. 2.9 Segmented layer printing

2.5.7 Contour Crafting

Contour crafting is a highly specialized form of 3D concrete printing that prioritizes the precision of surface finish and architectural detail. This technique uses a dynamically controlled extrusion nozzle capable of varying the print path and flow rate in real time, allowing for the production of smooth surfaces, sharp edges, and complex ornamental features.

Unlike conventional layered printing, contour crafting emphasizes sculptural finesse alongside structural functionality. Sophisticated slicing algorithms are employed to optimize toolpaths, minimize surface irregularities, and enhance dimensional accuracy. This makes contour crafting particularly suitable for printing decorative facades, customized moldings, and architecturally expressive components.

The method integrates both aesthetic and engineering objectives, often requiring advanced coordination between software, materials, and robotic control systems. The outcome is a printed product that closely mirrors the designer's vision, both visually and functionally (Fig. 2.10).

Fig. 2.10 Contour crafting

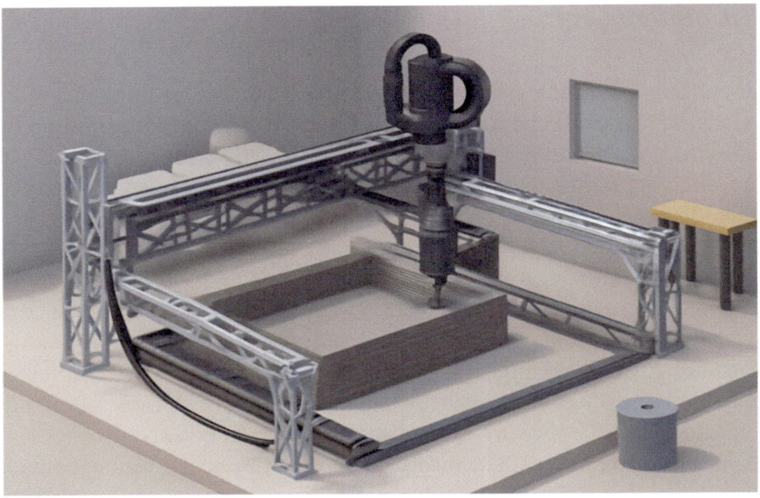

Fig. 2.11 Shotcrete 3D printing

2.5.8 Shotcrete 3D Printing

Shotcrete 3D printing, or spray-based concrete printing, diverges from traditional layered extrusion by employing high-velocity projection of a concrete mix onto a target surface. The technique allows free-form printing without the need for conventional flat-layer buildup, making it highly adaptable for irregular or vertical surfaces.

This method is especially useful for curved geometries, structural shells, retaining walls, and retrofit overlays. By adjusting nozzle orientation and spray dynamics, shotcrete printing can deliver concrete onto varied angles and contours, enabling novel architectural possibilities.

One of the key advantages of this technique is its reduced dependence on formwork. As the sprayed mix adheres and sets rapidly, it supports subsequent layers or coats in quick succession. The process also lends itself to high build rates and structural integrity, particularly when reinforced with steel fibers or mesh during application (Fig. 2.11).

2.5.9 D-Shape Printing

D-shape printing is a large-scale additive manufacturing technique based on binder jetting, in which a powdered material (typically sand or crushed stone) is selectively bonded using a binding agent. Unlike extrusion-based methods, this process

Fig. 2.12 D-shape-printed house

builds structures by depositing successive layers of powder and solidifying them in predetermined patterns.

The D-shape printer is capable of creating intricate geometries and surface textures by precisely controlling the binder application. It is especially well-suited for producing architectural components, sculptures, complex formwork, and urban furniture with organic or nonlinear shapes.

This method also allows for the inclusion of voids, lightweight structures, and internal detailing without the need for complex supports. Since the powder acts as a self-supporting medium during printing, geometries that would otherwise collapse in extrusion-based methods can be successfully realized (Fig. 2.12).

2.5.10 Hybrid Construction Printing

Hybrid construction printing represents an integrative approach that combines automated 3D printing technologies with conventional construction methods. Typically, a robotic or gantry-based system is used to print the primary structural framework, while skilled laborers perform tasks such as embedding reinforcements, laying conduits, and installing mechanical, electrical, and plumbing (MEP) systems.

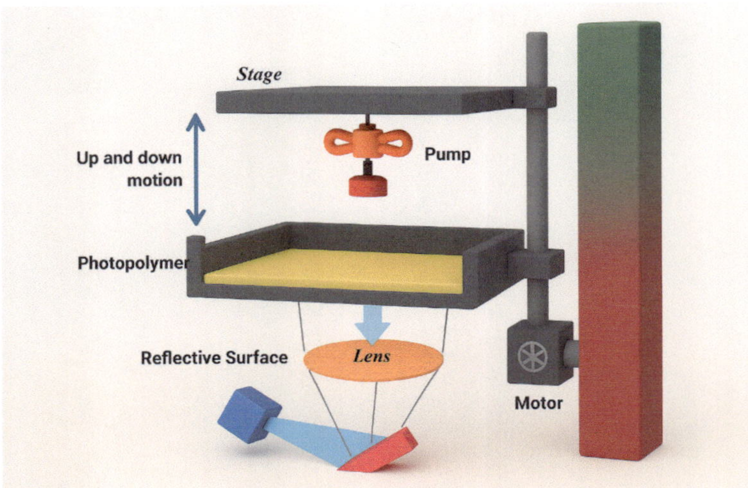

Fig. 2.13 Hybrid construction printing

This method capitalizes on the speed and precision of automation for the structural envelope, while leveraging human expertise for intricate or adaptive tasks that are difficult to automate. It allows for enhanced build complexity, functional integration, and compliance with safety standards and local building codes.

Hybrid construction printing is especially valuable for large-scale buildings where automation can accelerate the shell construction, and conventional trades can follow up with detailing, finishing, and systems integration (Fig. 2.13).

2.5.11 Concrete Co-Extrusion Printing

Concrete co-extrusion printing involves the simultaneous deposition of two or more different materials through a shared or multi-channel nozzle. This enables the creation of layered or composite components where structural concrete, insulation, fiber reinforcement, or even decorative mortar are integrated within the same printed layer.

The ability to vary material properties within a single print path introduces new functional possibilities. For instance, one can design walls with a high-strength outer shell and a thermally insulating core, all produced in a single pass. This approach enhances material efficiency, energy performance, and construction speed.

Co-extrusion also facilitates graded structures and multi-property elements, which are especially valuable for climate-responsive or sustainable building envelopes.

2.5.12 Digital Light Processing (DLP) 3D Printing for Concrete

Digital light processing (DLP) 3D printing, traditionally used in polymer-based additive manufacturing, is an emerging frontier in concrete printing. This technique uses a light projection system to selectively cure layers of a photo-reactive concrete mixture.

In DLP printing, concrete is mixed with photosensitive binders or resins, and a patterned light is projected to harden specific regions, allowing extremely fine resolution and high-detail geometry. Though current applications are mostly experimental or small-scale, research is progressing toward adapting this method for larger structural components and precision architectural detailing.

This technique opens the door to high-resolution printing of concrete components with intricate features, potentially suitable for heritage restoration, façade elements, and fine art in architecture (Fig. 2.14).

Each of the above 3D concrete printing methods has distinct strengths and suitability depending on the structural requirements, project scale, material strategy, and design goals. From rapid monolithic construction to intricate artistic fabrication, the spectrum of available techniques reflects the growing versatility of 3D concrete printing in the built environment.

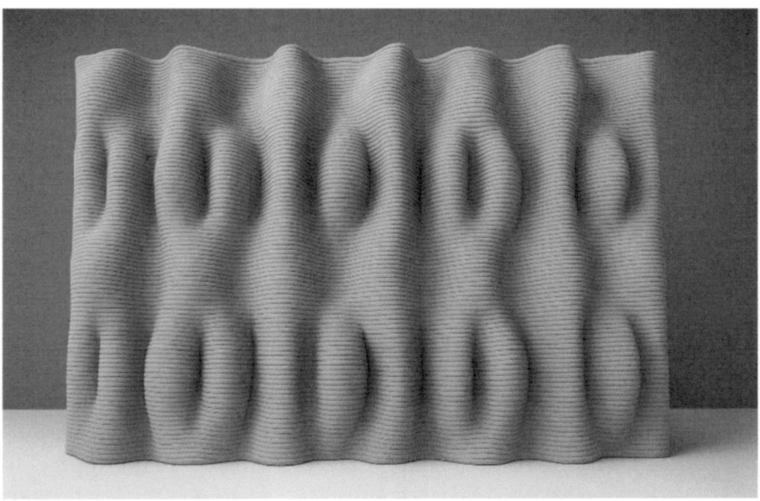

Fig. 2.14 Digital light processing (DLP) concrete printing

2.6 Structural Integrity in 3D Concrete Printing

The structural integrity of any built structure is a non-negotiable aspect that determines its strength, safety, durability, and serviceability over time. In the context of 3D concrete printing (3DCP), structural integrity becomes even more critical due to the novel nature of the construction process, which eliminates formwork, relies on layer-by-layer deposition, and introduces new variables such as material rheology, interlayer bonding, and digital control systems. Ensuring structural integrity in 3DCP requires a comprehensive understanding of various interacting parameters—ranging from material composition to environmental conditions and printing precision.

This section presents an in-depth discussion of the principal factors influencing structural integrity in 3D concrete printing. It also explores the current challenges and emerging strategies being adopted to overcome these issues in both laboratory and field-scale applications.

2.6.1 Key Parameters Influencing Structural Integrity

2.6.1.1 Material Properties and Their Role in Printed Concrete

The properties of the concrete used in 3D printing play a foundational role in shaping the strength and reliability of the structure. Unlike traditional concrete that is poured into molds and later vibrated or compacted, 3D-printed concrete is extruded in layers without any formwork or manual finishing. Hence, the mix must be capable of being pumped and extruded smoothly, retaining its shape immediately upon deposition, and also developing adequate mechanical properties after setting and curing.

The concrete mix must achieve an optimal balance between workability (for extrusion), buildability (to support multiple layers), and structural strength. Special admixtures, such as superplasticizers, are frequently incorporated to improve the flow of concrete without increasing its water content. Accelerating agents may be added to regulate setting time, allowing the next layer to be printed before the previous one has fully hardened, thus promoting better adhesion.

Another critical aspect is interlayer bonding [3]. The interface between two successive layers can be a potential point of weakness, especially if the gap between layer depositions is too long or if the surface has already begun to dry. In such cases, the new layer may not bond properly, leading to delamination or reduced shear resistance. This issue can be mitigated through strategies such as reducing time intervals between layers, moistening the surface before deposition, or applying chemical bonding agents. The formulation of the mix itself also influences the quality of adhesion.

Curing, too, plays a vital role in the structural performance of printed elements. Inadequate curing can result in incomplete hydration, lower strength gain, and a porous microstructure that is more prone to cracking. Controlled curing methods, such as maintaining a consistent moisture level around the structure or applying curing compounds that prevent rapid evaporation, are critical in preserving strength and durability.

2.6.1.2 Influence of Printing Parameters on Structural Behavior

The printing process itself introduces a range of parameters that influence structural performance. These include the height and width of each deposited layer, the speed at which the print head moves, and the rate at which concrete is extruded. Each of these parameters must be precisely controlled to ensure consistent layer deposition, shape retention, and bonding.

If the **layer height** is too large, the structure may lose its intended geometry or experience weak bonding between layers. On the other hand, very small layers can improve surface finish and bonding but significantly slow down the printing process. Hence, finding an optimal layer dimension is crucial.

Similarly, **print speed** directly affects the build quality. If the print head moves too quickly, the concrete may not have enough time to adhere to the underlying layer. If it moves too slowly, excess material may accumulate, distorting the shape. The **extrusion rate** must also be precisely aligned with the print speed; any mismatch may lead to material overflow, gaps, or inconsistency in layer thickness. These variables are typically managed through automated control systems, which use feedback from sensors to adjust parameters in real time and maintain printing fidelity.

2.6.1.3 Environmental Conditions and their Impact on Print Quality

Environmental factors such as ambient temperature, humidity, wind speed, and exposure to sunlight significantly influence the setting, curing, and strength development of concrete. In outdoor printing scenarios, high temperatures may cause rapid moisture loss, leading to shrinkage and surface cracking. Low humidity can accelerate drying, preventing proper bonding between layers.

Wind and sun exposure can lead to uneven curing, resulting in differential shrinkage or warping. To counter these effects, protective enclosures or printed shelters are often used to maintain a controlled micro-environment around the print area. By regulating humidity and temperature, these enclosures help ensure a uniform and predictable curing process.

2.6.1.4 Design and Geometric Complexity

3D concrete printing allows for highly creative and nonlinear architectural forms that are difficult to achieve using conventional techniques. However, complex geometries—such as overhangs, inclined planes, or cantilevered sections—pose structural challenges during printing. These features may require temporary supports or highly optimized print paths to prevent collapse or deformation during construction.

The internal geometry of the structure also affects stress distribution and long-term performance. For example, thin vertical walls may be easy to print but may not be adequate to bear loads unless properly reinforced. Hence, the design phase must incorporate structural considerations such as load transfer, center of gravity, and lateral stability. Software tools are now available to simulate and optimize designs for structural efficiency and printability simultaneously.

2.6.1.5 Load-Bearing Considerations

Ultimately, the printed structure must safely support imposed loads throughout its service life. These may include its own weight (dead load), occupancy loads (live loads), wind, seismic forces, or any additional service-related loads. Load transfer paths must be clearly defined, and the printed walls or elements must be capable of resisting compression, tension, shear, and bending forces.

Structural analysis tools, such as finite element modeling (FEM), are often employed during the design phase to simulate how loads will act on the printed components and how stresses are distributed throughout the structure. This analysis helps identify weak points and areas that require reinforcement or geometric refinement.

2.6.2 *Strategies for Ensuring Structural Integrity in 3DCP*

To address the above challenges and ensure the safe performance of 3D-printed concrete structures, researchers and engineers have developed several strategies:

2.6.2.1 Tailoring Concrete Mix Designs

Customized mix designs have been developed specifically for 3D printing applications. These are formulated to optimize the rheological properties, such as yield stress and viscosity [4–6], while still achieving desired compressive and tensile strengths. The use of supplementary materials like fly ash, silica fume, or nanomaterials further enhances the mix's performance in terms of durability and sustainability.

2.6.2.2 Improving Interlayer Bonding

Techniques such as moistening the surface before deposition, using bonding gels, or creating a rough surface texture between layers are effective in improving interlayer adhesion. In some experimental approaches, ultrasonic or vibratory techniques are being employed to increase bonding at the microstructural level.

2.6.2.3 Implementing Advanced Curing Techniques

Modern 3DCP systems are increasingly incorporating integrated curing systems that travel with the printer or envelop the printed area. These may use controlled mist sprays, insulated tents, or infrared heating to create the ideal curing environment, even under challenging conditions.

2.6.2.4 Real-Time Monitoring and Process Control

One of the biggest advancements in 3D concrete printing is the integration of real-time monitoring tools. Sensors embedded in the nozzle or print path can detect anomalies such as flow inconsistency, dimensional deviation, or nozzle clogging. These are fed into a control system that makes immediate adjustments to maintain print quality and structural consistency.

2.6.2.5 Environmental Control Measures

To mitigate the effects of wind, sun, or temperature fluctuations, many large-scale printing projects are conducted within temporary shelters or climate-controlled environments. These setups protect the printed structure and allow for continuous, uninterrupted curing.

2.6.2.6 Incorporating Structural Reinforcement

While 3DCP aims to reduce traditional formwork and manual labor, it does not eliminate the need for reinforcement. Reinforcing strategies include inserting steel rebars or meshes between layers, integrating synthetic or steel fibers within the concrete mix, or using robotic arms to place reinforcement in real time. These hybrid systems combine the speed of additive manufacturing with the proven performance of reinforced concrete.

2.6.2.7 Designing with Printability in Mind

An important shift in practice is the emphasis on "design for additive construction" (DfAC). This involves collaborative design thinking, where structural engineers, architects, material scientists, and automation experts work together from the outset.

Such an integrated approach helps produce structures that are not only aesthetically innovative but also structurally optimized and feasible to print.

2.6.2.8 Rigorous Quality Testing and Control

Before any printed structure is approved for use, it undergoes strict quality control checks. These include compressive and flexural strength testing of printed specimens, dimensional accuracy checks using 3D scanning, and non-destructive testing methods like ultrasonic or X-ray scanning to detect internal defects or voids.

Maintaining structural integrity in 3D concrete printing requires a deep, multidisciplinary understanding of materials, processes, environmental dynamics, and structural behavior. Unlike traditional construction, where many adjustments can be made manually, 3DCP relies on precision, automation, and predictive design to achieve the same outcomes. As the technology matures, continuous improvements in material formulations, printing systems, and design strategies will further enhance the structural reliability of 3D-printed buildings. Ultimately, the convergence of digital fabrication, smart monitoring, and engineering science promises to make 3D concrete printing a mainstream, safe, and robust construction method for the future.

2.7 Finishing in 3D Concrete Printing

The finishing phase in 3D concrete printing (3DCP) is pivotal in transforming a raw printed structure into a refined, aesthetically appealing, durable, and functional final product. Unlike conventional concrete construction, where molds and vibration-based compaction typically provide smooth surfaces and uniform finishes, 3DCP results in a layer-by-layer texture that demands specialized post-processing. The finishing process addresses this challenge through systematic steps involving surface treatment, curing, and detailing. Each of these components plays a crucial role in ensuring the structural performance, visual quality, and serviceability of the printed structure.

2.7.1 Surface Treatment

The surface of a 3D-printed concrete structure inherently carries the horizontal ridges and contours created by sequential layer deposition. Surface treatment methods are employed to enhance both the visual and functional attributes of the structure. These include smoothing, polishing, surface coatings, and intentional texturing, depending on the design goals and environmental requirements.

2.7.1.1 Smoothing

Post-print smoothing is often required to eliminate roughness, reduce the prominence of layer lines, and prepare the surface for further treatments such as painting or cladding. This may involve manual or mechanical methods such as sanding, grinding, or using surface finishing machines. For example, in residential 3D-printed structures, external and internal walls may be smoothed to create a uniform appearance, similar to conventional plastered walls. Unlike traditional cast-in-form concrete, 3D-printed surfaces are not inherently smooth and thus require more rigorous post-processing.

2.7.1.2 Polishing

Polishing is utilized in applications where a glossy or reflective finish is desired, particularly for floors or architectural features in commercial buildings. This process involves using progressively finer abrasives to remove imperfections and refine the surface to a high sheen. While polishing is commonly used in conventional concrete floors, it is more labor-intensive for 3D-printed concrete due to its rougher initial texture. Nonetheless, with proper surface preparation, equally high-quality finishes can be achieved.

2.7.1.3 Surface Coatings

Surface coatings serve both protective and aesthetic functions. Protective coatings may include water repellents, anti-carbonation barriers, or UV-resistant finishes that prolong the service life of the structure. Decorative coatings, such as stains and paints, enhance the visual appeal and allow for a broader palette of design options. In the case of 3D-printed infrastructure, such as pedestrian bridges, coatings help prevent degradation due to environmental exposure and heavy usage. Due to the uneven texture of printed layers, coating application in 3DCP often requires additional care and potentially multiple layers to ensure full coverage.

2.7.1.4 Texturing

Beyond smoothing and polishing, surface textures may be intentionally introduced to serve specific functions. Anti-slip textures can be incorporated into walking surfaces, while decorative patterns or engravings can enhance façades. The layer-wise printing approach in 3DCP provides opportunities to integrate custom surface textures during the printing process itself or post-print through mechanical means. This level of customization is significantly more complex and costly in conventional construction, making it a unique advantage of 3DCP.

2.7.2 Curing

Curing is an essential phase in the lifecycle of concrete, determining its strength development, durability, and dimensional stability. In 3DCP, curing must be carefully managed to address the unique characteristics of printed elements, such as high surface area exposure and rapid setting mixes. Uniform and controlled curing is vital to avoid early-age cracking, incomplete hydration, and compromised structural performance.

2.7.2.1 Controlled Curing Environment

One of the most effective approaches in 3DCP is the use of controlled environments—curing tents or enclosures that maintain optimal temperature and humidity levels. This is particularly important for large printed elements like walls, columns, or building envelopes. Controlled curing ensures consistent strength development across all layers and minimizes differential shrinkage or thermal gradients. In contrast, conventional curing methods often rely on periodic water spraying or wet coverings, which may not be feasible or efficient for complex 3D-printed geometries.

2.7.2.2 Curing Compounds and Membranes

Surface-applied curing compounds form a thin film that retains internal moisture by reducing evaporation. These compounds are especially useful in outdoor or arid environments where rapid drying poses a risk. Membrane-forming compounds or plastic sheets can be applied immediately after printing to shield the surface from premature drying. Although similar techniques are employed in traditional concrete construction, their precise application is more critical in 3DCP due to the layered build-up and potential for non-uniform curing.

2.7.2.3 Steam Curing

In industrial-scale 3D printing or precast applications, steam curing can be used to accelerate the setting and early strength gain of concrete. Steam is applied in controlled chambers to raise the temperature and humidity, thereby enhancing hydration kinetics. This method is highly beneficial for producing load-bearing elements or components that require rapid deployment. Steam curing is less common in on-site conventional construction but finds parallel applications in precast yards.

2.7.3 *Detailing*

Detailing is the final step that converts a printed structure into a fully functional built environment. It involves integrating all building services, fixtures, and aesthetic enhancements while aligning with the design intent.

2.7.3.1 Fixtures and Fittings

Installation of fixtures and systems such as windows, doors, plumbing, and electrical conduits is carefully planned during the design stage of 3DCP. Openings for these elements are pre-programmed into the print model, allowing for precise placement. This contrasts with conventional construction, where adjustments are often made manually on-site. In a 3D-printed house, for instance, service lines can be embedded or routed through predefined channels, reducing labor and minimizing errors.

2.7.3.2 Finishing Touches

Additional treatments such as painting, wallpapering, or applying decorative panels are conducted based on the intended interior or exterior design. These finishing touches enhance the visual quality and user experience of the built space. However, they may require additional surface preparation to address the texture of printed concrete. Special primers or undercoats are sometimes necessary to ensure adhesion and durability of finishes.

2.7.3.3 Functional Enhancements

To improve energy performance and comfort, insulation materials, soundproofing layers, and thermal barriers may be added post-printing. The geometry and precision of 3DCP enable seamless incorporation of such enhancements, either within cavity walls or as external layers. In contrast, conventional construction often retrofits these features post hoc, which can compromise thermal bridges or introduce inefficiencies.

2.7.3.4 Landscaping and Site Work

The finishing stage may also include site development such as laying pathways, creating garden spaces, and integrating outdoor utilities. For 3D-printed community facilities or public buildings, coordinated landscaping reinforces the architectural intent and enhances usability. As with traditional construction, this phase involves collaboration between civil, architectural, and landscaping teams but benefits from the upfront planning enabled by digital design in 3DCP.

Fig. 2.15 3D-printed homes—examples of exterior and interior finishing

2.7.4 Case Examples of Finishing Applications

2.7.4.1 3D-Printed Residential Buildings

In residential projects, finishing begins with smoothing and cladding of external walls, followed by interior painting and fixture installation. Curing tents are often used to regulate conditions during the early phases, and detailing ensures integration of utilities and furniture. The result is a fully habitable structure that rivals conventional homes in comfort but exceeds in customization and material efficiency (Fig. 2.15).

2.7.4.2 3D-Printed Bridges

For infrastructure such as pedestrian bridges, surface treatments include the application of protective sealants to resist moisture ingress and corrosion. Steam curing is often adopted to accelerate strength gain, while detailing involves the installation of railings, lighting, and pathway markings. The ability to customize form and texture provides both engineering robustness and architectural distinction (Fig. 2.16).

2.7.4.3 3D-Printed Commercial Buildings

In commercial buildings, floors may be polished to achieve a premium look, while walls receive surface coatings to improve aesthetics and durability. HVAC systems, partitions, and elevator shafts are integrated during the detailing phase. Finishing in these structures is aligned with occupancy requirements and leverages the precision of 3DCP for quicker, more coordinated construction (Fig. 2.17).

Fig. 2.16 3D-printed bridges—finishing treatments and safety enhancements

Fig. 2.17 3D-printed commercial buildings—high-performance finishing features [7]

Finishing in 3D concrete printing is not merely a cosmetic step but a comprehensive process that influences the durability, aesthetics, and performance of the entire structure. Unlike conventional methods, 3DCP introduces unique challenges due to its textured, layered surfaces and rapid production cycles. However, it also opens up new opportunities for precision, customization, and design freedom. Effective surface treatment, controlled curing, and thoughtful detailing ensure that the final product meets the functional and visual standards expected of modern construction. As technologies continue to evolve, the scope and sophistication of finishing techniques in 3DCP will further expand, enabling the realization of more resilient, beautiful, and sustainable structures. The following table (Table 2.1) summarizes the comparison between finishing in 3DCP and conventional concrete construction [2].

Table 2.1 Comparative overview of finishing in 3DCP versus conventional concrete construction

Aspect	3D concrete printing (3DCP)	Conventional concrete construction
Surface texture	Naturally layered; rough without treatment	Smooth (mold-cast); minimal treatment needed
Smoothing methods	Requires sanding/grinding due to ridges	Often not required due to formwork finish
Polishing	Requires more preparation due to surface texture	Easier to polish due to flat cast surfaces
Surface coatings	Application requires attention to layer lines	Uniform coating due to flat surfaces
Texturing	Custom patterns integrated during printing	Needs manual imprinting or formwork design
Curing control	Needs careful moisture retention per layer	Typically cured with wetting or membranes
Use of curing compounds	Essential to prevent rapid surface drying	Applied commonly, but with more margin for error
Steam curing	Widely used for early strength in 3DCP elements	Mostly limited to precast environments
Integration of fixtures	Predefined cavities allow precise integration	Manual chiseling and cutting often required
Functional enhancements	Easier to embed insulation/sound barriers	Often added post-construction
Landscaping & Site Work	Seamless due to digital planning	Conventional planning post-structure completion
Aesthetic customization	High flexibility in decorative forms/textures	Limited to mold shapes and surface rendering

References

1. Rangel B, Guimarães AS, Lino J, Santana L (2023) 3D printing for construction with alternative materials. Springer
2. Chourasia A, Pal B, Kapoor A (2025) Influence of printing direction and interlayer printing time on the bond characteristics and hardened mechanical properties of agro-industrial waste-based 3D printed concrete. J Mater Civ Eng ASCE 37:04025065. https://doi.org/10.1061/JMCEE7/MTENG-18196
3. Kruger J, Zeranka S, van Zijl G (2020) A rheology-based quasi-static shape retention model for digitally fabricated concrete. Constr Build Mater 254:119241
4. Jayathilakage R, Rajeev P, Sanjayan JG (2020) Yield stress criteria to assess the buildability of 3D concrete printing. Constr Build Mater 240:117989
5. Mahmoodzadeh F, Chidiac SE (2013) Rheological models for predicting plastic viscosity and yield stress of fresh concrete. Cem Concr Res 49:1–9
6. World's first 3D-printed office building completed in Dubai (n.d.). https://newatlas.com/3d-printed-office-dubai-completed/43522/. Accessed 3 Dec 2025
7. Sanjayan JG, Nazari A, Nematollahi B (2019) 3D concrete printing technology: construction and building applications. Butterworth-Heinemann

Chapter 3
Materials for 3DPC

3.1 Overview

The optimal performance of 3DPC technology is contingent upon the proper mix proportioning of cement-based materials, aggregates, admixtures, and other materials with good properties to achieve fluidity, ease of extrusion, form a robust build, and quick set to satisfy process demands, along with sufficient strength, stiffness, and durability. Consequently, the choice of material in 3DPC should be based on the fact that the developed mix satisfies the requirements of both rheological and printing performance and hardened mechanical and durability properties. This chapter introduces the various materials that could potentially be used in fabricating the mix to be used for 3DPC. Understanding the role of each of the constituents and their interactions plays a pivotal role in designing the mix for 3DPC. The various employed materials in 3DPC are illustrated in the subsequent sections. The choice of material depends on the specific requirements of the construction project, such as load-bearing capacity, environmental conditions, sustainability considerations, and architectural design.

3.2 Cement

Cement serves as the primary component in 3D printable concrete (3DPC) due to its essential role as a binding agent. For 3DPC applications, any material capable of functioning as a cementitious binder can be employed—ranging from conventional ordinary Portland cement (OPC) of various grades (e.g., 43 and 53 grades) to newly developed, more sustainable alternatives. Such greener binder is basically made of various agro-industrial waste ranging from fly ash (FA) [1], ground granulated blast furnace slag (GGBFS) [2], ladle furnace slag (LF) [3], lead–zinc tailings [4], etc., to bagasse ash (BA) [5], rice husk ash (RHA) [6], hazelnut shell ash [7], etc. Details

A. Chourasia et al., *Advances and Challenges in 3D Printed Concrete (3DPC)*, Materials Horizons: From Nature to Nanomaterials, https://doi.org/10.1007/978-981-95-8557-1_3

on the utilization of various such agro-industrial waste in 3DPC and their role are discussed in detail in Chap. 9. When any aforementioned cementitious material reacts with water and a suitable chemical activator, hydration or polymerization-type reaction occurs that results in the development of time-dependent rheology or mechanical properties required for 3D printing construction. Through the reaction mechanism, CSH or polymeric-gel-type products are formed, which helps in achieving the desired fluidity and hardened properties. Further, the kind of raw material employed as cementitious material (e.g., chemical, mineralogical, and morphological characteristics) and curing technique utilized have a big impact on the 3D-printed object's overall strength. Some of the key aspects with regard to the role of cement in 3DPC are illustrated below:

i. **Binding Agent**: Cement acts as the primary binding agent that holds the aggregate and other components of the concrete mix together, giving the printed structure its strength and durability. Modifications in the cement mix, such as adding admixtures, can enhance the final product's strength, durability, and resistance to environmental factors.
ii. **Workability**: The properties of the cement used can significantly affect the workability of the concrete mix. For 3D printing, the mix needs to be extrudable and flowable enough to pass through the printer nozzle yet maintain its shape after deposition.
iii. **Setting Time**: The setting time of the cement is critical in 3DPC. The mix needs to be set quickly enough to support the weight of subsequent layers without deforming, but should be slow enough such that it does not clog the printer or cause interruptions in the printing process.
iv. **Adhesion Between Layers**: Cement helps ensure good adhesion between the layers of printed material, which is essential for the structural integrity of the final construction. Proper adhesion prevents delamination and ensures the building can withstand loads and stresses.
v. **Customization**: Different types of cement and additives can be used to customize the properties (e.g., strength, durability, and resistance to environmental factors) of the printed concrete for specific applications. For instance, high-performance cement can be used for structures requiring extra strength, while eco-friendly cements can be used to reduce the environmental impact.
vi. **Compatibility with Additives**: Cement in 3DPC must be compatible with various additives and admixtures, such as fibers for reinforcement, superplasticizers for improved flow, and retarders or accelerators for controlling setting time.

Overall, cement is integral to the success of 3DPC, influencing the mechanical properties, printability, and longevity of the structures created through this innovative construction method.

3.3 Sand

Another major constituent of 3DPC, which forms a significant portion of the volume of concrete, is sand. It plays an essential role in improving the 3D-printed object's stability and robustness. In addition to ensuring that the concrete mix is densely and solidly packed, well-graded sand particles prevent material from clogging the hose pipe. Furthermore, the strength of a 3D-printed object is also significantly influenced by the choice of an appropriate sand-to-binder ratio. Here are the key aspects of sand's role in 3DPC:

i. **Consistency and Stability**: The inclusion of sand in the concrete mix improves its stability and consistency. It prevents excessive shrinkage and cracking by providing a more balanced and cohesive mix.
ii. **Workability**: The presence of sand in the mix improves its workability, making it easier to extrude through the 3D printer nozzle. Shape, size, gradation, and right proportion of sand ensure a smooth, consistent flow, which is essential for creating uniform and precise layers during printing.
iii. **Strength and Durability**: Sand contributes to the overall compressive strength of the printed concrete. It also plays a role in enhancing the durability of the structure by providing resistance to weathering and environmental factors.
iv. **Economical Filler**: Sand is an inexpensive and readily available material, making it a cost-effective component of the 3D printing mix. Its use helps to reduce the overall cost of the concrete while maintaining the desired mechanical properties.
v. **Optimizing Mix Design**: The grain size and shape of the sand can be tailored to optimize the concrete mix design for 3D printing. For example, well-graded sand with a good distribution of particle sizes can improve the packing density and reduce voids in the printed material.
vi. **Surface Finish**: Sand helps achieve a smoother surface finish on the printed layers. Finer sand can lead to a finer surface texture, which is important for both aesthetic and structural reasons.

In summary, sand is an essential component in 3DPC concrete mixes, contributing to the material's workability, strength, surface finish, and overall performance. Its role is integral to ensuring the success and viability of 3D-printed construction projects.

3.4 Water

In 3D printed construction (3DPC), water plays a vital role in the concrete mix, impacting its properties and the overall success of the printing process. Here are the key aspects of water's role in 3DPC:

i. **Hydration/Reactivity of Cementitious Material**: In the case of OPC, water initiates the chemical reaction known as hydration, which is essential for the hardening and strength development of cement. Without adequate water, the cement cannot properly hydrate, leading to weak or incomplete bonding. Whereas in the case of an agro-industrial waste-based alkali system, water acts as a transport medium of alkali ions that result in the initiation of the polymerization reaction between the constituents and consequently impacts the development of overall mechanical and printing characteristics.
ii. **Workability**: The amount of water in the mix affects its workability and fluidity. For 3DPC, the mix needs to be fluid enough to be extruded through the nozzle but stiff enough to hold its shape after deposition. The water-to-cement ratio is carefully controlled to achieve this balance.
iii. **Setting Time**: Water influences the setting time of the concrete mix. The right amount of water ensures that the concrete sets within a desirable timeframe, allowing for efficient layer-by-layer construction without premature hardening or excessive delay.
iv. **Consistency**: Consistent water content throughout the mix is crucial for uniformity in the printed structure. Variations in water content can lead to inconsistent layer thicknesses and weaker bonds between layers.
v. **Adhesion Between Layers**: Water helps in maintaining good adhesion between successive layers of printed material. Proper moisture levels ensure that each new layer bonds well with the previous one, contributing to the overall structural integrity.
vi. **Temperature Control**: Water can help in controlling the temperature of the concrete mix, especially in hot conditions. It aids in preventing the mix from drying out too quickly, which can cause cracks and reduce strength.
vii. **Additive Activation**: Many admixtures and additives used in 3DPC, such as superplasticizers or accelerators, require water to be effective. The right water content ensures these additives work as intended to enhance the properties of the concrete mix.
viii. **Surface Finish**: The presence of water affects the surface finish of the printed concrete. Properly balanced water content helps achieve a smooth, even surface on the printed layers, which is important for both aesthetic and structural reasons.

Hence, it can be stated that water is also a critical component in 3DPC, influencing the mix's hydration, workability, setting time, strength, consistency, and overall performance. Careful control of water content is essential for achieving the desired properties and successful outcomes in 3D-printed construction projects.

3.5 Admixtures

Admixtures are essential in 3DPC for optimizing the concrete mix's properties, ensuring workability, controlling setting time, enhancing strength and durability, improving layer adhesion, and achieving the desired surface finish. They enable the production of high-quality, reliable, and efficient 3D-printed structures. The precise combination and dosage of these admixtures are tailored to optimize the performance of the concrete mix for 3D printing applications.

Admixtures are chemical or natural substances added to the concrete mix to modify its properties and make it more suitable for the 3D printing process. Several specific admixtures are commonly used in 3D printed construction (3DPC) to enhance the properties of the concrete mix. Here are some of the key types of admixtures commonly employed in 3DPC along with their key roles.

3.5.1 Superplasticizer

Superplasticizer (SP), also called high-range water reducer (HRWR), is a necessary ingredient in 3DPC mix that regulates workability and plays a crucial role in making the mix pumpable and extrudable without compromising its strength or stability. SP helps to reduce the viscosity of the mix and thereby increases the flowability and extrusion of 3DPC. It also improves the rheological properties of the concrete mix. In 3DPC, the commonly employed SPs are polycarboxylate ether (PCE) superplasticizers, sulfonated melamine formaldehyde (SMF), and sulfonated naphthalene formaldehyde (SNF). Some of the key roles of superplasticizers in 3DPC are illustrated below:

- **Precision**: The improved flow properties and workability with the addition of SP ensure that the concrete can be extruded in thin, precise layers, which is essential for achieving the desired geometries in 3D-printed structures.
- **Structural Integrity**: By reducing the water-to-cement ratio, superplasticizers help in maintaining the strength and durability of the printed structure. Lower water content also reduces shrinkage and cracking. Further, some superplasticizers can be tailored to adjust the setting time of the concrete. This allows for better control over the curing process, which is crucial for building up layers quickly without deforming the lower layers.
- **Speed**: Faster setting times and improved flow allow for quicker printing processes, which can speed up construction timelines. In 3DCP, the concrete mixture needs to be pumped through hoses to the printer head. Superplasticizers help in maintaining a consistent flow, preventing blockages and ensuring a continuous printing process.
- **Material Savings**: Better workability and flow can reduce the need for additional materials to achieve the desired consistency, leading to cost savings.

3.5.2 Retarders

Retarders, also known as set-retarding admixtures, play a crucial role in 3D concrete printing (3DCP) by controlling the setting time of the concrete mix. This is particularly important in 3DCP, where the balance between workability and buildability needs to be carefully managed. Among various types of retarders, phosphatases, lignosulfonates, borates are commonly used in 3DPC. Here is an overview of their role and benefits in 3DCP:

- **Enhanced Precision**: The extended workability provided by retarders allows for more precise control over the extrusion process, leading to higher accuracy in the printed structure. Retarders slow down the hydration process of cement, providing a longer working time for the concrete mix. This extended workability is essential for ensuring the smooth and continuous flow of concrete through the printer, especially in large-scale or complex projects.
- **Improved Structural Integrity**: By delaying the setting time, retarders help ensure that successive layers of printed concrete bond effectively. This prevents cold joints and ensures a monolithic structure with consistent strength throughout.
- **Operational Efficiency**: Reduced risk of blockages and increased flexibility in construction scheduling contribute to more efficient and reliable 3D printing operations. In the absence of retarders, the concrete mix might set too quickly, leading to blockages in the pumping system or the printer nozzle. Retarders help mitigate this risk by keeping the mix workable for a longer period. Consequently, if there are delays or interruptions in the printing process, the extended setting time ensures that the mix remains workable and ready for use.

3.5.3 Accelerators

Accelerators, also known as set-accelerating admixtures, are used in 3D concrete printing (3DCP) to speed up the setting and hardening process of the concrete mix. This can be particularly beneficial in applications where rapid strength development is required, or where faster layer buildup is essential to support the structure during the printing process. This is particularly important for early load-bearing capacity and for projects that require rapid progression. Various commonly utilized accelerators are calcium chloride, sodium thiocyanate, and calcium nitrate. Here is an overview of their role and benefits in 3DCP:

- **Improved Buildability**: Accelerators help ensure that each layer of printed concrete sets quickly enough to support the subsequent layers, reducing the risk of deformation or collapse during the printing process. Thus, accelerators enhance the buildability of 3D-printed structures by providing quicker strength gain, which helps in maintaining the shape and stability of the layers.

- **Improved Integrity**: Accelerators reduce the initial and final setting times of concrete. This allows for faster layer buildup, which is crucial in 3DCP to maintain the shape and integrity of the printed structure.
- **Adaptability**: Accelerators can be particularly useful in cold-weather conditions where the natural setting and hardening process of concrete would be slowed down.
- **Increased Efficiency**: Faster setting times allow for a reduction in the overall printing time, increasing efficiency and productivity in construction projects.

3.5.4 Air-Entraining Agents

Air-entraining agents (AEAs) are admixtures used in concrete to introduce and stabilize microscopic air bubbles within the mix. In the context of 3D concrete printing (3DCP), these agents can offer specific benefits, although their use must be carefully managed to avoid potential drawbacks. Vinsol resin, wood resins, synthetic detergents, and fatty acids are some of the commonly employed air-entraining agents in 3DPC. Here is an overview of their role, benefits, and challenges in 3DCP:

- **Improved Extrusion**: Air-entraining agents improve the workability of the concrete mix by increasing its plasticity. This is particularly beneficial in 3DCP, where a highly workable mix is necessary for smooth extrusion and layer deposition. Increased workability and cohesion can lead to more precise and stable extrusion, resulting in higher-quality printed layers and overall structure.
- **Consistency**: The presence of air bubbles can improve the cohesion of the concrete mix, reducing segregation and bleeding. This ensures a more uniform and consistent mix during the printing process.
- **Reduced Density**: Air-entrained concrete has a lower density, which can be advantageous for certain applications where reduced weight is desirable.
- **Durability**: The microscopic air bubbles introduced by AEAs provide spaces for water to expand upon freezing. This reduces internal stresses and helps prevent cracking, which is important for the durability of 3D-printed structures in cold climates. Enhanced freeze–thaw resistance leads to longer-lasting structures, especially in environments subject to cyclic freezing and thawing.

3.5.5 Silica Fume

Silica fume, also known as micro silica, is a by-product of silicon and ferrosilicon alloy production. It is a fine amorphous powder with particles approximately 100 times smaller than the average cement particle. It is comprised of silicon dioxide, a primary component of sand and most rocks. It is commonly used as a pozzolanic material in concrete to improve strength, reduce permeability, and enhance durability. Incorporating silica fume in 3DPC boosted the plastic viscosity and yield stress while decreasing the flowability of cementitious mixes. Therefore, it can serve as an inorganic viscosity modifier in concrete applications particularly for underwater concreting and shotcrete. Here is a detailed look at its role and benefits:

- **Structural Integrity**: The fine particles of silica fume improve the bond between the cement paste and aggregate, as well as between successive layers in 3D printing. This ensures a monolithic structure with fewer weak interfaces.
- **Improved Mechanical and Durability Performance**: Silica fume significantly increases the compressive, tensile, and flexural strength of concrete. Its fine particles fill the voids between cement grains, leading to a denser and stronger matrix. Further, the addition of silica fume enhances the durability of concrete by reducing permeability. This makes the concrete more resistant to chemical attacks, chloride ingress, and other environmental factors. The increased strength and reduced permeability result in more robust and durable structures, capable of withstanding greater loads and harsh environments.
- **Improved Printing Characteristics**: While silica fume can make the mix more cohesive, it may reduce workability. However, when used in combination with superplasticizers, it helps achieve a balance between workability and stability, essential for smooth extrusion and buildability in 3DCP.
- **Precision**: Silica fume improves the cohesiveness of the concrete mix, reducing the risk of segregation and bleeding. This is particularly important in 3DCP, where a consistent mix is crucial for layer-by-layer deposition. The improved cohesiveness and reduced segregation help maintain the desired shape and dimensions of the printed layers, leading to higher precision and accuracy in the final structure.
- **Surface Finish**: Silica fume can contribute to a smoother surface finish, reducing the need for additional post-processing.
- **Material Efficiency**: The pozzolanic reaction of silica fume with calcium hydroxide, a by-product of cement hydration, contributes to the overall strength and durability of the concrete. This can reduce the amount of cement needed, leading to cost savings and environmental benefits.

3.5.6 Shrinkage-Reducing Admixtures

Shrinkage-reducing admixtures (SRAs) are additives used in concrete to mitigate the shrinkage that occurs during the curing process. In the context of 3D concrete printing (3DCP), where precise dimensional stability is crucial, SRAs can offer significant benefits. Polyethylene glycol (PEG) and calcium sulfoaluminate are a few of the generally employed shrinkage-reducing admixtures in 3DPC. Here is an overview of their role, benefits, challenges, and considerations in 3DCP:

- **Dimensional Stability**: SRAs reduce the water content needed for a given workability, which in turn decreases the drying shrinkage of concrete. This is critical in 3DCP to prevent dimensional changes and maintain accuracy in printed structures leading to higher-quality and more accurate printed components.
- **Enhanced Performance**: Further, by reducing shrinkage, SRAs help to mitigate cracking, which can compromise the durability and structural integrity of 3D-printed components.
- **Improved Printing Quality**: Some SRAs can improve the workability of concrete mixtures, allowing for better flow and extrusion properties during the printing process. This contributes to smoother layer deposition and reduced risk of printing defects.

3.5.7 Rheology Modifiers

Rheology modifiers are additives used in concrete to control its flow properties, viscosity, and workability. In the context of 3D concrete printing (3DCP), where precise control over material flow and deposition is crucial, rheology modifiers play a significant role in optimizing printability and achieving high-quality printed structures. Viscosity-modifying admixture (VMA) adjusts the viscosity of the mix by absorbing water from it. Additionally, it improves the extrusion-based mix's buildability. Some of the employed rheology modifiers are hydroxypropyl methylcellulose (HPMC), xanthan gum, and welan gum. Here is an overview of their role, benefits, challenges, and considerations in 3DCP:

- **Improved Precision**: Rheology modifiers alter the viscosity and rheological properties of the concrete mix, ensuring that it flows smoothly and consistently during the printing process. Thus, rheology modifiers enable more precise control over material flow and layer deposition, resulting in accurate geometries and dimensional consistency in printed structures.
- **Reduced Print Failures**: By improving the workability and extrudability of the concrete, rheology modifiers facilitate better control over the printing process. This reduces the risk of print failures, such as clogging or uneven deposition, and enhances overall print quality.
- **Optimized Performance**: By optimizing the rheological characteristics of the concrete, these additives contribute to better overall performance, including strength development and durability of printed components.

3.5.8 Bonding Agents

In the context of 3D concrete printing (3DCP), bonding agents are crucial additives that enhance the adhesion and bonding between successive layers of printed concrete. Bonding agents enhance the adhesion between layers of printed concrete, minimizing the risk of delamination or separation. This is critical for maintaining the structural integrity and overall strength of the printed structure. Effective bonding agents help distribute stresses more evenly across the interface between layers, reducing the likelihood of cracking and improving the durability of printed components. Further, by promoting strong interlayer bonding, bonding agents contribute to the overall stability, load-bearing capacity, and long-term performance of 3D-printed structures. Styrene-butadiene rubber (SBR) and acrylic polymers are some of the bonding agents utilized in 3DPC. Here is an overview of their role, benefits, challenges, and considerations in 3DCP:

- **Increased Structural Integrity**: Bonding agents ensure that successive layers adhere well to each other, resulting in a monolithic structure with enhanced strength and resistance to external forces.
- **Improved Print Quality**: Strong interlayer bonding leads to smoother and more consistent layer deposition, reducing the risk of print defects and enhancing the overall quality of printed components.
- **Enhanced Durability**: Minimizing layer separation and improving bonding help in achieving greater durability and longevity of 3D-printed structures, particularly in demanding applications.

3.5.9 Water Retaining Agents

Water-retaining agents (WRAs) are additives used in concrete to maintain the moisture content within the mixture, especially during the initial curing stages. In 3D concrete printing (3DCP), where precise control over hydration and setting times is crucial, WRAs play a significant role in ensuring optimal workability, hydration, and strength development. Cellulose ethers and starch ethers are a few of the water-retaining agents that can be used in 3DPC. Here is an overview of their role, benefits, challenges, and considerations in 3DCP:

- **Enhanced Printability**: WRAs help to retain water within the concrete mix, preventing rapid evaporation and maintaining adequate moisture levels for hydration. This is particularly important in 3DCP to ensure consistent flow properties and workability throughout the printing process. Hence, WRAs contribute to smoother extrusion and deposition of concrete layers by maintaining consistent flow characteristics and preventing premature drying or stiffening.

Further, by slowing down the drying process, WRAs extend the workable time of the concrete mix. This allows for better control over the printing process, reducing the risk of clogging or interruption during layer deposition.

- **Reduced Shrinkage and Cracking**: By controlling moisture loss, WRAs help minimize shrinkage and cracking in printed components, ensuring dimensional stability and improving overall structural integrity.
- **Uniform Quality**: Ensuring uniform moisture distribution throughout the printing process helps achieve consistent material properties and surface finish in 3D-printed structures.
- **Improved Mechanical Properties**: Maintaining optimal moisture conditions promotes proper hydration of cement particles, which enhances the strength and durability of printed structures. This is critical for achieving desired mechanical properties in 3DCP.

3.5.10 Anti-washout Admixtures

Anti-washout admixtures (AWAs) are additives used in concrete to enhance its cohesiveness and prevent segregation or separation of ingredients when placed underwater or in situations where excessive water might cause material loss. In the context of 3D concrete printing (3DCP), where precise control over material consistency and deposition is critical, AWAs can play a significant role in ensuring stable printing conditions and high-quality printed structures. Polyvinyl alcohol (PVA) and polyethylene oxide (PEO) are some of the anti-washout admixtures employed in 3DPC. Here is an overview of their role, benefits, challenges, and considerations in 3DCP:

- **Optimized Printability**: AWAs enhance the pumpability and extrudability of the concrete, facilitating smoother material flow through the printing nozzle and ensuring continuous and reliable layer deposition. AWAs contribute to stable and uniform material flow, reducing the risk of nozzle clogging or interruptions during the printing process.
- **Enhanced Structural Integrity**: By improving cohesion and reducing water segregation, AWAs help achieve stronger interlayer bonding and overall structural integrity in printed components.
- **Minimized Material Waste**: Preventing washout and material loss ensures efficient use of materials and reduces the need for rework or adjustments during printing.

3.5.11 Corrosion Inhibitors

Corrosion inhibitors are additives used in concrete to protect embedded reinforcement (such as steel bars or mesh) from corrosion caused by exposure to aggressive environments, such as moisture, chlorides, and carbonation. In 3D concrete printing (3DCP), where the durability and long-term performance of printed structures are critical, corrosion inhibitors can play a vital role in extending the service life and maintaining the structural integrity of the components. Calcium nitrite, amines, and phosphates are a few of the corrosion inhibitors that are commonly employed in 3DPC. Here is an overview of their role, benefits, challenges, and considerations in 3DCP:

- **Extended Service Life**: Corrosion inhibitors protect embedded reinforcement from corrosion, significantly extending the service life of 3D-printed structures.
- **Reduced Maintenance**: By minimizing the need for repairs and replacements due to corrosion damage, inhibitors help reduce maintenance costs and downtime for printed components.
- **Improved Sustainability**: Enhancing the durability of concrete through corrosion inhibition promotes sustainable construction practices by increasing the longevity of structures and reducing material waste.
- **Challenges**: Corrosion inhibitors should be compatible with other admixtures and materials used in the concrete mix, ensuring that they do not adversely affect setting time, workability, or mechanical properties.

3.6 Summary of the Role of Various Constituents

- **Workability Enhancement**: Admixtures like superplasticizers or water reducers improve the workability and flowability of the concrete mix without increasing the water content. This ensures that the mix can be easily extruded through the 3D printer nozzle and form precise layers.
- **Setting Time Control**: Retarders or accelerators are added to control the setting time of the concrete mix. Retarders slow down the setting process, which is useful in hot climates or for large-scale prints. Accelerators speed up the setting time, which helps in faster layer bonding and quicker construction.
- **Strength Improvement**: Certain admixtures, such as silica fume or fly ash, can enhance the mechanical properties of the concrete, including compressive strength, tensile strength, and durability. This is crucial for ensuring the structural integrity of the printed elements.

- **Durability Enhancement**: Admixtures like air-entraining agents can improve the durability of the concrete by introducing tiny air bubbles. These bubbles help the concrete withstand freeze–thaw cycles and reduce the risk of cracking.
- **Adhesion Between Layers**: Bonding agents or other specialized admixtures can improve the adhesion between layers of the printed concrete. This ensures that each new layer bonds well with the previous one, maintaining the overall strength and stability of the structure.
- **Rheology Modification**: Rheology-modifying admixtures adjust the viscosity and thixotropy of the concrete mix. This helps maintain the mix's shape after extrusion and prevents it from collapsing or deforming, ensuring accurate layer deposition.
- **Shrinkage Reduction**: Shrinkage-reducing admixtures minimize the risk of shrinkage cracks during the drying process. This is important for maintaining the dimensional stability and aesthetic quality of the printed structure.
- **Water Retention**: Some admixtures help retain water within the concrete mix, which is beneficial for the curing process. Proper curing enhances the strength and durability of the printed structure.
- **Surface Finish Improvement**: Admixtures can also influence the surface finish of the printed concrete. For example, additives that reduce surface tension can help achieve a smoother and more uniform surface.
- **Eco-Friendliness**: Using admixtures like fly ash or slag, which are industrial by-products, can make the concrete mix more sustainable and environmentally friendly by reducing the need for cement.

Table 3.1 provides a summary of studies that utilized ordinary Portland cement (OPC) as the primary binder in combination with various admixtures for the development of 3D printable concrete. Table 3.1 also includes details on the types of materials used, categories of tests performed (such as mechanical, printability, and durability assessments), and key performance metrics of the developed binders. This overview of selected studies not only highlights the feasibility of using cement as a core component in 3DPC but also offers insights into its current capabilities and areas for potential enhancement within the construction sector.

Table 3.1 Summary of the studies employing OPC as primary binder in 3DPC

Mix composition	Different studies conducted and key observations			References
	Mechanical test	Rheological test	Durability test	
• OPC-40% • Calcium aluminate cement (CAC)-60% • Lithium carbonate-4.5% of binder (OPC + CAC) • Water cement ratio-0.31	• Maximum compressive strength = 8.26 MPa	–	• Porosity test-lowest porosity observed at 170% saturation for both shell and core	[8]
• Cement (C)-70% • River sand-1355 kg/m^3 • Limestone (L)-24% • Metakaolin-6% • SP-0.9% of binder (C + L) • VMA-0.15% of binder (C + L) • Water cement ratio-0.35	• Compressive strength = 21–66 MPa • Flexural strength = 11.47–20 MPa • Interlayer adhesion = up to 5.06 MPa		• Void index = 4.32–16.6%	[9]
• Cement-100% • FeO-5% of binder (cement) • Water cement ratio-0.35	• Mechanical testing of 3D-printed steel rebar = yield strength was ~20% lower than conventional steel	• Storage modulus test = 27.6 to 82.2 kPa • Rheology test, shear stress = 200 Pa Interlayer bond strength depends on time gap; leads to anisotropy		[10]

(continued)

Table 3.1 (continued)

Mix composition	Different studies conducted and key observations			References
	Mechanical test	Rheological test	Durability test	
• Cement-50% • Limestone-25% • Kaolin clay-25% • Water cement ratio-0.41 • SP-0.3% of cement		• Initial yield stress 4 kPa • Yield stress increased linearly for ~40 min, then exponentially • Buildability test and failure stress of a layer under vertical load = 4.15–5.60 kPa • Surface fracture monitoring = fractures appeared faster at higher build rates; no fracture at 1.1 m/h rate		[11]
• Cement-30% • Sand-60% • Water cement ratio-0.31 • PVA fiber-0.1% of binder	• Compressive test = 60.4 to 62 Mpa • Bond test = Best bond strength observed for ≤15 min interlayer time gap • Poor bonding after 30–60 min gap	• Slump flow test = 190–200 mm for printability • Open time: 15 min • Crack resistance test = best result with 0.1% PVA (no cracks or clogging)		[12]
• Rapid hardening portland cement (RHPC)-97% • Polyvinyl alcohol (PVA)-3% • Water cement ratio-0.2	• Flexural strength = After 1-day water curing: 0.8 ± 0.1 MPa • After 26 days water curing: 2.2 ± 0.2 MPa • Curing at 80 °C was less effective	• Best green strength with RHPC/PVA ratio of 97:3		[13]

(continued)

Table 3.1 (continued)

Mix composition	Different studies conducted and key observations			References
	Mechanical test	Rheological test	Durability test	
• Cement (C)-96% • Metakaolin (M)-2% • Silica fume (SF)-2% • High-range water reducer (HRWR)-0.024% of binder (C + M + SF) • Thickening agent-0.0125% of binder (C + M + SF) • Sand: binder-1:2 • Water cement ratio-0.35 • PPF fiber-0.4% of binder (C + M + SF)	• Compressive strength (28 days) = 52–64 Mpa • Flexural strength = 7–10.5 MPa		• Freeze–thaw resistance = dynamic modulus: 82.8–92.2% and remaining • Weight loss: 0.3–1.9% • Chloride ion penetration = $3.6 \times 10-^{12}$ m^2/s • Sulfate attack resistance = 3DPC maintained strength for longer, reduction started after 60–90 cycles • Carbonation depth-9–13 mm • Porosity test = total pore volume: 11.2% • Surface area: 5.62 m^2/g Void analysis = Porosity: 5.1%	[14]
• Cement-40% • Sand-60% • Water cement ratio-0.40 • Isopropyl alcohol-10% of binder (cement) • Methyl cellulose (MC)-1.75% of binder (cement)	• Compressive strength = 25 MPa	• Shape Retention = Best shape retention at w/c = 0.26 and 1.0–1.75% MC • Buildability-Arch with 50 mm span successfully printed		[15]

References

1. Shah SFA, Chen B, Oderji SY, Haque MA, Ahmad MR (2020) Improvement of early strength of fly ash-slag based one-part alkali activated mortar. Constr Build Mater 246:118533
2. Reddy MS, Dinakar P, Rao BH (2018) Mix design development of fly ash and ground granulated blast furnace slag based geopolymer concrete. J Build Eng 20:712–722

3. Singh SK, Vashistha P (2021) Development of newer composite cement through mechano-chemical activation of steel slag. Constr Build Mater 268:121147
4. Deng P, Zheng Z (2023). Mechanical properties of one-part geopolymer masonry mortar using alkali-fused lead-zinc tailings. https://doi.org/10.1016/j.conbuildmat.2023.130522
5. de Sande VT, Sadique M, Bras A, Pineda P (2022) Activated sugarcane bagasse ash as efficient admixture in cement-based mortars: mechanical and durability improvements. J Build Eng 59:105082
6. Jittin V, Bahurudeen A (2022) Evaluation of rheological and durability characteristics of sugarcane bagasse ash and rice husk ash based binary and ternary cementitious system. Constr Build Mater 317:125965
7. Yurt Ü, Bekar F (2022) Comparative study of hazelnut-shell biomass ash and metakaolin to improve the performance of alkali-activated concrete: a sustainable greener alternative. Constr Build Mater 320:126230
8. Shakor P, Sanjayan J, Nazari A, Nejadi S (2017) Modified 3D printed powder to cement-based material and mechanical properties of cement scaffold used in 3D printing. Constr Build Mater 138:398–409
9. Barbosa MS, dos Anjos MAS, Cabral KC, Dias LS (2022) Development of composites for 3D printing with reduced cement consumption. Constr Build Mater 341:127775
10. De Schutter G, Lesage K, Mechtcherine V, Nerella VN, Habert G, Agusti-Juan I (2018) Vision of 3D printing with concrete—technical, economic and environmental potentials. Cem Concr Res 112:25–36
11. Perrot A, Rangeard D, Pierre A (2016) Structural built-up of cement-based materials used for 3D-printing extrusion techniques. Mater Struct 49:1213–1220
12. Jo JH, Jo BW, Cho W, Kim J-H (2020) Development of a 3D printer for concrete structures: laboratory testing of cementitious materials. Int J Concr Struct Mater 14:1–11
13. Gibbons GJ, Williams R, Purnell P, Farahi E (2010) 3D Printing of cement composites. Adv Appl Ceram 109:287–290
14. Zhang Y, Zhang Y, Yang L, Liu G, Chen Y, Yu S, Du H (2021) Hardened properties and durability of large-scale 3D printed cement-based materials. Mater Struct 54:1–14
15. Robayo-Salazar R, de Gutiérrez RM, Villaquirán-Caicedo MA, Arjona SD (2023) 3D printing with cementitious materials: challenges and opportunities for the construction sector. Autom Constr 146:104693

Chapter 4
Fresh State Characteristics and Rheology of 3D-Printed Concrete

4.1 Introduction

The successful employment of 3D printed concrete (3DPC) in construction relies not only on its hardened characteristics but also has a strong dependence on the fresh-state behavior of the constituents' materials. In contrast to conventional concrete, where it is placed into formwork and permitted to set, 3DPC is deposited following a layer-by-layer strategy in an open environment, without the need of molds or any type of supporting elements. Consequently, the employed material for 3DPC is required to be designed to simultaneously satisfy two key performance requirements: appropriate rheological performance during the printing process and sufficient mechanical integrity after deposition.

This two-fold requirement has led to the development of "printability" concept, a multi-dimensional terminology encompassing workability, extrudability, buildability, and shape stability of a material. Workability describes the ease with which the mix can be pumped and conveyed through the delivery system. Extrudability signifies the ability of any mix to make a smooth and continuous filament after the exits of the nozzle. Buildability reflects its capability to sustain successive layers without collapsing. Whereas, shape retention represents the mix's capacity to sustain geometric precision after the material extrusion.

Attaining the optimal balance among the fresh state and hardened properties necessitates meticulous design of the mix, which includes a suitable combination of various constituents, e.g., cementitious binders, fine aggregates, supplementary cementitious materials (SCMs), chemical admixtures (superplasticizers, viscosity-modifying agents) along with a carefully selected water-to-binder ratio. Mixes that are too watery in nature may extrude quite easily but possess risk of slump or losing of shape, whereas an excessively stiff material may resist the deformation and retain shape but poses challenges in pumping and extrusion of the material.

In this regard, rheological parameters, such as yield stress, viscosity, and thixotropy, play a pivotal role. These properties govern the flow characteristics of a

A. Chourasia et al., *Advances and Challenges in 3D Printed Concrete (3DPC)*, Materials Horizons: From Nature to Nanomaterials, https://doi.org/10.1007/978-981-95-8557-1_4

mix under the given stress, including its response at rest and during motion along with the strength development of the mix with time. This chapter presents a comprehensive discussion of various such fresh-state characteristics and their role on the printability and structural performance of 3DPC.

4.2 Fresh-State Properties Governing 3DPC Printability

4.2.1 Static and Dynamic Yield Stress

The highest stress required to get a material to flow out of its state of rest is termed as static yield stress. However, below the yield stress, the material does not flow and behaves elastically. During printing, as layers are placed one on top of the other, a rise in compressive stress occur in the lowest layer. The printed layer needs to have a sufficient yield stress so that it can sustain the weight of the concrete layers placed on top of it with negligible deformation. The concrete behavior in 3DPC is typically described by Perrot model and the non-Newtonian fluids based Bingham model [1]. The above models are described through Eqs. (4.1–4.2), as follows:

$$\tau_y = \tau_{y0} + \mu \gamma \tag{4.1}$$

$$\tau_y(t) = A_{thix} t_c \left(e^{t/tc} - 1\right) + \tau_{y0} \tag{4.2}$$

Here, τ_y, γ, and μ stands for the shear stress, shear rate, and plastic viscosity of the material. τ_{y0} represents the initial value of yield stress, i.e., the value corresponds to the first printed layer (Fig. 4.1a). Yield stress (τ_{y0}) of the cementitious mix is a time-dependent function of the fresh state that evolves with the cement hydration (which results in the development of calcium silicate hydrate gel) and due to the flocculation [2], A_{thix} and t_c are the structuration rate and characteristic time, R_{thix} is the re-flocculation rate of 3DPC [3].

The yield stress required to keep a material flowing after it is in motion is called dynamic yield stress (Fig. 4.1b). A material's thixotropy index (I_{thix}) can be determined by dividing the static yield stress by the value of dynamic yield stress, [4]. These yield stress is strongly related with the buildability of the printed concrete, where a higher yield stress results in better buildability. To prevent plastic failure during printing, the yield stress at the bottommost layer after time t must be larger than the corresponding $c(t) + \sigma_n \tan(\phi)$, as per the Mohr–Coulomb failure criteria. Here, $c(t)$, σ_n, and ϕ represent the cohesion, stress acting normal to the plane, and the angle of friction [5].

Researchers have sought to examine the variability of static yield stress with different constituents. In the work of Mahaut et al. [6], it was reported that a cement paste's static yield stress rises when it contains larger particles. Additionally, the authors provided a mathematical relation to predict the static yield stress. Analogous

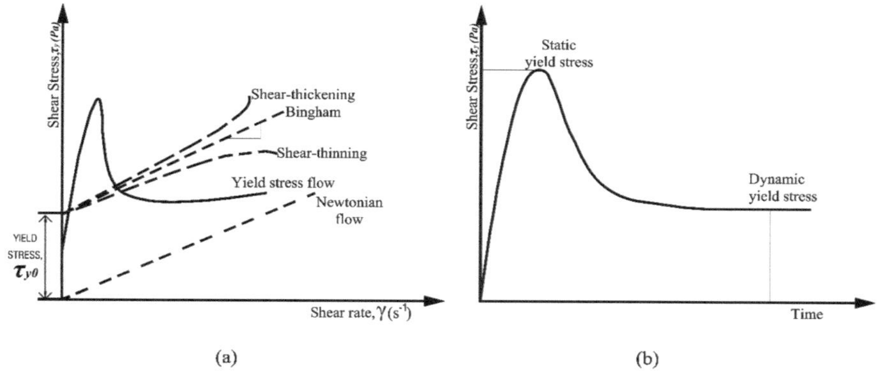

Fig. 4.1 **a** Shear stress vs shear rate of various fluids type, **b** Rheological characteristics of material at fixed value of shear rate

research has endeavored to establish a correlation between the yield stress value of mortar and cement binder [7, 8]. Weng et al. [9] used the Marson-Percy model and the Fuller-Thompson theory to show how silica sand can be used to arrive at a gradation of sand which improves mixture buildability. They consequently put up a mathematical formula for the production of hollow cylindrical columns by printing. In the context of 3D printed concrete (3DPC), the yield stress can change rapidly, with factors ranging from 25 to 150 [10], which are described in Eq. (4.3)

$$\frac{\tau_f}{\tau_i} = \frac{H}{h};\ 25 \leq \frac{\tau_f}{\tau_i} \leq 150 \tag{4.3}$$

where h represents the printed layer's height, τ_i and τ_f stand for the initial and final yield stresses at bottom or first printed layer respectively. This expression, proposed in the context of layer-by-layer printing, relates the final and initial yield stress to the total height of the printed structure and the height of individual layers, providing a predictive tool for assessing build stability.

4.2.2 Penetration Resistance

Penetration resistance is a practical and widely adopted method for evaluating the early stiffness or rigidity of cementitious mixes in their fresh state. The test involves driving a standard needle or plunger into the fresh mix [11] and measuring the resistance encountered. This method is particularly useful in estimating the rate of structural build-up, a critical factor for 3DPC.

Penetration resistance is closely associated with static yield stress; as the material gains stiffness, both the resistance and yield stress increase proportionally. In earlier studies, penetration resistance was primarily investigated in the context of shotcrete; however, the technique has proven equally effective in assessing 3DPC behavior.

The addition of materials like attapulgite clay and calcium sulphoaluminate (CSA) cement has a strong impact on the penetration resistance. Especially, CSA acts as a rapid setting material along with improving the early-age stiffness characteristics and thus it is advantageous for vertical buildability.

Additionally, it has been observed that adding accelerators like aluminum sulfate leads to a corresponding increase in penetration resistance. It has also been found [12] that an increase in the dosage of various accelerators increases the early penetration resistance proportionally, thereby escalating the feasible printing window and decreasing the deformation risk among layers.

4.2.3 Green Strength

The term green strength denotes the compressive strength of the printed mix at early-state, specifically that is developed within first few hours after extrusion. This characteristics serves as a crucial factor in evaluating whether an extruded layer can sustain the weight of subsequent layers without buckling or deformation.

The measurement of green strength is usually performed via uniaxial unconfined compression tests, adapted from soil mechanics. Additionally, non-destructive techniques, such as ultrasonic pulse velocity (UPV), have been employed to track strength development over time. Studies have shown a strong linear relationship between the compressive strength measured destructively and the UPV readings within the first 90 min after mixing.

The green strength curve of a 3D-printed mix generally mirrors the development of static yield stress, suggesting that both properties are governed by similar hydration and flocculation kinetics. Additives such as nanoclay significantly improve green strength by promoting denser particle packing and stronger floc structures. The colloidal nature of nanoclay facilitates rapid water absorption and bridging between particles, leading to early strength gain and better interlayer adhesion.

Refined attapulgite clay has also been demonstrated to boost green strength. Its high surface area and thixotropic properties contribute to enhanced stiffness and resistance to layer collapse, especially under sustained load.

4.2.4 Slump and Slump Flow

Though originally developed for conventional concrete, the **slump test** remains a useful indicator of flow behavior in 3DPC, particularly during the preliminary stages

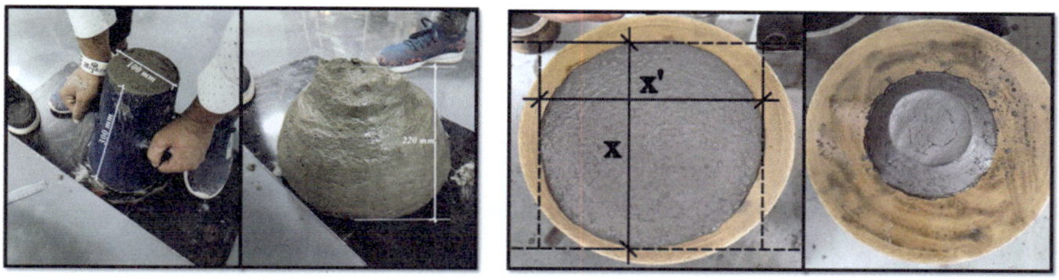

(a) Slump test (b) Flow table test

Fig. 4.2 Different tests used for flowability

of mix design. The slump test evaluates vertical deformation of the material once a standard cone is removed, while the **slump flow** test measures the lateral spread after 25 drops of the cone (Fig. 4.2).

In 3DPC, a high slump value typically indicates **excessive fluidity**, which may enhance pumpability but results in **poor shape retention** and risk of collapse during printing. Conversely, a low slump implies **stiff behavior**, which may impede extrusion but enhances buildability.

A detailed statistical investigation by Tay et al. [13] suggested that a **slump value between 4 and 8 mm** and a **slump flow between 150 and 190 mm** offer the most desirable combination of pumpability, surface smoothness, and shape retention. These ranges balance ease of printing with the mechanical requirements of layered construction.

Thus, the slump and slump flow tests, though simple, serve as valuable proxies for assessing the **static and dynamic yield stress** of the mix. They help identify the "printable zone" in terms of fluidity, reducing the need for more sophisticated rheometers in early-stage development.

The fresh-state behavior of 3D-printed concrete is fundamentally different from that of conventional concrete due to its unique flow and deposition requirements. In this regard, the rheological parameters that govern such behavior are extremely sensitive to mix constituents type and amount, ambient conditions, and time. With the evolution of research, along with the integration of **real-time monitoring of rheological characteristics**, **adjustment of the mix constituents**, and **smart control algorithms** for **printing** is likely to improve both performance and dependability. A deep understanding of these fresh-state properties would help the engineers and materials scientists in optimizing 3D printable mixes, that paves the way for formwork-free, and structurally sound construction of concrete structures.

4.3 Different Printability Parameters

4.3.1 Pumpability

Pumpability refers to the characteristics of the concrete mix to be sufficiently workable so that it can flow precisely and swiftly to the nozzle of printer via a pumping system under pressure without compromising on the original characteristics throughout the entire period. In other words, pumpability ensures the stability and quality of the printed object or structure and is essential for attaining accurate layer deposition.

The concrete mix needs to be workable and have the right consistency in order to guarantee that it passes through the printing equipment without experiencing problems like segregation or clogging. However, this presents a rheological contradiction: while the mix must possess high workability to support pumping, it must also retain enough stiffness to preserve the filament's shape post-extrusion. Therefore, achieving a balance between these opposing requirements is at the heart of designing pumpable concrete for 3D printing applications.

Several factors influence pumpability. These include the **hose diameter, pumping technique** (e.g., piston pump, rotor–stator system), and **length of the pumping circuit**. For instance, long-distance pumping in large-scale 3D printing introduces complexities absent in lab-scale setups, as the flow rate, time in transit, and pipe friction may alter the concrete's rheological behavior en route. Consequently, large-scale printing applications often demand a **longer setting time** to allow for transport without premature stiffening.

From a rheological standpoint, a mix with **moderate yield stress** and **low plastic viscosity** is desirable for pumpability. Yield stress ensures the material starts to flow under pressure, while low plastic viscosity allows it to maintain that flow with minimal energy. Moreover, an **adequate viscosity recovery rate** is crucial to stabilize the shape post-deposition, which often requires **higher pumping pressures**.

However, pumping also introduces specific challenges, such as **inner-wall smearing**—a phenomenon where larger aggregate particles accumulate near the inner pipe surface, creating a thin lubricating layer. If this lubricating layer is disturbed (e.g., due to excessive pressure or poor mix cohesion), **segregation** may occur, potentially causing **blockages**. Adjustments in **pumping speed, flow regulation at corners**, and **printing path angles** are thus essential to ensure uninterrupted flow, particularly at geometrically complex sections of the print.

Experimental investigations have proposed metrics to quantify pumpability. For instance, Tay et al. [13] introduced a **pumpability index**, defined as the ratio of a mix's volumetric flow rate to that of water under similar conditions. Their results revealed that the pumpability index is directly proportional to the mix's **slump flow** and inversely proportional to the **required pumping pressure**.

More precise techniques such as the **Sliding Pipe Rheometer (SLIPER)** have also been employed to assess pumpability. This method relates the **pressure at piston head to rate of flow**, allowing evaluation of the material characteristics at real time under various pumping conditions. Researchers utilizing SLIPER have validated that both the rotor-type and piston pumps are effective for pumping 3DPC [14, 15]. Among other alternative testing devices, e.g., **rheographs, viscometers**, and **tribometers**, also helps to understand various insights into concrete's rheological characteristics, particularly under the dynamic flow condition.

Thus, pumpability is impacted not only by the pumping equipment and applied pressure, but also strongly depends on the **mix design of concrete**, **aggregate-to-binder ratio, admixture dosage**, and the **thickness of printed layer**. Use of **plasticizers, SP**, and **VMA** can further modify such rheological properties, resulting in smoother and more consistent pumping along with maintaining the printing performance and structural integrity.

4.3.2 Extrudability

The term "extrudability" describes a printed mix's capacity to extrude smoothly and continuously out of the printhead nozzle while maintaining the desired quality of the printed layer. Here, the quality of the layer implies that it should be free from any kind of defects like discontinuity or void.

From a material science viewpoint, extrudability is primarily impacted by two rheological characteristics: **yield stress** and **plastic viscosity. A low yield stress** may cause material segregation during flow, especially under gravity or pumping pressure, whereas **excessively high yield stress** results in a mix that is too stiff to be extruded smoothly. Research by Rahul et al. [16] identified an optimal yield stress range of **1.5 to 2.5 kPa** as ideal for achieving smooth extrusion.

Key determinants of extrudability include

- **Mix Design Parameters: A low water-to-binder ratio, reduced sand content, and optimized fiber dosage** improve extrudability.
- **Rheological Modifiers**: The use of **water-reducing agents and superplasticizers** can reduce water demand, enhancing flow without increasing bleeding.
- **Aggregate Gradation and Size**: The largest aggregate size should not exceed **one-tenth of the nozzle diameter** to avoid blockage [17, 18].
- **Sand-to-Binder Ratio**: Le et al. [19] found that a 3:2 ratio yielded effective results with 2 mm aggregate for a 9 mm nozzle. Panda et al. [20] observed mix segregation beyond a 1.9 sand-to-binder ratio.
- **Water Content**: Lowering water from 232 to 216 kg/m^3 improved extrudability in Le et al.'s [19] study.

Nozzle design plays a pivotal role. The **shape** (circular, square, elliptical) and **size** of the nozzle influence both the **aesthetics** and **interlayer bonding**. Circular nozzles are effective for printing around curves and angles but may offer **less contact area** between layers. In contrast, square nozzles improve **surface finish** and **layer alignment** [21]. However, smaller nozzle openings, while enhancing control, reduce print speed and can cause blockages if the **aggregate-to-nozzle ratio** falls below 4.25 [22].

The **orientation** of the nozzle also matters—especially at curves or junctions—where **tangential positioning** along the path reduces the risk of twisting newly deposited filaments [23]. Furthermore, the **hydraulic conductivity** of the mix must be matched with the extrusion rate to prevent inconsistent flow and pressure build-up inside the nozzle.

Various testing methodologies have been developed to assess extrudability. Le et al. [19] used **a filament continuity test**, measuring whether a 4,500 mm filament could be deposited without breakage. Figure 4.3 shows a typical example of a successfully extrudable mix. Ma et al. [24] introduced an **extrudability coefficient** that ties extrusion rate to printing speed, optimizing these parameters together. Perrot et al. [25] developed a **ram extrusion test**, while others used **squeezing tests, penetration resistance tests, and electrical energy-based flow rate assessments**.

To replicate real printing conditions, Rushing et al. [17] designed **a Bailey Ceramics-based extruder** for laboratory testing. Such systems avoid the influence of geometric artifacts and provide a more reliable assessment of extrudability under practical constraints.

The **delivery-extrusion interface** also influences print quality. Systems may follow two configurations:

Fig. 4.3 Test on extrudability of cementitious mix

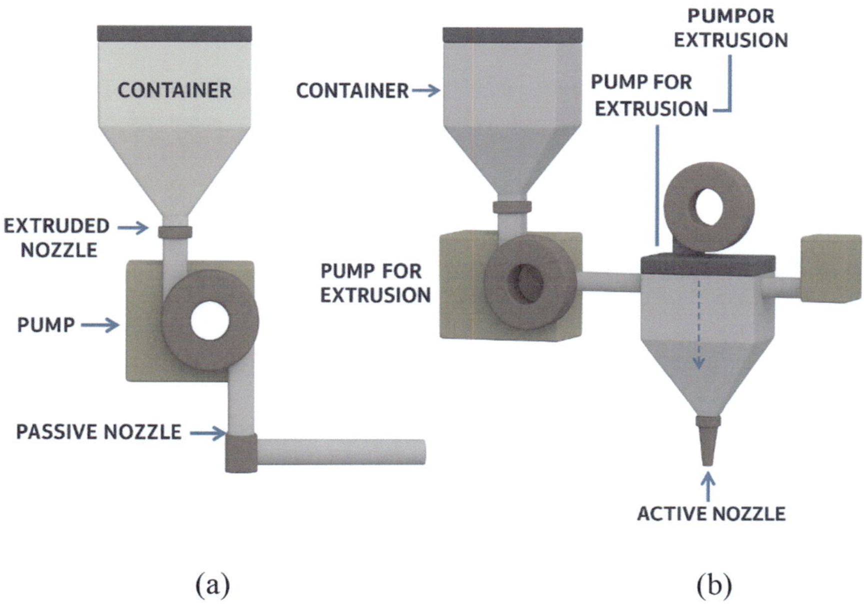

Fig. 4.4 Different processes of material delivery and extrusion of 3DPC: **a** combined process, **b** separated system

- **Combined Delivery and Extrusion** (Fig. 4.4a): The same pump delivers and extrudes the material, suitable for lab-scale applications where pumping distances are short and mix consistency can be tightly controlled.
- **Separated Delivery and Extrusion** (Fig. 4.4b): A delivery pump feeds material to a printhead where extrusion is managed independently. This is more appropriate for **large-scale** operations where **distance, flow control, and admixture integration** require flexible modulation.

Additionally, the Danish Technological Institute [23] categorizes robotic nozzles into

- **Passive nozzles**, which act as simple dispensers.
- **Active nozzles**, which incorporate advanced features such as **flow rate control, shape modulation**, and **on-site admixture dosing**, significantly improving extrusion consistency and adaptability. Thus, extrudability is a dynamic property governed by a complex interplay of material formulation, rheological control, and mechanical hardware—each of which must be tailored to the scale and application of the 3D concrete printing project.

4.3.3 Buildability

Buildability is one of the most critical fresh-state characteristics in 3D concrete printing (3DCP). It refers to the capacity of the freshly extruded concrete to sustain its own weight and the weight of the subsequent layers during the printing process without experiencing deformation or collapse. Unlike conventional concrete construction methods, which rely on formwork to support the structure until hardening occurs, 3D printing eliminates this external support system. Therefore, the printed material must inherently possess the mechanical stability to retain its shape layer by layer, a property governed by its early-age strength, yield stress, and viscosity.

In practice, a 3D printable mix must demonstrate the ability to form and maintain layers with minimal deformation or sagging. The first layer, that directly share interfaces with the printing base, should be strong enough to sustain the weight of the successive upper layers without collapse or lateral deformation. To achieve this, the fresh concrete should gain sufficient green strength, a terminology that describes the structural integrity of concrete before achieving the complete hardness. Without sufficient green strength, the bottommost layers could fail under weight of upper layers that leads to failure modes such as buckling or plastic collapse.

Yield stress and plastic viscosity serve as two vital rheological parameters that impact buildability. Yield stress governs the moment at which the mix starts to deform under a given stress, while plastic viscosity represents the resistance of a material to flow after the yielding starts. For a mix to be buildable, it must exhibit high static yield stress, which helps the layers remain rigid and self-supporting. At the same time, the concrete must be thixotropic—meaning it can regain its stiffness after being disturbed during pumping or extrusion—so that the shape of the extruded filament is maintained and interlayer bonding is not compromised.

Buildability is not just a function of material composition but is also deeply influenced by the timing of the layer deposition. If the time gap between printing successive layers is too short, the lower layer may not have gained adequate strength, leading to deformation. On the other hand, if the time gap is too long, cold joints or weak interlayer adhesion may form due to partial setting of the previously deposited layer. Therefore, achieving the right balance of setting time and interlayer interval is essential.

Several empirical and analytical methods have been proposed to evaluate the buildability of a printed mix. A common and practical approach is to assess the maximum vertical height that can be built without significant deformation or collapse. This method considers the number of successfully printed layers and observes the extent of deflection or distortion with increasing height. Typically, the buildability decreases as the vertical build rate increases because the lower layers are subjected to more stress within a shorter time frame, reducing the window for strength gain.

Experimental studies have categorized buildability failure into three primary modes: elastic buckling, plastic collapse, and mixed-mode failure (Fig. 4.5). Elastic buckling is primarily controlled by the stiffness of the material at early ages and occurs when the printed element bends under its own weight without permanent deformation. Plastic collapse, on the other hand, is governed by yield strength; it manifests when the stress in the lower layers surpasses their yield stress, causing permanent deformation or structural collapse. A mixed mode combines features of both elastic and plastic failure, often seen in taller or slender printed structures where both stiffness and strength become limiting factors.

To quantify these behaviors, researchers have devised several testing protocols. For instance, Le et al. [19] evaluated buildability by counting the number of printable layers before failure occurred. Zhang et al. [26] extended this method by printing square specimens with voids and measuring the tallest stable structure achieved by different mixes. Kazemian et al. [27] introduced two complementary tests: the first examined deformation between layers over time using photographic measurements, while the second—known as the cylinder stability test (see Fig. 4.6c)—subjected fresh concrete cylinders to incremental vertical loads to observe deformation under self-weight. Another relevant method involved calculating vertical strain and aspect ratio (the ratio of height to width of extruded layers) to monitor structural stability during the print.

In parallel, buildability has also been assessed using rheological and mechanical measurements. Green strength, often evaluated via unconfined uniaxial compression tests, provides a direct measure of the early-age load-bearing capacity. Yuan et al. studied this property by applying successive loads to printed layers at defined intervals, identifying the critical time at which the lower layer fails. Perrot et al. proposed a theoretical model based on yield stress evolution and derived expressions to predict the safe printing height and failure time of a printed element under increasing load. Their study suggested that mixes with higher initial yield stress offered greater buildability and less deformation, allowing for taller structures to be printed safely.

Thixotropy also plays a crucial role in buildability. A highly thixotropic mix stiffens rapidly after extrusion, helping layers retain their shape. This behavior can be characterized using hysteresis loop tests, which measure the area between increasing and decreasing stress–strain curves under shear—a larger area indicating higher

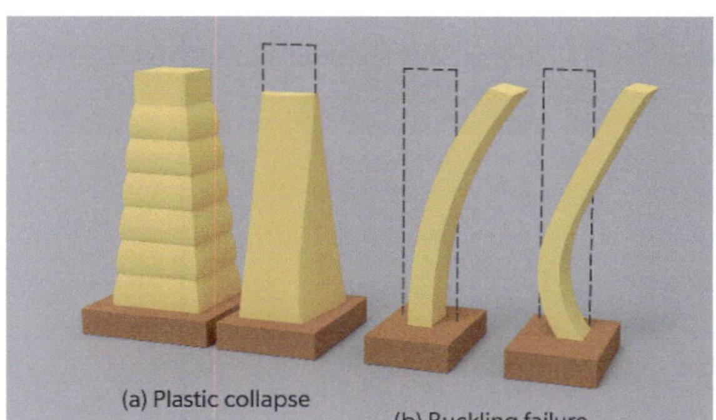

Fig. 4.5 Different types of failure mode during buildability in the case of 3DPC

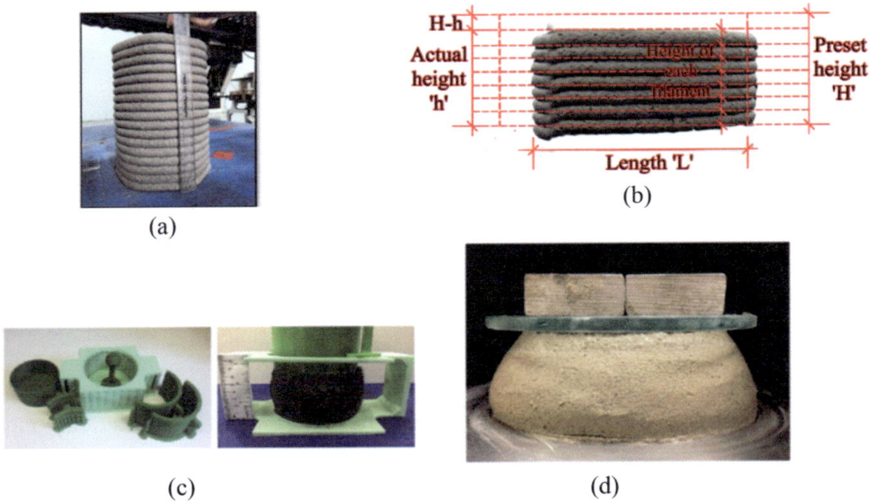

Fig. 4.6 Various test methods and processes employed for buildability: **a** buildability test, **b** measurement of deformation, **c** cylinder-based stability test [27], **d** shape retention test [28]

structural build-up. Incorporating additives and supplementary cementitious materials (SCMs) has shown promise in tuning this behavior (Table 4.1). For instance, the use of attapulgite nanoclay and calcium sulphoaluminate cement has been found to enhance early-age strength and reduce deformation. Accelerating admixtures can further improve buildability by expediting setting reactions, as shown in experimental studies where non-alkaline accelerators were applied to the surface post-printing to strengthen the structure in situ.

Material composition, particularly aggregate gradation, also influences buildability. Weng et al. demonstrated that mortar made with continuously graded sand performed better than that made with gap-graded sand, highlighting the role of particle packing and internal friction in structural retention. Finer aggregates or higher contents of fine materials, such as nano-silica, have been shown to improve shape retention, although they may reduce flowability. To address this trade-off, partial replacement of Portland cement with reactive materials like fly ash or metakaolin

Table 4.1 Influence of commonly employed supplementary cementitious materials on the fresh state properties of 3D-printed concrete [29]

Material	Primary composition	Effect on fresh properties
Silica fume	Si	Improved flowability and buildability
GGBFS	Ca–Si–Al	Better shape retention, i.e., lowered slump and buildability
FA	Si–Al/Si–Ca–Al	Increases flowability
Metakaolin	Si–Al	Higher agglomeration of particles
RHA	Si–C–K_2O	Reduces free available water and reduces workability

has been investigated. Nano-silica, in particular, has been found to reduce setting time while preserving initial flowability, thus improving both extrudability and buildability.

To predict and optimize buildability further, researchers have recently begun to employ numerical modeling. Using finite element methods based on the Mohr–Coulomb failure criterion and linear-elastic stress–strain relationships, simulations have been developed to replicate the layer-by-layer construction and determine failure thresholds. These models take into account the time-dependent strength gain and stress distribution across layers, offering valuable insight into mix design and printing parameters for complex geometries and large-scale applications.

When it comes to full-scale or large-format 3D concrete printing, the challenges of buildability become even more pronounced. In such cases, the concrete must be stored, pumped, and extruded over longer durations, during which material properties can change due to hydration, segregation, or moisture loss. The proper sequencing and timing of admixture additions become essential to ensure that the material maintains its fresh state properties throughout the printing operation. As large-scale 3D printing continues to evolve, there is a pressing need for standardized testing protocols and robust modeling frameworks that can capture the buildability performance under realistic, field-scale conditions.

Ultimately, buildability remains a multifaceted property, deeply interconnected with other fresh state characteristics such as pumpability, extrudability, and setting time. Striking a balance among these often-contradictory demands is critical for the success of any 3D printed concrete structure. The insights drawn from both laboratory tests and theoretical models lay the groundwork for designing advanced 3DPC mixes that can enable taller, more complex, and structurally sound constructions without the need for formwork (Fig. 4.7 and Table 4.1).

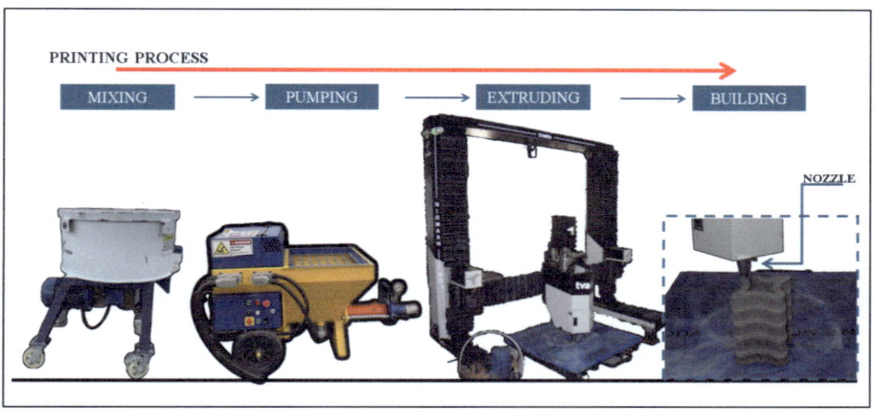

Fig. 4.7 Overview of the different steps involved in the printing processes of extrusion based 3DPC

4.4 Contradicting Requirements of Fresh State Properties

The development of a 3D printable concrete mix presents a unique challenge in balancing multiple fresh-state properties, each of which must satisfy a specific functional requirement during different stages of the printing process. These properties—such as pumpability, extrudability, buildability, open time, workability retention, and setting time—are often interdependent and, at times, mutually conflicting. Although each of rheological property impacts the printing quality and final structure, optimization of any one characteristics may unconsciously affects another. The current section investigates into such intrinsic contradicting characteristics and to determine the possible strategy to overcome such complex optimisation problem.

The primary conflict in the material characteristics of 3DPC lies in the fact that it should be fluid enough during transportation and pumping. In continuation to that it should transform into a solid-like, stable shape material immediate after the extrusion. For example, a concrete mix that is designed for greater pumpability typically shows low plastic viscosity so that it can travel through long pipes without getting clogged or segregated. Such types of mixes typically show a lower value of yield stress. In contrast, buildability (which represents the ability of a material to sustain its own weight along with retaining the shape of when successive layers are printed) requires exactly the contrasting material behavior, a higher value of static yield stress along with increased thixotropy characteristics in order to prevent deformation or collapse after material deposition.

The conflict becomes more evident when a mixture that can effortlessly be pumped but fails to hold its shape after extrusion. Conversely, a mix designed for a high value of yield stress along with greater buildability can become hard to pump the material, leading to blockages of line, usage of higher energy, or pump failure. Thus, achieving a balance in the mix design of 3DPC has become an extensive area of research for the scientific community to balance flow characteristics and stability.

Extrudability increases another type of intricacy in designing the material for 3DPC. As mentioned, pumpability signifies the flow performance of the mix through the pumping system, while, extrudability signifies a material's ability to come out from the nozzle and develop a continuous filament. Thus, a mix with a higher viscosity may create clogging inside the nozzle, i.e., the extruded filament may have poor quality in terms of material extrusion. In oppose to that, a well-designed mix with better extrusion quality could have poor quality of shape after printing that ultimately could hamper the interfilament bond. In this regard, pressure at the nozzle, rate of extrusion, and nozzle geometry also control such processes. Therefore, a very careful calibration of the mix and printing parameters are required so that the produced filaments remain dimensionally stable along with a well interfilament bond.

Moreover, concrete mixes pose additional challenges with respect to open time and workability. A mix that has a higher open time confirms uninterrupted printing over longer durations, a characteristic that is required in the case of large-scale structures. However, a higher open time typically involves delayed setting that hinders the growth of green strength. It could have an adverse effect on the interlayer bond and buildability, particularly when the interlayer time gap crosses certain threshold time. It results in the creation of cold joints, i.e., weak bond where mechanical and durability properties are compromised.

On other side, a mix that is designed for fast strength development along with improved buildability and interlayer bond might set too rapidly, thus the open time and relative workability. Thus, the printing process can be interrupted resulting in waste of material, and difficulty in restarting the operations. In fact, a mix that sets rapidly may also hydrate inside the pump or delivery system, causing early stiffening characteristics and potential damage of the equipment. Thus, it is very crucial to achieve a balance between the optimum early strength development and operational flexibility.

The situation is additionally complicated when different admixtures are used in the mix to alter various properties. For instance, inclusion of VMA such as hydroxypropyl methylcellulose can improve the buildability by increasing cohesion and static yield stress. However, an excess amount of VMA could reduce the pumpability and extrudability, resulting in difficulty in handling the mix. Similarly, thixotropic materials that help to recover yield stress after shear cessation may spoil the flow through the pump or cause clogging at bends in the delivery system.

Another level of interdependence is observed with the application of accelerators that imparts rapid strength gain and setting. Although its inclusion improves buildability and reduces workability, it can increase the risk of microcracking due to thermal effects, especially if higher amount of cement is used, ultimately impacting the structural performance of the 3DPC element.

The inconsistency also includes in the case of print resolution and interlayer bond. A mix with higher buildability and lower value of deformation might form well-defined edges and vertical profiles, refining the architectural view of the 3DPC element. However, a stiff mix may pose problem in terms of lower interlayer bond with the already printed layer, leading to drop in mechanical properties. In contrast, a slightly more fluid mix with extended setting time may ensure better interlayer diffusion and bonding but at the expense of shape precision and vertical load-bearing capacity.

Navigating these conflicting demands requires a holistic approach to mix design, where the material composition, rheological modifiers, admixture timing, and printing parameters are co-optimized. One common strategy is to develop time-dependent rheological profiles that match the material's evolution with the printing process. For instance, a mix may be tailored to have low viscosity and moderate yield stress during pumping, with a rapid buildup of static yield stress and green strength shortly after extrusion. Such behavior is achievable using thixotropic agents that exhibit time-dependent structural rebuilding or using dual-admixture systems where setting accelerators are introduced just prior to or after extrusion.

Process-based interventions can also mitigate conflicting demands. For example, adjusting the print speed or interlayer interval can compensate for slow strength gain or poor bonding. Similarly, temperature control during mixing and pumping can be used to modulate the setting behavior and rheological evolution of the mix. In some advanced systems, in-line rheometers or sensor-based feedback loops are

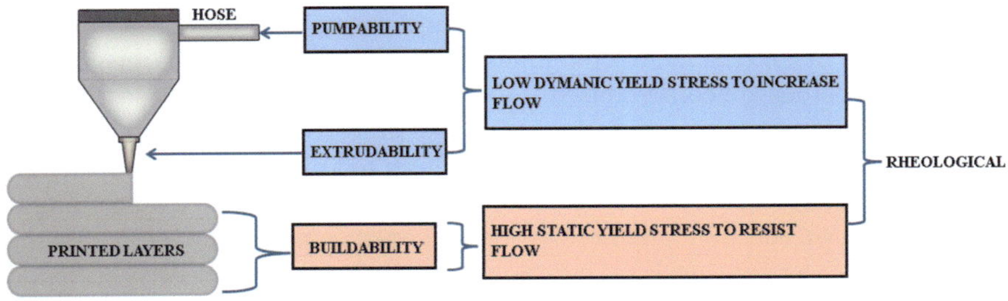

Fig. 4.8 Contradicting requirements of rheological characteristics for 3DPC

employed to monitor material behavior in real-time and adapt the printing parameters dynamically, ensuring a better balance among the competing requirements.

Recent advances in computational modeling have also helped to address these contradictions. Multi-objective optimization algorithms, informed by rheological, mechanical, and performance data, can help identify mix compositions and printing sequences that maximize overall print quality while satisfying critical thresholds for each fresh state property. These tools allow for the simulation of different scenarios and trade-offs before actual printing, saving time and material in the trial-and-error stage of mix development.

In conclusion, the fresh state behavior of 3D printable concrete is characterized by a complex interplay of contradictory requirements. Designing a mix that excels in all parameters simultaneously is virtually impossible; instead, the goal is to find a viable compromise that aligns with the specific demands of the intended application. Whether the priority is structural performance, architectural detail, or rapid printability, understanding these contradictions and managing them through integrated design and process control remains the cornerstone of successful 3D concrete printing. The above discussion on contradicting requirements of different rheological properties for 3DPC is presented in Fig. 4.8.

4.5 Open Time

In the domain of 3D concrete printing (3DCP), one of the most critical time-dependent parameters that governs print quality and operational continuity is open time. Open time refers to the duration for which the freshly prepared concrete mix remains extrudable through the nozzle without encountering clogging, filament breakage, or significant degradation in flowability. In simpler terms, it is the time window within which the concrete behaves as a printable material—capable of forming continuous, cohesive filaments upon extrusion. Once this period expires, the increasing yield stress and internal structural buildup of the material begin to inhibit smooth flow, marking the end of its printable phase.

For successful implementation of 3D printing in construction sector, the open time of a concrete mix should be higher than the total duration for material extrusion. If the condition is not met, the printing process may be disrupted, and filament discontinuities or nozzle blockages will be observed. The implications can further be impacted in large-scale structures, where long printing durations is required. Early stiffening of the concrete mix can lead to formation of cold joints, weaken the interlayer bond, and higher wastage of material, ultimately affecting the structural integrity and mechanical performance of 3DPC element.

The development of open time is primarily impacted by the rheological characteristics of the concrete mix, particularly the increment in yield stress. With the progress of hydration, the internal microstructure of the mix evolves over time, and the material transforms from a deformable to a stiff mix. Such a transformation, although helpful for early age strength development and improved buildability, however shortens the effective printable time window. Hence, monitoring and controlling the transition serves to be vital in ensuring the fact that the material remains sufficient flowability during the printing process.

To measure the open time at lab and field scale, researchers utilized a wide variety of rheological and flow-based tests various intervals after mixing. Such tests measure the variation in flowability of the mix over time and evaluate the time when the process of extrusion becomes unfeasible. Among the different types of methods, flow table test, Vicat needle test, V-funnel flow, slump cone test, and rheometer-based shear stress analysis are commonly employed to evaluate the flow characteristics of the mix.

For instance, Le et al. [19] employed the vane shear test to monitor the shear strength evolution of the mix over time. They concluded that the end of the open period is indicated by the concrete's shear strength increasing by 0.3 kPa with respect to its initial value, a level after which the material extrusion becomes discontinuous, i.e., internal resistance to flow became extreme. Likewise, the Vicat apparatus, generally employed for evaluating the setting time, was used by various researchers [30] to track the progressive loss in penetrability and to find the end of the open time following on the setting kinetics of the concrete mix.

In oppose to that, Chen et al. [31] developed a visually instinctive and process-based technique, where the open time is calculated based on the duration up to which uniform and uninterrupted extrusion of the material occurs. In the process, the filament geometry is continuously observed, and any visible contraction, tearing, or lack of continuity is considered as the end of the open time of the mix. This methodology closely allies with the real-world 3D printing construction, where visual quality of the filament serves as a practical indicator of the material performance.

Figure 4.9a presents the visual illustration process in which the filament width is reduced over the duration of printing. With the increase in time, the width of the extruded concrete layer gradually narrow down, demonstrating the drop in flowability and development of internal resistance. As a result, the filament becomes thin, irregular, or broken, specifying the end of open time. Figure 4.9b reveals another failure process, nozzle blockage, which result from the rise in yield stress. With the increase in yield stress, the material get stiffens and exert excessive backpressure

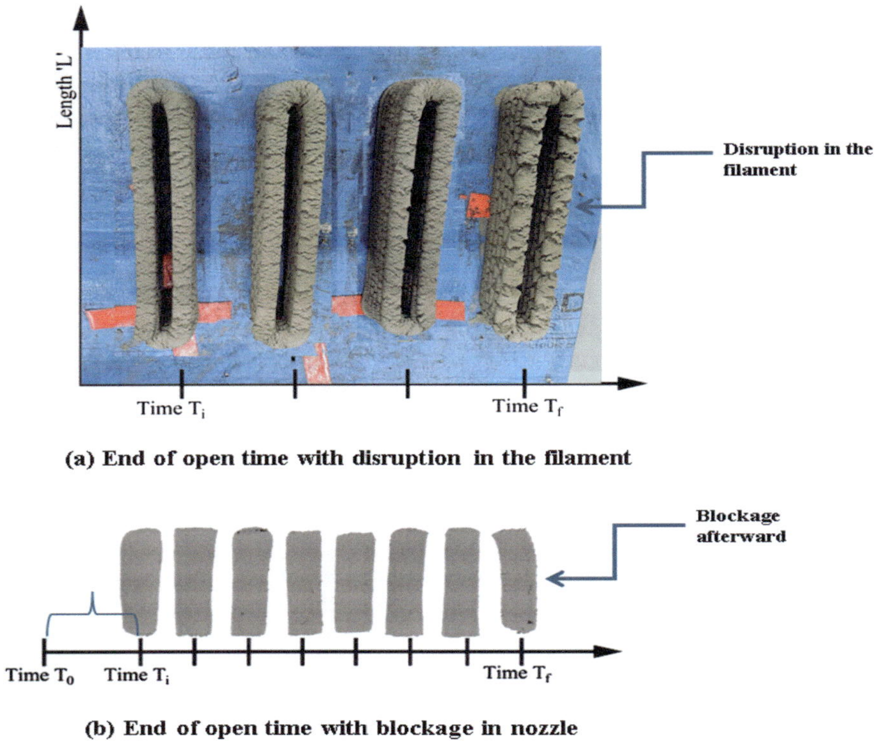

Fig. 4.9 Different processes to evaluate the open time of concrete mix

and flow resistance, eventually the process of extrusion stopped. Thus, the above-mentioned two figures together reveal the dual mechanisms when open time ends, either through degradation of filament slowly or a sudden extrusion type failure due to clogging of the material.

Here, it is required to be mentioned that open time depends on multiple factors, like ambient temperature, admixture dosage, types of binder, and loss of moisture. For instance, mixes that contain rapid-setting type binders or higher amount of calcium aluminate cement often show shorter open times owing to accelerated hydration. Likewise, environmental exposure increases the surface stiffening and thus the available time for printing decreases. To overcome these, researchers progressively started using retarders or moisture control techniques which allows real-time modification of the concrete properties during printing.

Eventually, the open time of a 3D printable concrete mix should be carefully designed to meet the desired size, complexity, and operational speed of printing. A mix that is well designed continues to remain stable and serve as extrudable throughout the entire printing duration. As a result, a consistent quality of filament, interlayer adhesion, and geometric fidelity can be ensured. Thus, a deep understanding of the open time along with its controlling factors serves as a crucial part of the rheological characteristics of printable concretes. The two figures (Fig. 4.9a–b) show the two distinct methods for figuring out when the material's open time ends. In Fig. 4.9a, filament width decreases with time, indicating diminishing flowability and onset of discontinuous extrusion. Whereas Fig. 4.9b shows nozzle blockage due to yield stress buildup over time, resulting in complete extrusion failure.

4.5.1 Experimental Analysis of Open Time

To understand the influence of chemical admixtures on the open time of 3D printable concrete mixes incorporating agro-industrial wastes, a systematic experimental investigation was carried out. The focus of the analysis was not on optimizing mix proportions for strength or durability—although the selected binary and ternary mix compositions were finalized only after iterative trials that ensured acceptable compressive strength—but rather on assessing how varying dosages of superplasticizer (SP), viscosity modifying agent (VMA), and retarder affect the flow retention or open time characteristics of such composites.

The compositions employed for the binary and ternary blends are presented in Table 4.2. In the binary system, fly ash was excluded, and 40% of cement was replaced with bagasse ash (BA). In the ternary system, cement was partially replaced by equal proportions of BA and fly ash (FA). For both cases, the sand-to-binder ratio was maintained at 1.4 and the water-to-binder ratio at 0.40 to ensure consistent baseline rheological properties across tests.

In both mix types, the dosage of admixtures was systematically varied to evaluate their impact on open time. Figure 4.10a–b shows the effect of different SP contents on the open time of the binary and ternary cementitious composites. Open time, in this context, is interpreted as the period during which the mix maintains a target flowability sufficient for consistent pumping, extrusion, and buildability. This threshold flow value is determined to be around 160 ± 10 mm, representing the range within which the material is deemed suitable for 3D printing operations.

The results clearly show a trend where open time increases with the addition of SP up to a dosage of 1%, beyond which it begins to decline. For both binary and ternary mixes, a peak open time of approximately 25 min was achieved at this 1% SP content. The initial increase can be attributed to the dispersing action of SP, which promotes deflocculation of cement particles, thereby delaying structural buildup and

Table 4.2 Mix-proportion of raw materials in binary and ternary cementitious composites

Mix type	Cement (%)	BA (%)	FA (%)	Sand/binder	Water/binder
Binary	60	40	0	1.4	0.40
Ternary	60	20	20	1.4	0.40

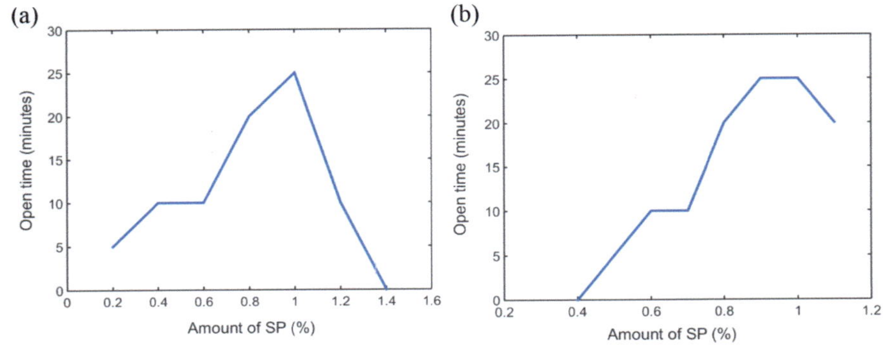

Fig. 4.10 Effect of SP on open time of agro-industrial waste-based **a** binary and **b** ternary cementitious composite

maintaining flowability. However, beyond this optimum level, the structural stability of the mix begins to deteriorate due to excessive dispersion, which undermines its thixotropic behavior and leads to faster collapse of the microstructure, reducing the effective open time.

Following the determination of the optimum SP content, the effect of VMA dosage was evaluated while maintaining SP at 1%. Figures 4.11a–b illustrate the impact of VMA on open time. Here, it is observed that an increase in VMA dosage up to 0.5% significantly improves the open time, extending it to approximately 35 min in both binary and ternary systems. This is primarily due to the VMA's ability to enhance cohesion and minimize bleeding, which stabilizes the internal matrix and retards structural stiffening without compromising flow. As a result, VMA addition complements the dispersing effect of SP by providing stability, which translates to prolonged open time.

Building upon the SP-VMA combination, the influence of retarders was then studied by adding different dosages while holding SP at 1% and VMA at 0.5%.

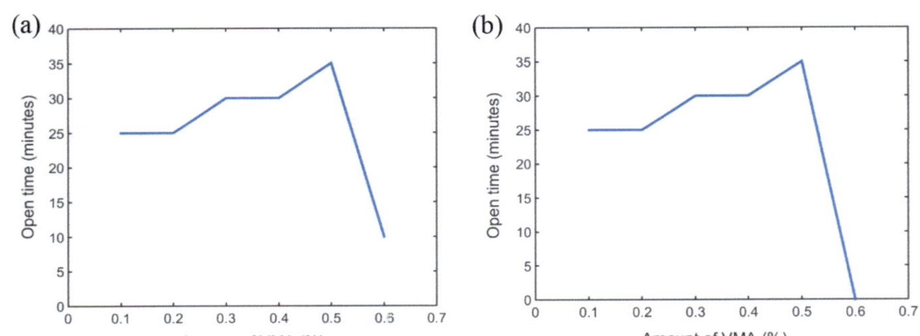

Fig. 4.11 Effect of VMA on open time of agro-industrial waste-based **a** binary and **b** ternary cementitious composite

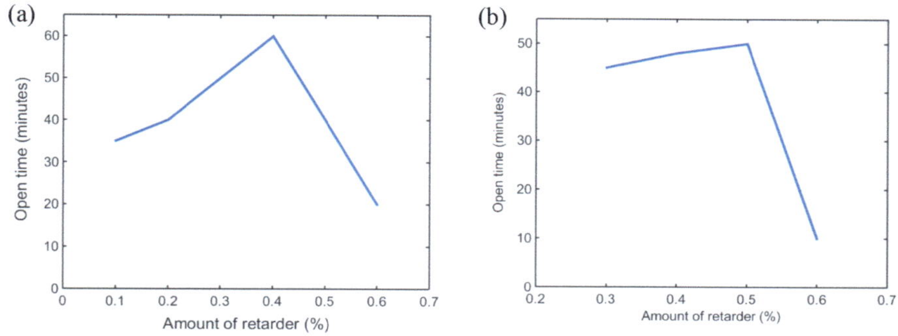

Fig. 4.12 Effect of retarder on open time of agro-industrial waste-based **a** binary and **b** ternary cementitious composite

Figure 4.12a–b shows that with the addition of 0.4% retarder, the open time of the binary mix increases dramatically to nearly 60 min. In the case of the ternary mix, the highest open time of about 50 min is achieved at a slightly higher dosage of 0.5% retarder. The action of the retarder delays the hydration of cementitious phases, thus preserving the flowability and workability over a prolonged duration. Additionally, some retarders have a dual effect, acting as mild water reducers, which further improve the rheological characteristics without increasing water demand.

These findings suggest that the synergy between SP, VMA, and retarder must be carefully calibrated for each mix type, especially when using agro-industrial wastes with variable pozzolanic activity and water demand. The binary mix, with only bagasse ash, responded more strongly to retardation, potentially due to lower reactivity compared to the ternary mix. Nonetheless, both systems demonstrated significant improvement in open time, confirming the efficacy of admixture optimization in enhancing the printability window of sustainable cementitious composites.

4.5.2 Determination of Open Time from Flowability

To validate the experimental determination of open time and its relevance to actual printing conditions, flowability-based assessments were conducted for the optimized binary and ternary mixes. In these tests, flow measurements were performed at regular time intervals to identify the range during which the mixes sustained the target flowability required for smooth printing operations.

Figure 4.13a–b displays the results of this investigation. In both mixes, the flowability was monitored over time using a standardized flow table, and the duration over which the flow values remained within the desired range of 160 ± 10 mm was used to define the open time.

In the case of the binary mix, the flowability remained within the acceptable range for approximately 10 to 70 min (Fig. 4.13a). This indicates a wide operational

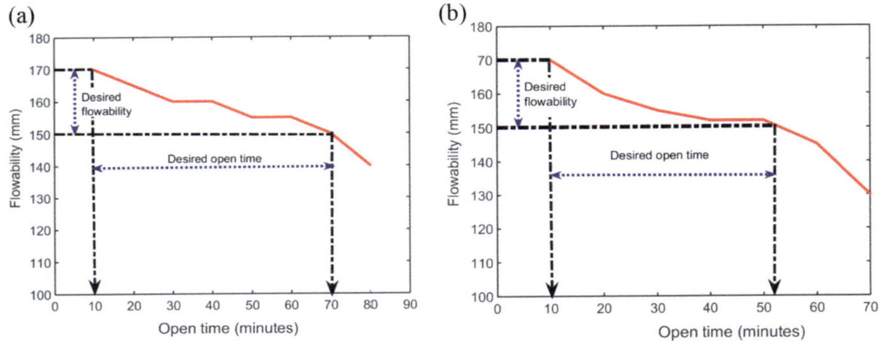

Fig. 4.13 Identification of optimum range of open time from desired flowability for **a** binary (B40:C60) and **b** ternary (B20:F20:C60) cementitious composite

window for extrusion and buildability, making it well-suited for medium-duration prints. The enhanced performance may be attributed to the stabilizing action of VMA and the extended setting time due to the optimized retarder content.

For the ternary mix, a slightly narrower window of 10 to 50 min was observed (Fig. 4.13b). Despite the presence of fly ash, which generally improves flow characteristics, the ternary blend appeared to exhibit a faster buildup of structural strength, likely due to the combined pozzolanic reactions. This somewhat reduced the time for which flowability could be maintained within the desired limits, though still sufficient for practical printing operations.

These findings confirm that open time, as defined through rheological and visual filament-based observations, aligns closely with the practical flowability thresholds observed in printing. This reinforces the validity of using such flow-based criteria to determine the usable window for 3D printing with agro-industrial waste-based cementitious materials. Moreover, it emphasizes the importance of material characterization not only at the fresh state but throughout the early age evolution, to ensure uninterrupted, high-quality prints.

4.6 Closure

This chapter presented a comprehensive examination of the fresh state properties of 3D printed concrete (3DPC), with a particular emphasis on evaluating the parameters that govern the printability, structural stability, and process control of the material in its early age. Thus, an in-depth understanding of the fresh state characteristics of cementitious composites, that impacts the filament quality and mechanical integrity of printed concrete elements, serves as the crucial aspects in the field of 3DPC.

The discussion, in this chapter, started with an explanation of the basic rheological characteristics, e.g., yield stress, plastic viscosity, and thixotropy, which represents the ability of a concrete mix to pass through the nozzle and consequently maintain its shape after printing. Measurements of these rheological properties were described in conjunction with their influence in the context of printability of the material, i.e., pumpability, extrudability and buildability. Different experimental methodologies, comprising of rheometers and shear vane type devices, were emphasized as important tools in evaluating such time-dependent rheological properties.

Afterwards, the various concepts of printability, that forms the central part of 3DPC, were presented in detail through parameters like flowability, extrudability, and buildability. It was highlighted that attaining a optimal balance among various such parameters is essential in fabricating a structurally balanced and geometrically steady printing. In this regard, the test such as flow table test, V-funnel, and different customize nozzle extrusion based setups were found as practical means to measure these printability characteristics in a lab-scale or real field based 3DPC structures.

The chapter also explored into the setting time characteristics of 3DPC mixes, a critical factor that impacts interlayer bond and the management of time during the printing process. It was presented that how various additives, e.g., superplasticizers, VMA, and retarders can modify the hydration mechanism of the cementitious composite, ultimately impacting the setting behavior. Understanding this setting characteristics offered the foundation for identifying the effective buildability time period, the duration over which a printed layer maintains stability under the weight of subsequent filaments, a factor that directly depends on the thixotropic characteristics and early age structural build-up of 3DPC.

One of the most critical contributions of this chapter was the detailed evaluation of **open time**, defined as the period during which the mix remains extrudable and maintains the desired flowability without nozzle blockage or filament discontinuities. It was established that open time must be sufficiently longer than the actual printing duration to avoid cold joints and ensure monolithic bonding. Several experimental techniques—including the Vicat needle test, vane shear analysis, and filament observation—were reviewed to assess open time based on yield stress evolution and visual changes in filament geometry.

The latter part of the chapter focused on experimental investigations using agro-industrial waste-based cementitious composites, specifically binary (cement-bagasse ash) and ternary (cement-bagasse ash-fly ash) mixes. The influence of admixtures was systematically studied, revealing that a 1% dosage of superplasticizer resulted in the longest open time before structural degradation set in. Similarly, the addition of VMA at 0.5% improved the stability of the mix, while the inclusion of retarders significantly extended the open time—up to 60 min in binary and 50 min in ternary systems. Such observation emphasized the crucial role of various admixtures that controls early-age performance of the mix and established the viability of modifying fresh state properties for fabricating sustainable 3DPC structures.

Lastly, the investigation of open time behavior of the fabricated binary and ternary composite highlights the practical applicability of these mixes within the possible time period. The binary mix was observed to be appropriate for printing between 10 and 70 min. Whereas the ternary mix provides 10 to 50 min of effective time period for printing. Such understanding is of utmost importance especially for large-scale projects that involves critical planning for achieving an uninterrupted printing in a layer-wise deposition.

Thus, the current chapter has formed the basic scientific and practical application based foundation for an in depth understanding of the fresh state characteristics of 3DPC, especially the mixes that are made of agro-industrial by-products. By establishing the interrelation between the rheological characteristics, admixture dosage, and experimental observation, the present chapter indicates the potential for future research in the designing of concrete mix keeping in mind the fresh state properties that meets the desired performance and advances sustainable and improved performance based 3DPC structures.

References

1. Perrot A, Rangeard D, Pierre A (2016) Structural built-up of cement-based materials used for 3D-printing extrusion techniques. Mater Struct 49:1213–1220
2. Roussel N, Ovarlez G, Garrault S, Brumaud C (2012) The origins of thixotropy of fresh cement pastes. Cem Concr Res 42:148–157
3. Kruger J, Zeranka S, van Zijl G (2019) An ab initio approach for thixotropy characterisation of (nanoparticle-infused) 3D printable concrete. Constr Build Mater 224:372–386
4. Qian Y, De Schutter G (2018) Enhancing thixotropy of fresh cement pastes with nanoclay in presence of polycarboxylate ether superplasticizer (PCE). Cem Concr Res 111:15–22
5. Jayathilakage R, Rajeev P, Sanjayan JG (2020) Yield stress criteria to assess the buildability of 3D concrete printing. Constr Build Mater 240:117989
6. Mahaut F, Mokéddem S, Chateau X, Roussel N, Ovarlez G (2008) Effect of coarse particle volume fraction on the yield stress and thixotropy of cementitious materials. Cem Concr Res 38:1276–1285
7. Toutou Z, Roussel N (2006) Multi scale experimental study of concrete rheology: from water scale to gravel scale. Mater Struct 39:189–199
8. Noor MA, Uomoto T (2004) Rheology of high flowing mortar and concrete. Mater Struct 37:513–521
9. Weng Y, Li M, Tan MJ, Qian S (2018) Design 3D printing cementitious materials via Fuller Thompson theory and Marson-Percy model. Constr Build Mater 163:600–610
10. Wangler T, Lloret E, Reiter L, Hack N, Gramazio F, Kohler M, Bernhard M, Dillenburger B, Buchli J, Roussel N (2017) Digital concrete: opportunities and challenges. Rilem Technical Letters 1:67–75
11. Lootens D, Jousset P, Martinie L, Roussel N, Flatt RJ (2009) Yield stress during setting of cement pastes from penetration tests. Cem Concr Res 39:401–408
12. Bhattacherjee S, Santhanam M (2020) Enhancing buildability of 3D printable concrete by spraying of accelerating admixture on surface. In Second RILEM international conference on concrete and digital fabrication: digital concrete 2020 2. Springer, pp 13–22
13. Tay YWD, Qian Y, Tan MJ (2019) Printability region for 3D concrete printing using slump and slump flow test. Compos B Eng 174:106968
14. Mechtcherine V, Nerella VN, Kasten K (2014) Testing pumpability of concrete using Sliding Pipe Rheometer. Constr Build Mater 53:312–323
15. Nerella VN, Mechtcherine V (2018) Virtual Sliding Pipe Rheometer for estimating pumpability of concrete. Constr Build Mater 170:366–377
16. Rahul AV, Santhanam M, Meena H, Ghani Z (2019) 3D printable concrete: Mixture design and test methods. Cem Concr Compos 97:13–23

17. Rushing TS, Stynoski PB, Barna LA, Al-Chaar GK, Burroughs JF, Shannon JD, Kreiger MA, Case MP (2019) Investigation of concrete mixtures for additive construction. In: 3D concrete printing technology. Elsevier, pp 137–160
18. Rahul AV, Santhanam M (2020) Evaluating the printability of concretes containing lightweight coarse aggregates. Cem Concr Compos 109:103570
19. Le TT, Austin SA, Lim S, Buswell RA, Gibb AGF, Thorpe T (2012) Mix design and fresh properties for high-performance printing concrete. Mater Struct 45:1221–1232
20. Panda B, Tan MJ (2018) Experimental study on mix proportion and fresh properties of fly ash based geopolymer for 3D concrete printing. Ceram Int 44:10258–10265
21. Kwon H (2002) Experimentation and analysis of contour crafting (CC) process using uncured ceramic materials. University of Southern California
22. El Cheikh K, Rémond S, Khalil N, Aouad G (2017) Numerical and experimental studies of aggregate blocking in mortar extrusion. Constr Build Mater 145:452–463
23. Bos F, Wolfs R, Ahmed Z, Salet T (2016) Additive manufacturing of concrete in construction: potentials and challenges of 3D concrete printing. Virtual Phys Prototyp 11:209–225
24. Ma G, Li Z, Wang L (2018) Printable properties of cementitious material containing copper tailings for extrusion based 3D printing. Constr Build Mater 162:613–627
25. Perrot A, Rangeard D, Melinge Y, Estelle P, Lanos C (2009) Extrusion criterion for firm cement-based materials. Appl Rheol 19:53041–53042
26. Zhang Y, Zhang Y, Liu G, Yang Y, Wu M, Pang B (2018) Fresh properties of a novel 3D printing concrete ink. Constr Build Mater 174:263–271
27. Kazemian A, Yuan X, Cochran E, Khoshnevis B (2017) Cementitious materials for construction-scale 3D printing: laboratory testing of fresh printing mixture. Constr Build Mater 145:639–647
28. Nematollahi B, Vijay P, Sanjayan J, Nazari A, Xia M, Naidu Nerella V, Mechtcherine V (2018) Effect of polypropylene fibre addition on properties of geopolymers made by 3D printing for digital construction. Materials 11:2352
29. Nodehi M, Aguayo F, Nodehi SE, Gholampour A, Ozbakkaloglu T, Gencel O (2022) Durability properties of 3D printed concrete (3DPC). Autom Constr 142:104479
30. Tarhan Y, Şahin R (2021) Fresh and rheological performances of air-entrained 3D printable mortars. Materials 14:2409
31. Chen Y, Figueiredo SC, Li Z, Chang Z, Jansen K, Çopuroğlu O, Schlangen E (2020) Improving printability of limestone-calcined clay-based cementitious materials by using viscosity-modifying admixture. Cem Concr Res 132:106040

Chapter 5
Mechanical Properties of 3D-Printed Concrete

5.1 Introduction

Recently, 3D printing (3DPC) has seen increased attention in the construction sector [1], owing to its numerous advantages compared to the traditional concreting method. The requirement of a lower workforce, a faster construction process, and higher flexibility concerning the geometry of the structural element are a few of the essential characteristics of 3D printing in construction sector. Such advantages of 3DPC make it as an alternative in the construction field for fabricating a variety of structural and architectural elements that drives its acceptance among the engineers and designers.

However, before the large-scale applicability of 3D printing in the construction sector, there were several barriers that it had to cross [2]. The necessity for a 3D-printed concrete sample's precise mechanical qualities, which are mainly required for assessing the structural performance, grows significantly with the increased in the number of 3DPC projects. In contrast to traditional mold cast concrete, which is generally found to be homogeneous within formwork, 3DPC is placed following a layer-by-layer strategy without the need of any formwork. As a result of such layer-wise construction process, potential interlayer weaknesses arises, resulting in anisotropic mechanical behavior that may not follow to the assumptions of uniform strength as used in conventional design codes.

Moreover, the precise characterization of mechanical properties, such as compressive, tensile, and flexural strengths, is essential for evaluating the safety and load-bearing capacity of printed structures. These mechanical parameters form the foundation for structural design, durability assessments, and long-term performance predictions. However, the existing national and international standards, such as those outlined by the Bureau of Indian Standards [3], are primarily developed for conventionally cast concrete and may not adequately capture the unique behavior of 3D-printed materials. The absence of standardized guidelines for testing and evaluating 3DPC poses an additional barrier to its widespread adoption in structural engineering practice.

A. Chourasia et al., *Advances and Challenges in 3D Printed Concrete (3DPC)*, Materials Horizons: From Nature to Nanomaterials, https://doi.org/10.1007/978-981-95-8557-1_5

As the demand for 3D-printed concrete structures continues to grow, so too does the need for robust scientific understanding and standardized approaches to assess their mechanical performance. Addressing these challenges is crucial not only to advance the technological maturity of 3DPC but also to ensure the safety, reliability, and public acceptance of this emerging construction method.

5.2 Mechanical Properties

Concerning the mix-design of a 3D-printed binder, in addition to mechanical properties, the additional printability requirement makes it different from the traditional one [4]. Also, higher cement content is required in 3D-printed concrete than mold-cast specimens [5]. In the case of 3D printing, the additional cement takes the space of eliminated larger aggregate, which is necessary to maintain the required rheological properties of a 3D-printed mix for achieving quality extrusion and stability in shape in the layer-by-layer printing process.

However, utilization of such higher cement content creates a significant environmental challenge. It is already known that the production of 1-ton Ordinary Portland Cement (OPC) generates around 0.9 tons of CO_2; in total, 7% of the global CO_2 comes from the cement industries [6]. Further, the manufacturing of OPC is directly linked with the deployment of limestone natural reserves. Naturally, the utilization of OPC in 3D printing could undermine the ecological profits that this innovative technology seeks to offer.

Hence, it becomes essential to reduce the utilization of OPC, especially concerning to 3D printing, and to develop an alternate binder [7]. In response, in the current period, the development of greener construction materials employing agro-industrial wastes has attracted the attention of most civil engineering researchers. In the existing studies, different industrial wastes, e.g., fly ash (FA) [8], ladle furnace slag (LF) [9], ground granulated blast furnace slag (GGBFS) [10], and recycled glass powder [11], have gainfully been utilized in fabricating greener binder. Similar to industrial waste, various agricultural wastes, e.g., [12], hazelnut shell ash [13], and rice husk ash (RHA) [14] were also utilized in developing sustainable cementitious composite.

However, in the above-mentioned articles, the composite binary binder was prepared by partially replacing OPC with a specific waste material. In contrast to OPC based or partial OPC replaced agro-industrial waste-based binders, few studies have been found that employed geopolymer binders in 3D concrete printing. In one of the recent works, Pasupathy et al. [15] developed a 3D printable geopolymer-based binder employing GGBFS, FA, and grounded brick-based waste. The developed binder was then investigated for the mechanical properties of the samples extracted from the printed specimens from various printing directions. Likewise, Khan et al. [16] utilized GGBFS and construction-based wastes in developing a geopolymer binder that can be printed. Concerning the printing characteristics, the geopolymer binder was investigated for its buildability performance using an experimental process and a numerical approach [16]. Other investigations have incorporated steel slag [17], paper industry sludge [18], and glass powder [19] to develop printable binders. Yet, despite its potential, BA remains underexplored as a component in 3D printable ternary binder systems [20].

The choice of agro-industrial waste in binder development depends largely on local availability and chemical reactivity. In India, for example, around 259 million tons (Mt) of fly ash are produced annually [21]. Recognizing the established reactivity and favorable engineering properties of FA and GGBFS, the present study focuses on developing a 3D printable ternary geopolymer binder using these materials [22].

Furthermore, a number of researchers have already discovered that different printing factors, such direction of printing and printing interval between two layers, have a significant impact on the strength of a concrete that is 3D printed. Some of the main elements that cause a 3DPC specimen's mechanical properties to vary depending on the printing direction include the presence of gaps between two successive filaments, the relative displacement of an individual layer compared to another, and the layer-to-layer bond characteristics [23]. The greater time difference may result from the need for several printing sessions during large-scale construction in order to prevent the temperature of the printer's various components from rising or to position the reinforcement, which is one of the few. The impact of loading direction on the mechanical properties of additively constructed concrete, which serve as the foundation for assessing structural safety, must thus be investigated in relation to the interlayer gap duration [24, 25].

Globally, researchers have investigated the rheological behavior [26, 27], durability [28, 29], compressive strength [30], tensile strength [31], and flexural properties [32] of 3D-printed concretes subjected to loads in various directions relative to the printed layers, often comparing these results with those of conventionally cast specimens [33].

For instance, [34] reported that 3D-printed samples exhibited up to 45% lower compressive strength than mold-cast specimens. Conversely, [33] found cases where printed samples displayed 20% higher compressive strength compared to mold-cast samples. Similarly, the direction of loading relative to the printed layers can influence strength results: [35] observed reduced compressive strength when the load was applied parallel to the layers, whereas strength improved when loaded perpendicular. Conflicting findings also exist for flexural strength, with some studies [36, 37] reporting strong directional dependence, while others [38] found only minor effects on compressive strength but substantial impacts on flexural strength. In contrast, [39] observed minimal influence of loading direction on overall mechanical performance.

These discrepancies arise from complex, nonlinear phenomena, including variations in filament tensile and compressive strength, differences in interfacial bond strength across layers, and inconsistencies in filament size and geometry within the printed structure.

Motivated by these gaps, this study first examines the compressive strength of alkali-activated binary mixes composed of FA and GGBFS in various proportions to identify an optimal blend. The influence of superplasticizer (SP) content on compressive strength and open time is then evaluated, an important consideration to balance printability and mechanical performance. Subsequently, the effects of loading direction and interlayer time gap on the mechanical behavior (compressive, tensile, and flexural) of printed specimens are systematically investigated and compared to mold-cast samples.

This chapter would therefore be a major step toward comprehending the mechanical properties of a sustainable binder made of FA and GGBFS, which is ultimately crucial for assessing the structural safety of a structure that has been 3D printed.

5.3 Materials and Methods

5.3.1 *Materials*

In this study, the primary constituents that are used to fabricate the 3D printable concrete mixes include bagasse ash (BA) and Class-F type fly ash (FA) in compliance with ASTM C618 [40]. The choice of the above-mentioned materials reflects a focus on sustainability as well as performance, as each material contributes distinct chemical and physical properties essential for creating a robust, printable binder.

The employed fly ash was thoroughly analyzed for its particle size gradation, measuring d_{90} and d_{10} values of 50 μm and 1.25 μm, respectively. Such a type of particle size distribution indicates the presence of a wider range of particle sizes, that can impact packing density as well as rheological behavior, both of the above serve as critical factors in the case of extrusion-based printing. Furthermore, X-ray fluorescence (XRF) analysis shows that the employed FA in the current work contains approximately 52% silicon dioxide (SiO_2) and 25% aluminum oxide (Al_2O_3), highlighting its pozzolanic characteristics and appropriateness for the alkali activation.

In contrast, the utilized ground granulated blast furnace slag (GGBFS) contain 35.4% SiO_2 and a significantly higher amount of calcium oxide (CaO), approximately 42.2%. A higher value of CaO content makes GGBFS highly reactive, helping early-age development of strength properties when activated by alkaline solutions.

Mineralogical compositions of the employed raw materials are studied through the X-ray diffraction (XRD) method using a Rigaku X-ray diffractometer. A sample is scanned at a speed value of 1.5°/minute with the 2θ ranges between 5° and 80°. After that, the minerals are analyzed through X-pert High Score Plus software, and consequently, the XRD plots are developed in the Origin Pro software. XRD plots for the two employed waste materials such as FA and GGBFS are shown in Fig. 5.1. As shown in Fig. 5.1, the XRD patterns provide critical insight into the amorphous and crystalline phases present. Especially, the XRD plot for GGBFS shown in Fig. 5.1b indicates that GGBFS primarily contains a significant amount of reactive amorphous or glassy material phases, and during alkali activation, the alkaline solution (sodium hydroxide, for example) reacts with these glassy phases. The alkaline activator initiates dissolution and polycondensation reactions, leading to the formation

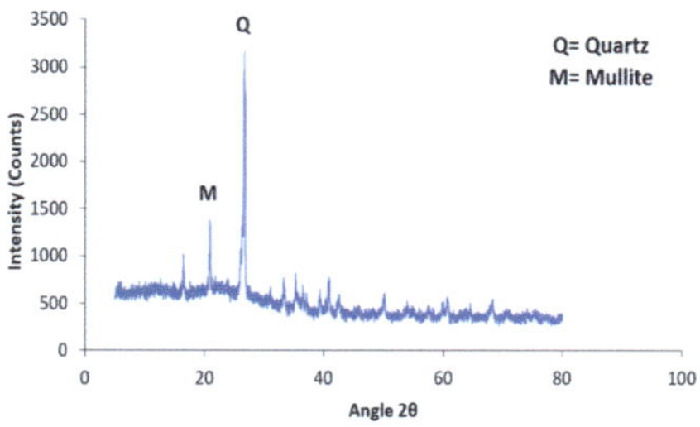

(a)

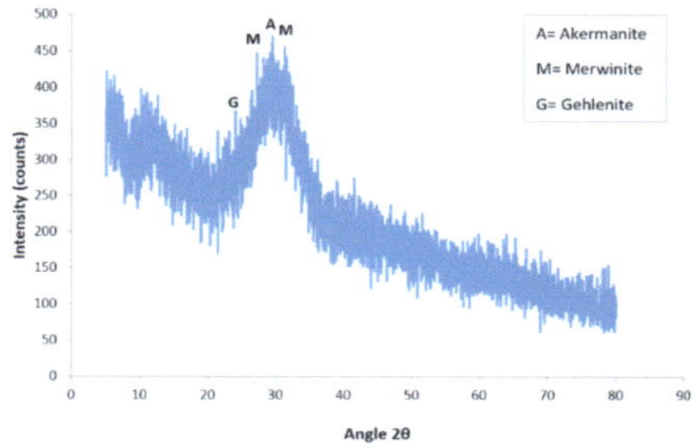

(b)

Fig. 5.1 XRD plots of **a** Fly ash and **b** GGBFS

of alkali alumino-silicate gel. The gel contributes to developing a three-dimensional cementitious matrix that provides strength and durability.

Similarly, the plot for FA shown in Fig. 5.1a demonstrates that fly ash is also rich in amorphous phases, and the alkali activator reacts with these phases. The reaction products include hydrated calcium aluminosilicates and other gel-like substances. The alkali activation of fly ash results in the forming of a binder phase that contributes to the strength of the material.

In addition to mineralogical analysis, morphological characteristics of the raw material are examined through Field Emission Scanning Electron Microscopy (FE-SEM) with TESCAN MIRA 3. Before imaging, the materials were carefully broken into smaller fragments and then oven-dried at 105 °C for 24 h to remove any residual moisture that could interfere with imaging. As illustrated in Fig. 5.2a, the SEM images of fly ash particles revealed a predominantly spherical shape with smooth surfaces. This spherical morphology is characteristic of Class-F fly ash and is known to enhance flowability and packing density—critical properties for achieving a uniform, continuous filament during 3D printing.

In contrast, the SEM images of GGBFS Fig. 5.2b displayed irregularly shaped, glassy particles with relatively smooth surfaces. Additionally, distinct crystalline phases such as gehlenite and merwinite were observed as more sharply defined and angular structures within the GGBFS. The interlocking nature of these particles can be advantageous, contributing not only to mechanical interlock within the hardened matrix but also to increased reactivity under alkaline conditions.

Complementing the binder system, fine aggregate was incorporated into the mix to ensure a balance of workability and mechanical performance. The fine aggregates used had a dry specific gravity of 2.62 and a fineness modulus of 2.4, both of which are suitable for producing a smooth and cohesive mortar suitable for extrusion through a nozzle.

To chemically activate the binder system, sodium hydroxide (NaOH) solutions of various molarities were used as the primary alkaline activator. The choice of NaOH and its concentration were carefully tailored to promote the dissolution of

(a) (b)

Fig. 5.2 SEM images **a** Fly ash and **b** GGBFS

aluminosilicate species in FA and GGBFS, thus facilitating rapid setting and strength gain required for 3D printing applications.

Additionally, a polycarboxylate-type superplasticizer was incorporated into the mixes to improve flowability without increasing water content. This admixture plays a critical role in enhancing printability by enabling smooth extrusion, reducing filament deformation, and ensuring layer buildability while maintaining the necessary mechanical strength in the hardened state.

The proportioning of these materials, along with the dosage of admixtures, was carefully optimized through iterative trial mixes to balance rheological properties, open time (essential for continuous printing), and final mechanical performance. The overall approach highlights a commitment to integrating sustainability, material science, and advanced construction technology—setting the stage for further investigations into the structural behavior and long-term durability of 3D-printed concrete elements.

5.3.2 Mix Proportions

In this study, a series of alkali-activated mortar mixes were systematically developed to achieve optimal mechanical and fresh properties tailored for 3D concrete printing applications. The primary binder system was formulated using fly ash (FA) and ground granulated blast furnace slag (GGBFS), two supplementary cementitious materials known for their high reactivity and environmental benefits.

The proportions of FA and GGBFS were carefully varied to understand their combined effects on key performance indicators such as strength development and workability. By adjusting the relative percentages of these two components, the mix design could be fine-tuned to balance early-age strength (favored by GGBFS) and long-term durability and flowability (associated with FA).

Alongside binder composition, the molarity of the sodium hydroxide (NaOH) alkali solution was systematically altered. This parameter is critical because the molar concentration of the activator governs the dissolution and subsequent geopolymerization reactions of the aluminosilicate phases in FA and GGBFS. A higher molarity typically accelerates setting and improves early strength, while lower concentrations may enhance open time and reduce premature stiffening—an essential requirement for continuous 3D printing.

Furthermore, a polycarboxylate-based superplasticizer was incorporated to refine the rheological behavior of the mix. The superplasticizer dosage was optimized to ensure smooth extrusion through the printer nozzle and to maintain adequate buildability of the printed layers, thus preventing undesirable deformations or collapse during printing.

Following the detailed evaluation of these mix-related parameters, each trial formulation was assessed for mechanical performance (primarily compressive strength) and open time. The optimal mix was identified based on its ability to achieve the required balance between printability and structural performance. Once established, this optimal mix was used to further study the influence of loading directions and interlayer time intervals on the mechanical behavior of printed samples—an investigation vital for understanding the anisotropic nature of 3D-printed concrete elements.

5.3.3 Printing Process and Printed Samples

This study employs a gantry type 3D printer that has a print volume capacity of 1.2 m × 1.2 m × 1.2 m. The average printing speed was kept at 60 mm/s, while the average extrusion volume was 27 cm^3/s. The material is 3D printed using a circular nozzle that has a 30 mm diameter. The printed filaments measure 30 mm in width and 15 mm in height. The above-mentioned dimensions of the printed filament were selected keeping in mind the standard size of the various cores that are to be extracted for different mechanical testing.

A vital part of the current work is to investigate the influence of interlayer time gaps on the various mechanical properties. Various samples have been printed with a layer-wise gap that was roughly 30 s and 60 min for a 60 mm/s printing speed. In the case of time gap of 30 s, the printing process was executed continuously, with no or minimal pause between consecutive layers. On contrary, for the 60-min gap, a deliberate delay was introduced and the next layer was printed after an hour had passed from the completion of the printing of previous layer. Such an approach mimics practical challenges that could be encountered on-site, e.g., machine downtime or the necessity for inserting reinforcements, managing the temperature of printer, etc.

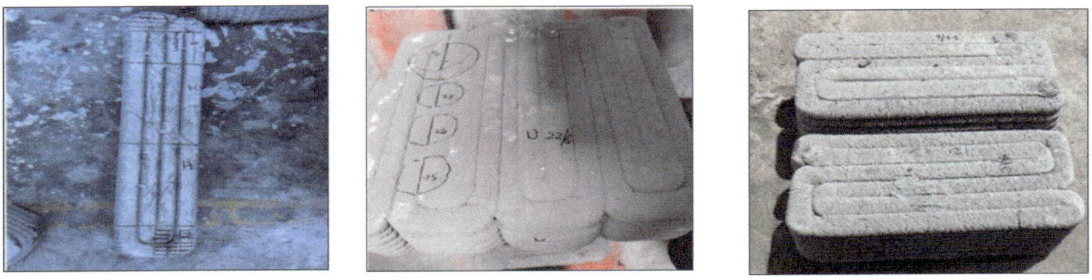

(a) (b) (c)

Fig. 5.3 Different lab-scale 3D printed samples: **a** 580 mm × 120 mm × 120 mm, **b** 420 mm × 420 mm × 120 mm, and **c** 400 mm × 150 mm × 150 mm

Specimens of different sizes are 3D printed in order to investigate how the printing time gap affects the anisotropic character of the printed samples under varied loading circumstances. The various 3D-printed samples made for various experimental experiments are displayed in Fig. 5.3. Beams measuring 580 mm in length, 120 mm in width, and 120 mm in height—composed of four horizontal and eight vertical layers along with printed bed of size 400 mm (length) × 150 mm (width) × 150 mm (height) were used for the test. The specifications of particular sample sizes for a given test type, using a particular nozzle that produces filament with dimensions of 30 mm in width and 15 mm in height, served as the inspiration for the selection of the different 3D-printed specimen sizes. Cores of the appropriate size are taken from samples that were printed and evaluated under various loading scenarios, which are covered in greater detail in Sect. 5.3.4. The thoughtful design of these specimens ensured that a wide range of mechanical properties could be assessed under different loading conditions, ultimately providing insights into the performance and reliability of 3D-printed concrete structures.

5.3.4 Evaluation Methods

To comprehensively understand the mechanical behavior of the printed samples, a series of rigorous evaluation methods were employed. Test specimens were precisely cut from the larger printed elements, taking into account different printing directions and interlayer time gaps. This allowed for a detailed assessment of anisotropic mechanical responses and comparison with conventionally mold-cast concrete samples. For reference and benchmarking, the compressive, split tensile, and flexural strengths of mold-cast specimens of identical dimensions were also determined.

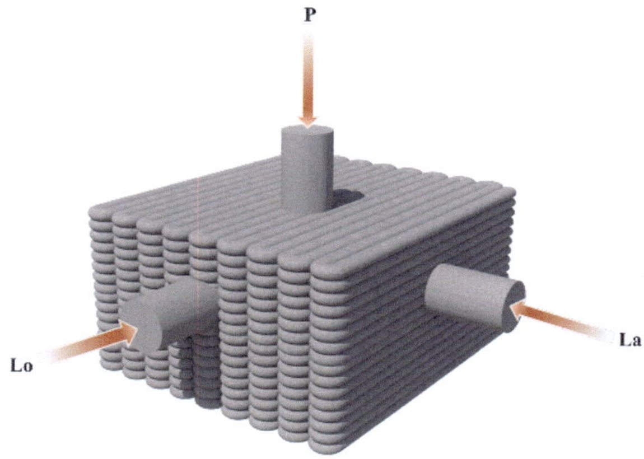

Fig. 5.4 Extraction of cores from printed sample in various direction for uniaxial compressive test (P-Load applied perpendicular to the printing direction, Lo-Load applied along the printing direction, and La-Load applied laterally to the printing direction)

5.3.4.1 Compressive Strength Test

Cylindrical cores with a diameter of 50 mm were extracted from the printed samples, following ASTM C42 [41] guidelines, as shown in Fig. 5.4. These cores were taken in three orientations to evaluate the directional dependence of compressive strength: along the longitudinal axis of printing (Lo), lateral to the printing direction (La), and perpendicular to the layers (P). Each core was tested under a loading rate of 0.5 $N/mm^2/min$ to capture the influence of the print-induced anisotropy. Furthermore, additional cores were taken from beds printed with different interlayer time gaps to understand the effects of construction interruptions on compressive performance.

5.3.4.2 Split Tensile Strength Test

For split tensile testing, cores measuring 75 mm in diameter and 150 mm in height were extracted, again considering the three principal directions relative to the print layers, as depicted in Fig. 5.5. The tests were carried out according to ASTM C496 [42] standards. Similar to compressive testing, the tensile strength of mold-cast control specimens was also evaluated for comparison. Additional tests on samples from different interlayer time gaps allowed for a detailed analysis of how these delays influence tensile properties, particularly important for understanding the material's crack resistance under service loads.

5.3.4.3 Flexural Strength Test

Flexural performance was assessed using beam specimens measuring 500 mm × 100 mm × 100 mm, as shown in Fig. 5.6. The tests followed ASTM C78 [43] standards and involved four-point bending to ensure a uniform moment distribution along the center span. The impact of interlayer time gaps on flexural strength was

Fig. 5.5 Extraction of cores from printed sample in various direction for split tensile test (P-Load applied perpendicular to the printing direction, Lo-Load applied along the printing direction, and La-Load applied laterally to the printing direction)

also investigated to determine the continuity and integrity of printed filaments across successive layers.

5.3.4.4 Split Prism-Based Tensile Bond Strength Test

Recognizing that weak interlayer bonds can significantly compromise the performance of 3D-printed elements, the tensile bond strength was examined using a split prism test, similar to the split tensile method and conforming to ASTM C496 [42] standards. Figure 5.7 illustrates the loading and boundary conditions adopted during this test. Here, a steel plate with an attached bar was placed along the interlayer plane, and the load was applied at a rate of 0.5 $N/mm^2/min$ until failure occurred. The specimens used for this test were cubic blocks measuring 120 mm on each side, extracted from the larger 580 mm × 120 mm × 120 mm printed beams. This evaluation was crucial for understanding how interlayer adhesion influences the overall integrity and reliability of 3D-printed concrete structures.

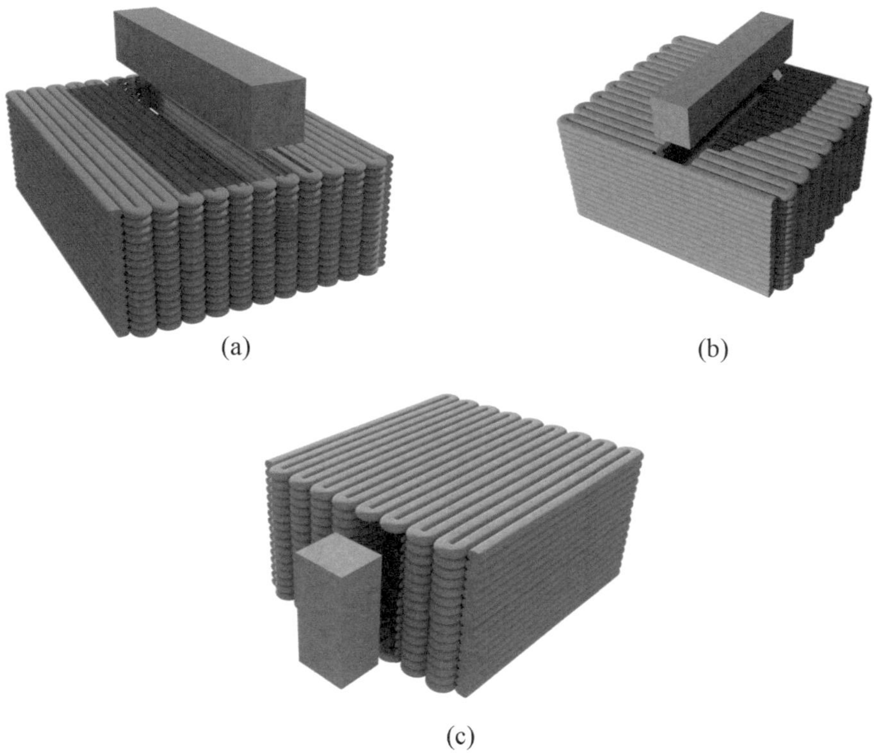

Fig. 5.6 Extraction of samples from different directions for flexure test

Fig. 5.7 Loading and boundary conditions for tensile bond strength test

5.4 Results and Discussion

This section comprehensively presents and analyzes the mechanical performance of the developed 3D-printed concrete specimens. Emphasis is placed on the influence of printing direction, interlayer time gap, and mix parameters such as Fly Ash (FA) to GGBFS ratio and superplasticizer content. By evaluating compressive, tensile, flexural, and bond strengths, the anisotropic behavior and interfacial properties of the 3D-printed concrete are systematically discussed.

5.4.1 Effect of FA to GGBFS Ratio and Activator Molarity on Binder Strength

The initial phase of this investigation focused on optimizing the mix design by studying the influence of different FA to GGBFS ratios and varying molarities of alkali activator solution (6 M, 8 M, 10 M, and 12 M) on the compressive strength of the binder. Figure 5.8 clearly illustrates the trend: as the GGBFS content increases, the compressive strength also increases across all molarities. This improvement can be attributed to the higher reactivity of GGBFS, which is rich in calcium and silica. When activated by alkalis, these components lead to the formation of additional cementitious phases, resulting in a denser and more robust microstructure.

A distinct increase in compressive strength is noted between 20 and 30% GGBFS replacement, with values peaking around 41.2 MPa for both 8 and 10 M activator solutions. Considering performance and cost-effectiveness, the 8 M solution is chosen

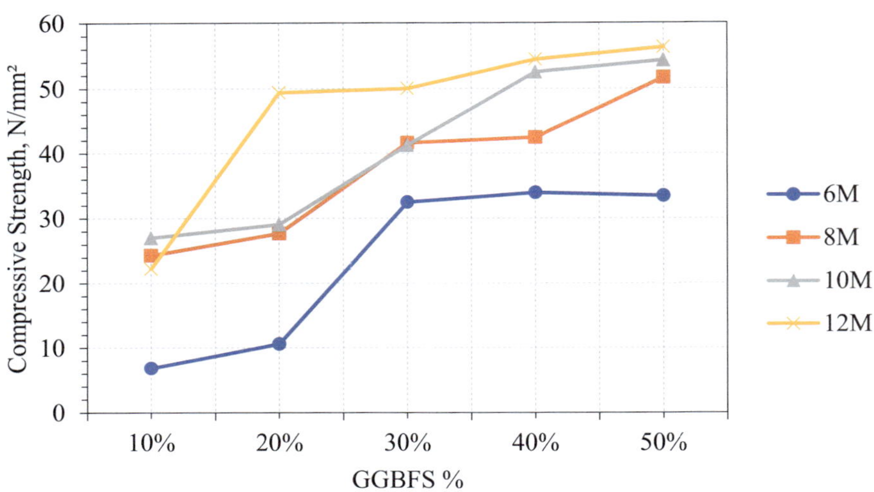

Fig. 5.8 Effect of GGBFS inclusion on compressive strength at various molarities

as optimal for subsequent studies. Nevertheless, the GGBFS content could be further tuned depending on project-specific strength requirements. The ultimate goal here is to design a printable mix capable of achieving M30 grade concrete with a target strength of 38.25 MPa, without compromising extrudability and buildability.

5.4.2 Influence of Superplasticizer Content on Strength and Open Time

The influence of superplasticizer (SP) content on the compressive strength and open time was investigated using the optimized FA:GGBFS ratio of 70:30 and 8 M alkali solution. Figure 5.9 shows that compressive strength increases progressively with SP content up to 1.2%, achieving a peak value of 45.2 MPa at 1% SP dosage. Beyond this point, the mix tends to lose structural integrity due to excessive flowability, leading to a reduction in open time. The increase in open time up to 1% SP dosage can be attributed to enhanced deflocculation and improved dispersion of binder particles, which delay setting and allow longer workable periods—a critical requirement in layer-by-layer printing.

Thus, the final optimized mix (FA:GGBFS = 70:30, 8 M alkali solution, 1% SP) balances strength, workability, and open time, serving as the baseline for further analysis of directional and time gap effects.

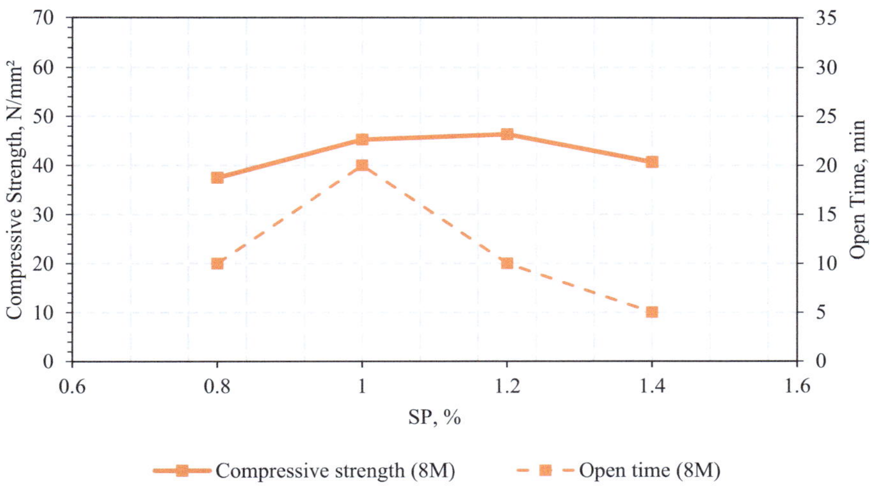

Fig. 5.9 Effect of SP dosage on compressive strength and open time (Fly ash = 70% and GGBFS = 30%)

5.4.3 Influence of Printing Direction and Interlayer Time Gap on Mechanical Properties

5.4.3.1 Compressive Strength Analysis

Figure 5.10 illustrates the compressive strengths measured in three principal loading directions—perpendicular, longitudinal, and lateral—for specimens printed with an interlayer gap of about 30 s. For comparison, the strength of mold-cast specimens is also presented.

Printed specimens exhibit reduced compressive strength compared to mold-cast counterparts, which aligns with findings from Singh et al. [44] and [45]. The reduced strength is largely due to weak interfilament bonds and internal pores that develop between deposited layers. Among the directions tested, the perpendicular orientation consistently shows the highest compressive strength, with only a 3.8% reduction relative to mold-cast specimens. In contrast, longitudinal and lateral directions show reductions of 14.5% and 19.6%, respectively. Similar trends have been reported in the literature [23, 46], emphasizing that compressive performance is strongly governed by layer orientation and interfacial quality.

Figure 5.11 further demonstrates the impact of different interlayer time gaps. Up to a 10-min gap, compressive strength remains relatively stable (within 1% of the 30-s benchmark), suggesting that the fresh mix maintains adequate cohesion within this timeframe. However, strength gradually decreases beyond 10 min, with marked reductions observed at 20 and 40-min gaps. At 40 min, reductions reach up to 9.6% in lateral, 7.3% in longitudinal, and 15.4% in perpendicular directions. The decreasing trend can be explained by the progressive setting of the previously printed layer, which leads to poorer adhesion with subsequent layers. Based on these observations,

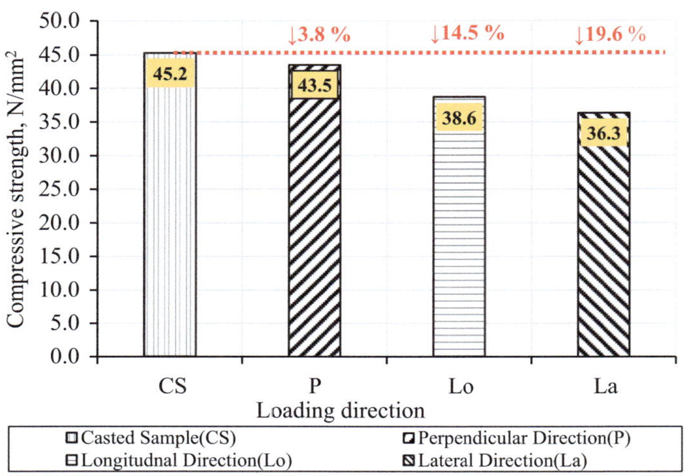

Fig. 5.10 Compressive strength in various loading directions

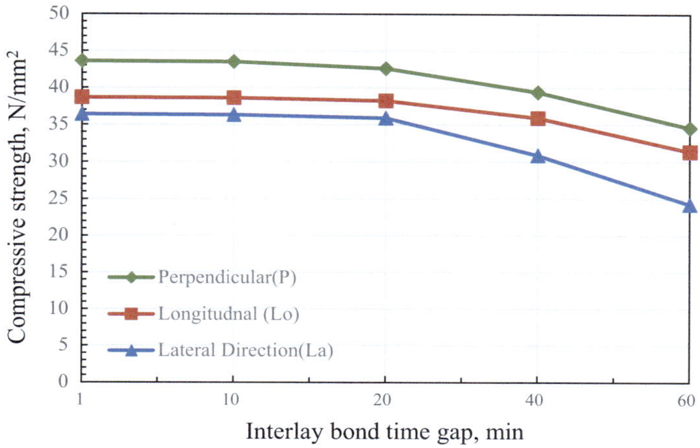

Fig. 5.11 Effect of interlayer time gap on compressive strength

a 20-min maximum time gap is recommended to minimize strength degradation while maintaining practical flexibility during printing.

5.4.3.2 Split Tensile Strength Analysis

Figure 5.12 presents the split tensile strength results across different directions for a 30-s interlayer gap, along with mold-cast specimen results. As expected, tensile strengths in all printed directions are lower than in mold-cast specimens.

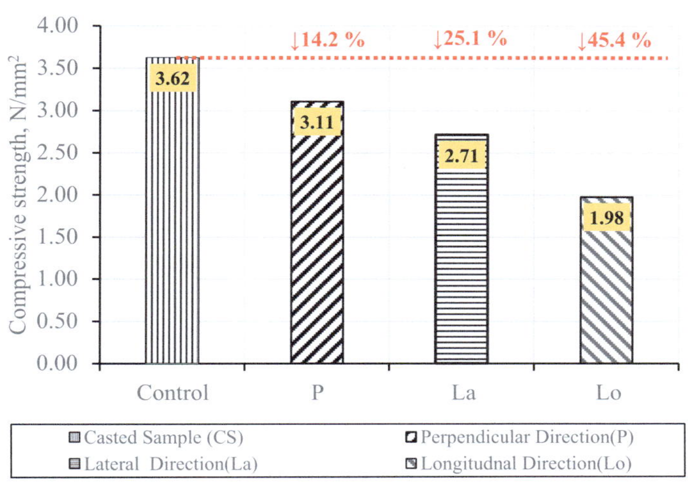

Fig. 5.12 Effect of printing in various directions on split tensile strength

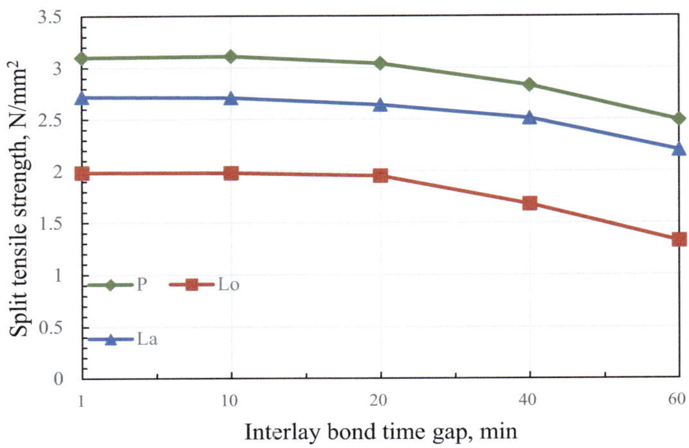

Fig. 5.13 Effect of interlayer time gap on split tensile strength

Notably, the longitudinal direction exhibits the highest reduction (approximately 45%), while the perpendicular direction maintains relatively higher tensile strength. This behavior highlights the crucial role of interlayer bond quality under tensile loading conditions. Figure 5.13 shows the effect of varying interlayer time gaps on tensile strength. Up to 10 min, tensile strength reductions are negligible. However, at 40 min, reductions approach 15% across directions, and at 60 min, reductions increase further—especially in the longitudinal direction (about 32%). These findings underline the importance of controlling time gaps to preserve tensile performance, which is critical for preventing early-age cracking and ensuring overall structural integrity.

5.4.3.3 Flexural Strength Analysis

The flexural strength results presented in Fig. 5.14 further corroborate the anisotropic nature of printed specimens. Printed samples show a more pronounced reduction in flexural strength compared to compressive and split tensile strengths, which can be attributed to greater sensitivity to interlayer weaknesses. As shown, the perpendicular orientation again demonstrates the highest flexural capacity, while the longitudinal direction exhibits the lowest strength. Such trends are consistent with observations from [45, 47].

Figure 5.15 illustrates typical damage patterns under four-point bending. In the perpendicular direction, failure initiates as a central V-notch crack, whereas in the longitudinal direction, cracks tend to form closer to the quarter-span locations, similar to mold-cast behavior. These patterns highlight the need for carefully planned printing paths and potential retrofitting strategies to mitigate specific failure modes.

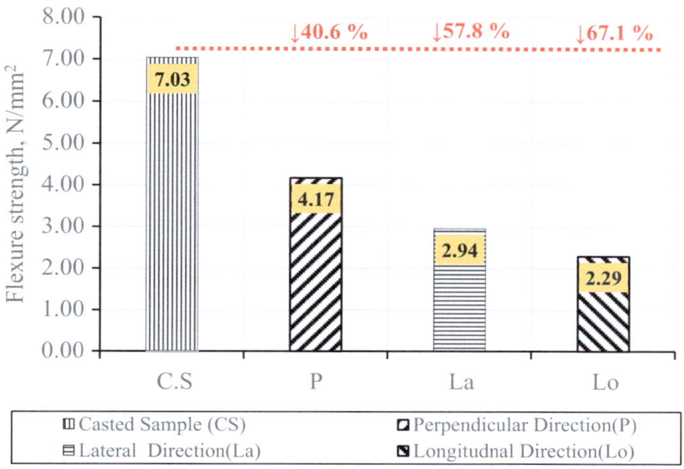

Fig. 5.14 Flexural strength in various directions

Figure 5.16 demonstrates that flexural strength remains relatively stable up to a 20-min interlayer gap, with reductions below 20%. Beyond this, strength degradation accelerates, reaching almost 35% at a 60-min gap. These insights are critical for on-site decision-making regarding pause durations and layer scheduling.

5.4.4 *Interlayer Bond Strength Analysis*

Finally, the interlayer tensile bond strength results are summarized in Fig. 5.17. Here, the split prism method was employed to evaluate adhesion strength, as detailed by [35]. The results confirm that the longitudinal direction exhibits the lowest bond strength, corresponding with lower compressive, tensile, and flexural strengths discussed earlier. Conversely, the perpendicular direction shows the highest bond strength, likely due to enhanced interlocking between filaments. Figure 5.18 depicts the typical failure through the interlayer region, emphasizing the significance of interface quality. Figure 5.19 shows the degradation of bond strength with increasing interlayer time gaps. As the time gap grows, the previously deposited layer stiffens, reducing cohesive interaction with the next layer.

At larger time gaps, bond strength reductions follow similar trends as other mechanical properties, reinforcing the importance of controlling layer scheduling. Although bond strength values vary widely in the literature, the results here align with ranges reported by [48], emphasizing the need for continued research into optimizing bond performance in printed concrete.

(a)

(b)

Fig. 5.15 Crack patterns in printed beams under flexure. **a** Perpendicular direction **b** Longitudinal direction

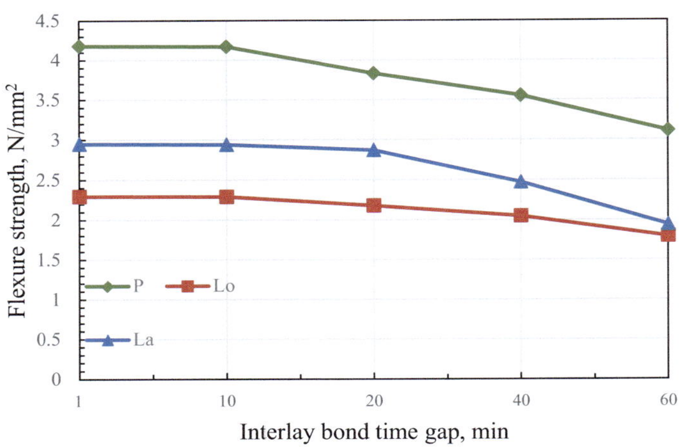

Fig. 5.16 Effect of interlayer time gap on flexural strength

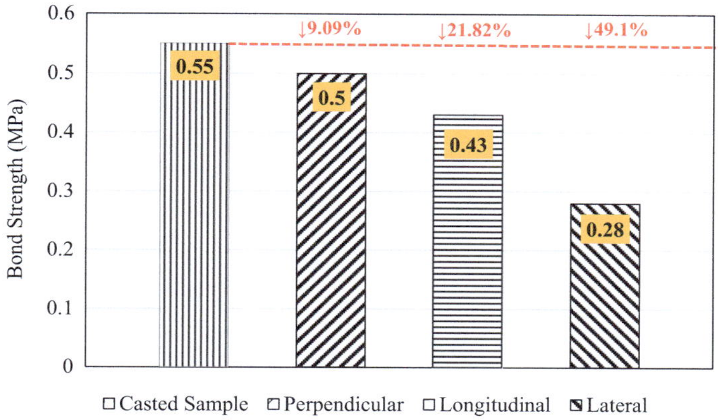

Fig. 5.17 Comparison of tensile bond strength of printed specimen (cores extracted from three direction: perpendicular, longitudinal and lateral)

Fig. 5.18 Debonding along interlayer region

5.4.5 Effect of Interfilament Pores on Compressive Strength of 3D-Printed Concrete

In this section, the impact of pore distribution and size—especially those present at different heights within printed specimens—on compressive strength is systematically investigated. To this end, cylindrical cores with a diameter of 50 mm and a height of approximately 60 mm were extracted from both the top and bottom regions of the printed beds Fig. 5.20. The cores were then separated into two groups to

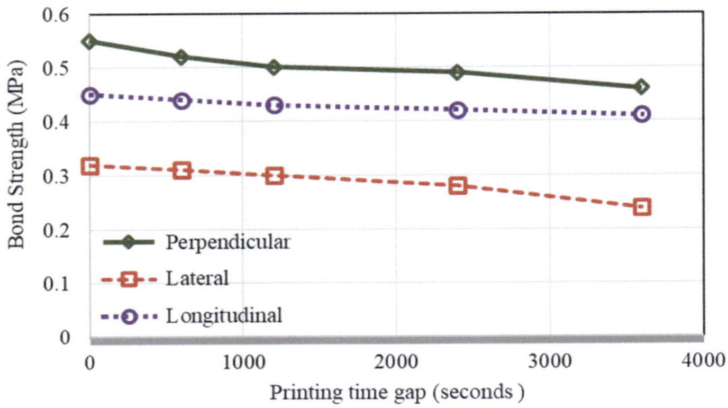

Fig. 5.19 Effect of interlayer time gap on tensile bond strength in different direction

evaluate how the vertical position and associated porosity affect their compressive behavior.

Visual inspection Fig. 5.20 reveals that the cores extracted from the top regions exhibit noticeably larger and more frequent interfilament gaps compared to those from the bottom. This observation aligns with the inherent characteristics of layer-by-layer extrusion in 3D printing, where gravitational settling and compaction during early layers help minimize void formation at the bottom, whereas top layers are more prone to reduced compaction and greater pore development as printing progresses.

The extracted cores were subjected to uniaxial compressive loading, and their strength values were corrected using standard size correction factors. As expected,

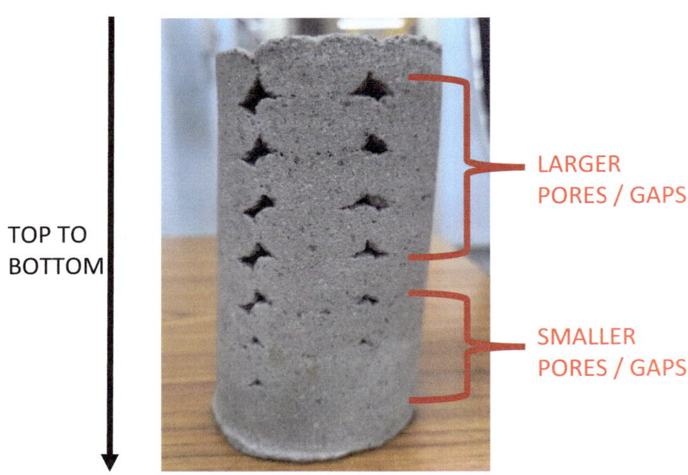

Fig. 5.20 Extracted cylindrical cores of diameter 50 mm showing higher interlayer pores at top and lower pores at bottom layers of printed specimens

cores extracted from the top layers exhibit a lower average compressive strength of **36.4 MPa**, whereas those from the bottom layers achieve a higher average strength of **38.5 MPa**. The reduction in strength for top cores can be directly attributed to the presence of larger interfilament pores, which act as stress concentrators and initiate early microcracking under load.

These findings underscore the critical role of layer compaction and pore control during printing. Improved strategies for minimizing top-layer porosity—such as adjusting extrusion parameters, modifying rheology, or using additional compaction aids—could lead to more homogeneous and robust printed concrete elements.

5.5 Closure

The study of various mechanical properties presented in the current chapter offers a detailed understanding of the performance of 3D printed concrete (3DPC) under different types of mechanical load, highlighting the effect of material composition, loading direction, interfilament time gap, and microstructural characteristics such as interfilament pores.

At first, the study demonstrated the advantages of optimizing the mix design, representing that a binder made with a fly ash (FA) and ground granulated blast furnace slag (GGBFS) in the ratio of 70:30 and activated by an alkali solution of 8 M along with 1% superplasticizer provides the desired compressive strength and maintains the satisfactory printability performance. The obtained mix design, thereafter, served as a baseline for analyzing the mechanical behavior of 3DPC.

An extensive investigation of anisotropy in mechanical properties due to layer-wise printing highlights that 3DPC displays direction dependent mechanical performance. In this regards, compressive, tensile, and flexural strengths were found to be higher when the load was applied perpendicular to the printing direction and lowest in the longitudinal direction. Such a dependency of mechanical properties on the loading direction with respect to printed layers could be due to the variation in interlayer bond along with the presence of interfilament pores that inherently impacts the mechanical properties differently in different directions.

Moreover, the interfilament time gap was identified as one of the crucial factors that affect the mechanical behavior. A shorter time interfilament gaps up to around 10 min had negligible effect on different strength values. Whereas, a larger time interval, beyond around 20 min, results in significant reductions in properties. The weakened bond among the consecutive layers in the case of larger time gap underlines the need for controlling the onsite printing process parameter carefully.

The investigation of split tensile and flexural strengths of 3DPC further strengthened the important role of interfilament bond. It is observed that split tensile strength has a stronger dependency on interlayer bond than compressive strength. This highlights the potential weaknesses in a loading condition where tensile or flexural stresses govern. Likewise, drop in flexural strength were more severe than compressive tests, emphasizing that flexural behavior is more susceptible to interlayer bond.

Assessment of tensile bond strength through a split prism-based test process gives crucial insights about interlayer adhesion. The test reveals that the bond strength in the longitudinal direction was lowest, whereas the perpendicular direction gives the highest strength. The sequential drop in the bond strength values with the increasing interfilament time gap underscored the importance of keeping the construction delays as low as possible to uphold the structural integrity.

Further, the influence of interlayer pores on the compressive strength of 3DPC specimen highlighted the variability of the strength properties across the printed samples spatially. Top layers, often characterized by larger pore sizes and higher porosity, exhibited lower compressive strengths compared to denser bottom layers. This observation suggests that continuous monitoring and optimization of extrusion and compaction parameters are essential to achieving homogeneous mechanical performance throughout the printed element.

In summary, the current chapter highlighted that the various printing process related entities (e.g., interlayer time gap) and mix design of the material plays an important role in the performance of 3DPC element under different types of mechanical loadings. Thus, without an optimal choice of the above-mentioned parameters, the immense potential of 3DPC such as customized, fast, and efficient utilization of material in construction could be sidelined. The insights obtained in the present chapter provide a strong groundwork for further developments of 3D printing in construction sector, allowing more reliable material and process design, safer applications at the structural level, and advancement toward large-scale adoption.

References

1. Kaikai X, Yadong G, Qiang Z (2023) Comparison of traditional processing and additive manufacturing technologies in various performance aspects: a review. Arch Civil Mech Eng 23:1–28
2. El-Sayegh S, Romdhane L, Manjikian S (2020) A critical review of 3D printing in construction: benefits, challenges, and risks. Arch Civil Mech Eng 20:1–25
3. BIS (2019) Concrete mix proportioning-guidelines. BIS, New Delhi, India, IS 10262: 2019
4. Xu Z, Zhang D, Li H, Sun X (2023) Effects of the distribution of solid particles on the rheological properties and buildability of 3DPM fresh pastes with different FA/GGBFS content. Arch Civil Mech Eng 23:82
5. Samudrala M, Mujeeb S, Lanjewar BA, Chippagiri R, Kamath M, Ralegaonkar RV (2023) 3D-printable concrete for energy-efficient buildings. Energies (Basel) 16:4234
6. Dinh HL, Doh J-H, Liu J, Lu L, Song H, Park D (2023) Comprehensive assessment of geopolymer concrete mechanical and environmental performance with glass cullet fine aggregates. J Build Eng 107094
7. Hansted FAS, Mantegazini DZ, Ribeiro TM, Gonçalves CEC, Balestieri JAP (2023) A mini-review on the use of waste in the production of sustainable Portland cement composites. Waste Manage Res 41:828–838
8. Shah SFA, Chen B, Oderji SY, Haque MA, Ahmad MR (2020) Improvement of early strength of fly ash-slag based one-part alkali activated mortar. Constr Build Mater 246:118533
9. Singh SK, Vashistha P (2021) Development of newer composite cement through mechano-chemical activation of steel slag. Constr Build Mater 268:121147

10. Reddy MS, Dinakar P, Rao BH (2018) Mix design development of fly ash and ground granulated blast furnace slag based geopolymer concrete. J Build Eng 20:712–722
11. Abdelli HE, Mokrani L, Kennouche S, de Aguiar JLB (2020) Utilization of waste glass in the improvement of concrete performance: a mini review. Waste Manage Res 38:1204–1213
12. de Sande VT, Sadique M, Bras A, Pineda P (2022) Activated sugarcane bagasse ash as efficient admixture in cement-based mortars: mechanical and durability improvements. J Build Eng 59:105082
13. Yurt Ü, Bekar F (2022) Comparative study of hazelnut-shell biomass ash and metakaolin to improve the performance of alkali-activated concrete: a sustainable greener alternative. Constr Build Mater 320:126230
14. Jittin V, Bahurudeen A (2022) Evaluation of rheological and durability characteristics of sugarcane bagasse ash and rice husk ash based binary and ternary cementitious system. Constr Build Mater 317:125965
15. Pasupathy K, Ramakrishnan S, Sanjayan J (2023) 3D concrete printing of eco-friendly geopolymer containing brick waste. Cem Concr Compos 138:104943
16. Khan SA, İlcan H, Aminipour E, Şahin O, Al Rashid A, Şahmaran M, Koç M (2023) Buildability analysis on effect of structural design in 3D concrete printing (3DCP): an experimental and numerical study. Case Stud Constr Mater 19:e02295
17. Yu Q, Zhu B, Li X, Meng L, Cai J, Zhang Y, Pan J (2023) Investigation of the rheological and mechanical properties of 3D printed eco-friendly concrete with steel slag. J Build Eng 72:106621
18. Chen M, Li H, Yang L, Wang S, Zhao P, Huang Y, Lu L, Yue G, Li Q (2022) Rheology and shape stability control of 3D printed calcium sulphoaluminate cement composites containing paper milling sludge. Addit Manuf 54:102781
19. Deng Q, Zou S, Xi Y, Singh A (2023) Development and characteristic of 3D-printable mortar with waste glass powder. Buildings 13:1476
20. Jesus M, Teixeira J, Alves JL, Pessoa S, Guimarães AS, Rangel B (2022) Potential use of sugarcane bagasse ash in cementitious mortars for 3D printing. In: Materials design and applications IV. Springer, pp 89–103
21. Annual Report on fly ash generation at coal/lignite based thermal power stations and its utilization in the country for the year 2021–22, Central Electricity Authority, Ministry of Power, Government of India., (n.d). https://cea.nic.in/wp-content/uploads/tcd/2022/08/Fly_ash_Generation_and_utilisation_Report_2021_22-1.pdf.
22. Siddique R, Bennacer R (2012) Use of iron and steel industry by-product (GGBS) in cement paste and mortar. Resour Conserv Recycl 69:29–34
23. Luo S, Lin Q, Xu W, Wang D (2023) Effects of interval time and interfacial agents on the mechanical characteristics of ultra-high toughnessnn cementitious composites under 3D-printed technology. Constr Build Mater 374:130936
24. Ghosh D, Anleu PB, Le Pape Y, Ma ZJ (2023) Effect of interlayer time-lapse and workability retention on printed concrete performance. J Mater Civ Eng 35:4023362
25. Tarhan Y, Şahin R (2024) The physicomechanical behavior and microstructure of air-entrained 3D printable concrete. J Mater Civ Eng 36:4023489
26. Comminal R, da Silva WRL, Andersen TJ, Stang H, Spangenberg J (2020) Modelling of 3D concrete printing based on computational fluid dynamics. Cem Concr Res 138:106256
27. Muthukrishnan S, Kua HW, Yu LN, Chung JKH (2020) Fresh properties of cementitious materials containing rice husk ash for construction 3D printing. J Mater Civ Eng 32:4020195
28. Nodehi M, Aguayo F, Nodehi SE, Gholampour A, Ozbakkaloglu T, Gencel O (2022) Durability properties of 3D printed concrete (3DPC). Autom Constr 142:104479
29. Şahin HG, Mardani A (2023) Mechanical properties, durability performance and interlayer adhesion of 3DPC mixtures: a state-of-the-art review. Struct Concr
30. Liu C, Wang Z, Wu Y, Liu H, Zhang T, Wang X, Zhang W (2023) 3D printing concrete with recycled sand: the influence mechanism of extruded pore defects on constitutive relationship. J Build Eng 68:106169

31. Chen W, Pan J, Zhu B, Ma X, Zhang Y, Chen Y, Li X, Meng L, Cai J (2023) Improving mechanical properties of 3D printable 'one-part'geopolymer concrete with steel fiber reinforcement. J Build Eng 107077
32. Deng Z, Jia Z, Zhang C, Wang Z, Jia L, Ma L, Wang X, Zhang Y (2022) 3D printing lightweight aggregate concrete prepared with shell-packing-aggregate method-printability, mechanical properties and pore structure. J Build Eng 62:105404
33. Murcia DH, Genedy M, Taha MMR (2020) Examining the significance of infill printing pattern on the anisotropy of 3D printed concrete. Constr Build Mater 262:120559
34. Kruger J, van Zijl G (2021) A compendious review on lack-of-fusion in digital concrete fabrication. Addit Manuf 37:101654
35. Zareiyan B, Khoshnevis B (2017) Effects of interlocking on interlayer adhesion and strength of structures in 3D printing of concrete. Autom Constr 83:212–221
36. Sanjayan JG, Nematollahi B, Xia M, Marchment T (2018) Effect of surface moisture on interlayer strength of 3D printed concrete. Constr Build Mater 172:468–475
37. Ma G, Li Z, Wang L, Wang F, Sanjayan J (2019) Mechanical anisotropy of aligned fiber reinforced composite for extrusion-based 3D printing. Constr Build Mater 202:770–783
38. Hambach M, Rutzen M, Volkmer D (2019) Properties of 3D-printed fiber-reinforced Portland cement paste. In: 3D concrete printing technology. Elsevier, pp 73–113
39. Wolfs RJM, Bos FP, Salet TAM (2019) Hardened properties of 3D printed concrete: the influence of process parameters on interlayer adhesion. Cem Concr Res 119:132–140
40. ASTM C618–15 (2015) Standard specification for coal fly ash and raw or calcined natural Pozzolan for use in concrete. ASTM International, West Conshohocken, PA, USA
41. ASTM (2013) Standard test method for obtaining and testing drilled cores and sawed beams of concrete. ASTM C42/C42M-13, West Conshohocken, PA
42. ASTM C 496–96 (1996) Standard test method for splitting tensile strength of cylindrical concrete specimens American standard test method. West Conshohocken, Penn, USA
43. ASTM C 78/C 78M (2010) Standard test method for flexural strength of concrete (using simple beam with third-point loading). West Conshohocken (PA)
44. Singh A, Liu Q, Xiao J, Lyu Q (2022) Mechanical and macrostructural properties of 3D printed concrete dosed with steel fibers under different loading direction. Constr Build Mater 323:126616
45. Xiao J, Liu H, Ding T (2021) Finite element analysis on the anisotropic behavior of 3D printed concrete under compression and flexure. Addit Manuf 39:101712
46. Pasupathy K, Ramakrishnan S, Sanjayan J (2021) Formulating eco-friendly geopolymer foam concrete by alkali-activation of ground brick waste. J Clean Prod 325:129180
47. Liu C, Zhang R, Liu H, He C, Wang Y, Wu Y, Liu S, Song L, Zuo F (2022) Analysis of the mechanical performance and damage mechanism for 3D printed concrete based on pore structure. Constr Build Mater 314:125572
48. Van Der Putten J, De Schutter G, Van Tittelboom K (2018) The effect of print parameters on the (micro) structure of 3D printed cementitious materials. In: RILEM international conference on concrete and digital fabrication. Springer, pp 234–244

Chapter 6
Durability Properties of 3DPC

6.1 Overview

Although 3D concrete printing is widely regarded as a promising approach for producing innovative structural elements with multiple advantages, the long-term durability and performance of these printed components depend on a set of time-sensitive parameters. These parameters play a critical role since the structures are likely to face exposure to external environments. In this context, the present chapter reviews durability aspects of 3DPC sections, focusing on the influence of printing conditions, mixture composition, and material properties with respect to shrinkage behavior, porosity, and the interconnection of pores, freeze–thaw resistance, resistance to chemical and acid, along with moisture absorption. The analysis indicates that the longevity of 3D-printed sections is strongly influenced by printing time gap and shrinkage potential, both might substantially modify porosity and pore connectivity, particularly at interlayer bonding zones. Recommendations for enhancing the durability of 3DPC elements under varying environmental exposures are also discussed at the end of the chapter.

For 3D printing, the concrete mix generally incorporates a higher binder content and follows a different placement mechanism (pumping instead of formwork) compared to traditional cast concrete. Consequently, the durability performance of 3DPC differs from conventional practices. Figure 6.1 illustrates, in schematic form, a four-phase deterioration process that begins with the hardening stage of the printed material. Moreover, due to its layer-by-layer construction, 3DPC exhibits anisotropy in durability characteristics, similar to its mechanical properties. Figure 6.2 schematically highlights these anisotropic features in relation to durability performance, showing that varying influences from different orientations cause differing levels of mechano-chemical durability loss in 3D-printed concrete.

A. Chourasia et al., *Advances and Challenges in 3D Printed Concrete (3DPC)*, Materials Horizons: From Nature to Nanomaterials, https://doi.org/10.1007/978-981-95-8557-1_6

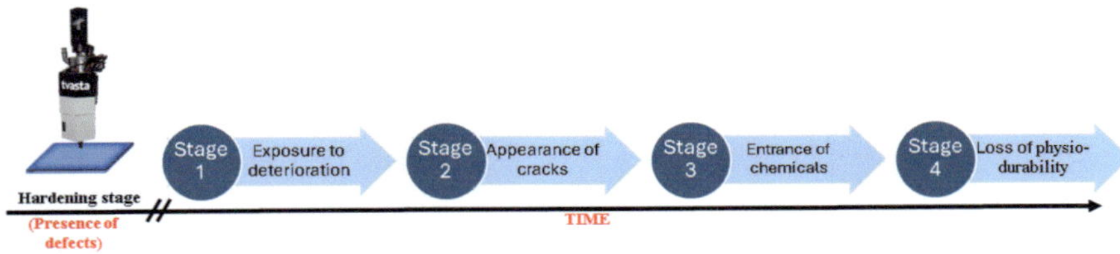

Fig. 6.1 Deterioration process of 3D printed concrete (3DPC)

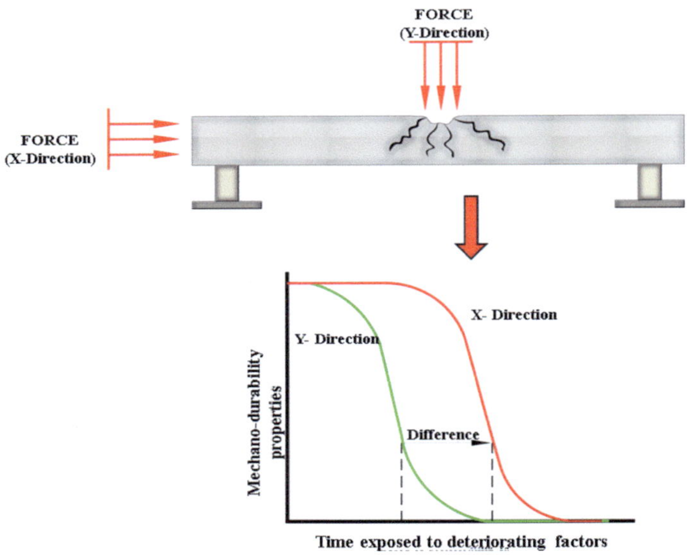

Fig. 6.2 Schematic diagram representing the directional dependency of durability of 3DPC

6.2 Durability Analysis

The advancement of 3D printed concrete (3DPC) is still at an early stage, with most efforts directed toward developing a printable mix and optimizing the material's early-age behavior. Because the longevity of 3D-printed structures is strongly tied to their durability, it must remain a primary focus. The durability of 3DPC largely depends on the performance of the concrete itself and the surrounding environmental conditions. Key environmental factors influencing durability include chloride-induced corrosion, carbonation, freeze-thaw cycles, and fire exposure [1]. Additionally, shrinkage and cracking are considered the most critical durability-related concerns [2]. Several of these influencing factors that govern the durability of 3DPC are illustrated in Fig. 6.3.

In recent years, research has primarily emphasized controlling rheological behavior, accelerating the additive manufacturing, and improving the mechanical

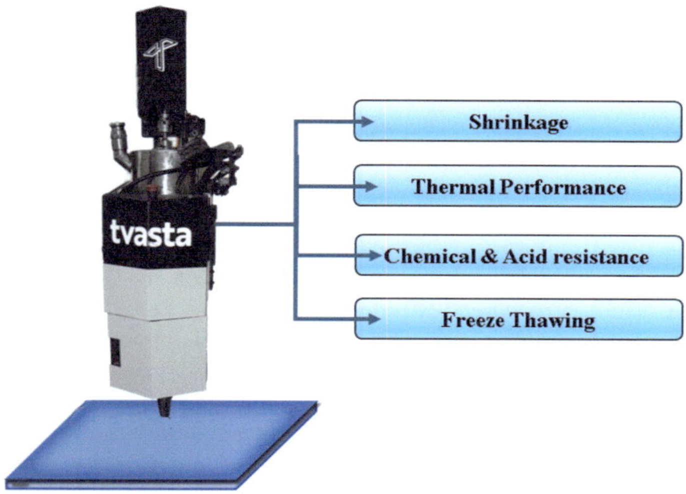

Fig. 6.3 Various durability properties of 3DPC

properties of 3DPC, while durability studies have received comparatively less attention. The impact of critical rheological parameters—such as dynamic yield stress, static yield stress, and plastic viscosity—on printing performance (including extrudability) is represented in Fig. 6.4. A comprehensive understanding of these parameters, as outlined in Chapter 4, provides insights into the durability characteristics of 3DPC specimens. In this context, durability performance of 3DPC has been investigated by a limited number of research groups, and their contributions are summarized in the following subsections.

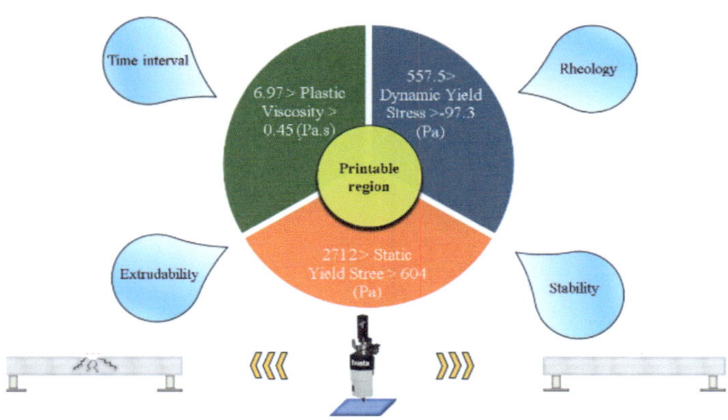

Fig. 6.4 Factors influencing overall durability of 3DPC

6.2.1 Porosity

Porosity is one of the key factors governing both the durability and mechanical performance of concrete structures, ultimately determining their service life—regardless of whether the concrete is cast in molds or produced through 3D printing. Higher porosity can facilitate the ingress of aggressive agents such as acids and chemicals, leading to deterioration of the concrete and necessitating costly repair and rehabilitation measures [3]. In extrusion-based 3DPC, the presence of interlayer pores can further contribute to an overall increase in porosity. Concrete pores are generally classified as capillary, dead-end, air-void, and gel pores. They range in size from roughly 10 nm to 2 mm, primarily gel or air voids [4]. Micropores and larger size air-voids are common faults found in the printed layers of 3D-printed concrete because of its buildability and relatively high shape retention (Fig. 6.5).

Such porosity characteristics have been attributed to variations in the rheological behavior of individual layers [5], the absence of vibration during the printing

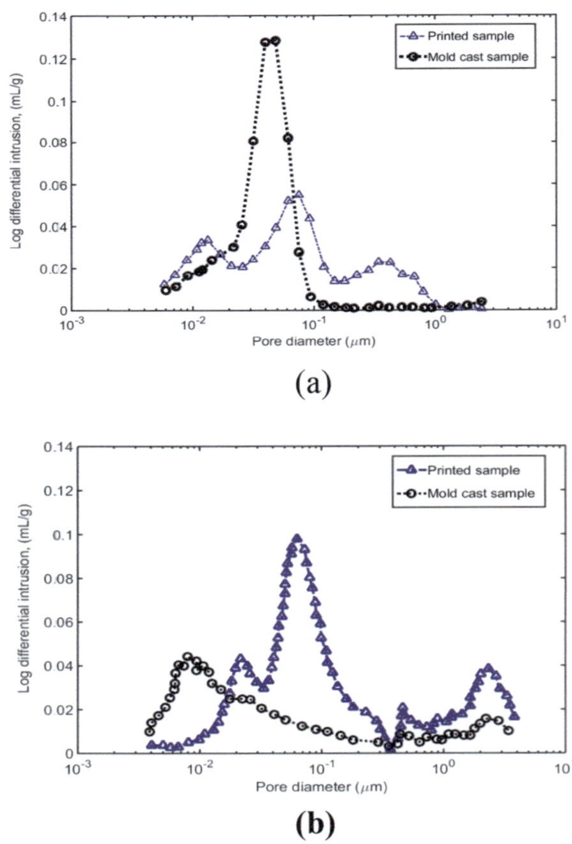

Fig. 6.5 Distribution of pore size printed and non-printed samples from two different studies [5, 15]

process [6], reduced adhesion between successive layers caused by time gaps [7, 8], and possible surface moisture loss during the deposition of subsequent layers [9, 10]. These findings have also been confirmed in Refs. [5, 11], which reported that, unlike conventionally cast concrete, pores in 3D-printed sections tend to be more widely distributed, interconnected, and continuous. A study by Ref. [12] further demonstrated that not only mixture design parameters-such as water-binder ratio and binder content- but also printing characteristics, e.g., printing speed (a higher speeds leading to greater pore volume [13]), interlayer time gap, and orientation of printing, significantly influence the pore network of 3D printed concrete. Meanwhile, Ref. [14] highlighted that curing conditions, particularly relative humidity, exert the most significant influence on porosity, with water curing shown to effectively reduce pore content in printed elements. However, since 3D printing is typically intended for on-site applications, ambient curing is the most common practice, and applying controlled curing regimes is often impractical. To investigate both surface and internal pore structures, techniques such as X-ray CT scanning and Mercury Intrusion Porosimetry (MIP) are frequently employed. Figure 6.5 further illustrates the differences in reported porosity between conventionally cast and 3D-printed concrete sections.

6.2.2 *Water Absorption and Its Rate of Water or Sorptivity*

Water absorption serves as an indicator of concrete permeability, as it reflects the amount of water entering the concrete surface. Figure 6.6 presents the water absorption behavior of both 3D-printed and mold-cast specimens. For the printed concrete, two types of mixes were evaluated: a control mix (containing only OPC) and a composite binder system incorporating 60% OPC, 20% BA, and 20% FA. The results (Fig. 6.6) show that the inclusion of BA and FA reduces water absorption compared to the control mix. This decrease can be attributed to the pore-filling effect, since BA and FA particles are smaller in size than OPC.

At 7 days, the water absorption values for printed concrete 0BA0FA100C (control), 20BA20FA60C, and cast 20BA20FA60C were 8.5, 7.3, and 6.9 respectively. At 28 days, the corresponding values decreased to 6.1, 5.4, and 4.3. The reduction in water absorption with longer curing durations is linked to the transformation of portlandite (lower density) into surplus C-S-H gel (higher density) due to the pozzolanic activity of SCBA. Furthermore, for the mold-cast specimens with the same mix design, water absorption was consistently lower than that of the printed specimens. This difference arises from the compaction process—cast samples were consolidated in molds using a vibrator, thereby eliminating voids and reducing permeability.

Sorptivity, on the other hand, is used to evaluate the ability of concrete to absorb and transport water through capillary suction. A decrease in sorptivity strongly reflects an improvement in durability. During the sorption process, water progressively rises into the interconnected capillary pores as the exposure time to moisture increases. Figure 6.7 illustrates the sorptivity, or rate of water absorption, for both

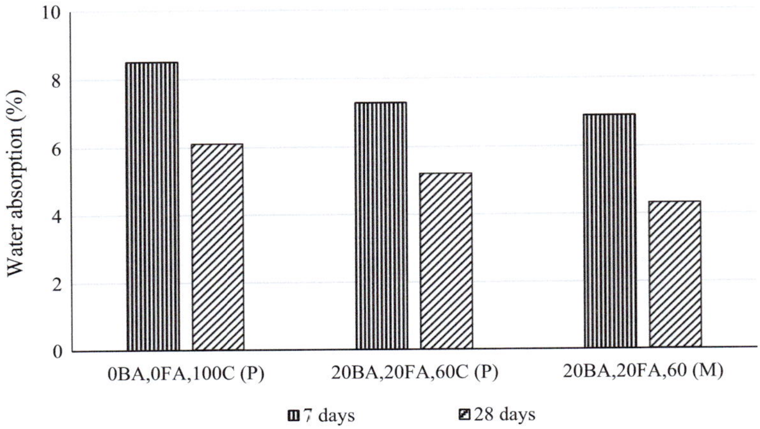

Fig. 6.6 Comparison of water absorption of 3D printed and mold cast sample with different mix design

printed and mold-cast concrete mixes. It is evident that the 3D-printed specimen without BA exhibits the highest water absorption rate among the three mixes shown in Fig. 6.7. The incorporation of BA in the printed sample, however, leads to a noticeable reduction in sorptivity, which can be attributed to the pore-filling effect of BA particles, as they are relatively finer than OPC. Moreover, the mold-cast specimen containing BA shows an even greater reduction in sorptivity. This improvement is likely due to the presence of a smaller proportion of coarse pores in printed samples compared to mold-cast ones, as previously discussed in Sect. 6.2.

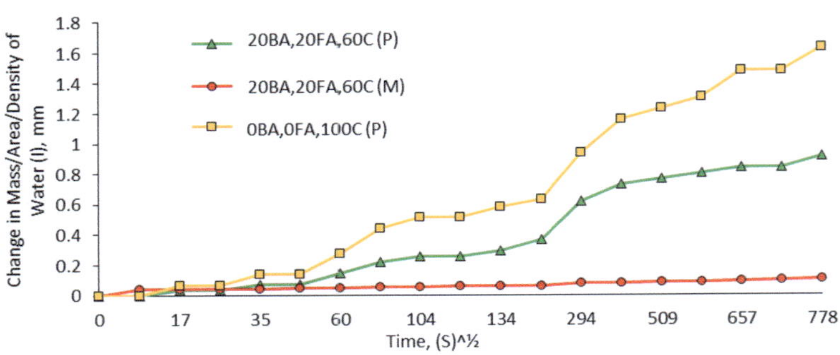

Fig. 6.7 Rate of water absorption of three different mixes

6.2.3 Chloride Attack

The pore connectivity at the interfaces of 3D-printed specimens enhances their vulnerability to chloride attack [11]. Putten et al. [9] examined chloride ion penetration in 3DPC sample employing the NT Build method [16] and compared the results with conventionally cast concrete. It was found that chloride ingress was greater in printed concrete, and the depth of chloride penetration increased as the interlayer time gap widened. Blaakmeer and Lobo [17] also examined penetration of chloride ions and sorptivity in 3DPC specimens, reporting that capillary water absorption in the external layer was around 0.1 kg/m^3, greater than the inner layer, attributed to the rougher and more porous structure network of 3DPC. The NT Build 492 test [18] indicated a chloride diffusion coefficient of 4×10^{-12} m^2/s. Similarly, Weger et al. [19] detected that longer time gaps (0, 10, and 60 min) between layers increased the rate of chloride ingress. Putten et al. [8] further assessed water absorption in printed specimen with a 15-s layer interval and for varying print speeds using neutron radiography, which showed that sorptivity decreased with higher printing speeds. The radiographic results also revealed no preferential water penetration along interlayers when the print surface was exposed to moisture.

- ***RCM Coefficient Test for Rapid Chloride Migration***

Cylindrical specimens of 100 mm diameter and 50 mm in height were utilized for the RCM test following three days of curing at 20 °C and at a humidity of 95%. The test employed water which is deionized for solution preparation, with 0.3 mol/L NaOH along with 10% NaCl solutions (by weight) serving as the anode and cathode electrolytes, correspondingly. A potential difference of 60 V was applied across the two chambers for 6 h, and to control the overall charge flowing through the item, the current was measured. Following the test, the sample surface was washed, dried, and then split axially by a cutting machine. To determine the depth of penetration of the white chromogenic zone, a 0.1 mol/L solution of silver nitrate was spattered onto the cut profile. The RCM coefficient was then computed using Eq. (6.1) [20].

$$D_{RCM} = \frac{(0.0239X(273+T)L}{(U-2)t}\left(X_d - 0.0238\sqrt{\frac{(273+T)LX_d}{U-2}}\right) \tag{6.1}$$

where U denotes the applied voltage (V); D_{RCM} is the RCM coefficient (m^2/s); T is the mean of the starting and ending temperatures; X_d is the mean chloride penetration depth (mm); L is the specimen thickness (mm); and t is the test duration (s).

Concrete's durability (including impermeability and resistance to frost) is directly impacted by its permeability, and the RCM coefficient can be used to assess the depth of chloride ion penetration [21]. In an investigation [20] on the RCM coefficient of 3D printable composite cement made with various amounts of slag and calcium sulphoaluminate cement, it was reported that RCM coefficient originally reduced and then increased with higher slag powder replacement. The RCM coefficient decreased from

7.28×10^{-12} to 3.27×10^{-12} m^2/s when the slag amount rose from 0 to 10%, indicating enhanced impermeability of the 3D-printed composite. This decrease in chloride penetration depth up to 10% substitution is attributed to the finer particle size of slag powder compared to cement, which reduces porosity and results in denser structures, thereby delaying chloride ion ingress. Therefore, adding slag powder is crucial to improving the characteristics of printed CSACMs. However, the drop in cementitious material becomes more noticeable if the slag powder percentage surpasses 10%. This results in reduced compactness and macroscopic cracking, which in turn makes it easier for chloride ions to penetrate 3D-printed CSACMs [22].

6.2.4 Chemical and Acid Resistance

Since 3DPC is designed for application in a wide range of environmental conditions, resistance to chemical and acidic attack is one of its key durability aspects. Although its composition is largely similar to conventionally cast concrete, differences in pore size distribution, greater binder content (especially Portland cement, which has low resistance to acid [23, 24]), and the heterogeneous deposition of layers make 3DPC more prone to chemical ingress [5].

Regarding acid exposure, H_2SO_4 and HCl are the two primary acids commonly present in aggressive environments. When specimens are exposed to these acids, they infiltrate pores and micro-cracks in the concrete. In this study, samples were submerged in acidic solutions with varying concentrations, and their compressive strength was observed after 7 and 28 days. Results (Fig. 6.8) revealed that strength loss was greater in specimens exposed to H_2SO_4 than in those exposed to HCl. Furthermore, surfaces immersed in hydrochloric acid exhibited a rusty-brown coloration, likely due to the formation of ferric chloride from the reaction of iron in cement with chloride ions. In contrast, for H_2SO_4 exposure, the interaction between free lime and sulfates was evident on the surfaces of the submerged specimens.

Within the available literature, only a limited number of studies have examined the chemical and acid resistance of 3DPC. For example, Baz et al. [5] investigated the

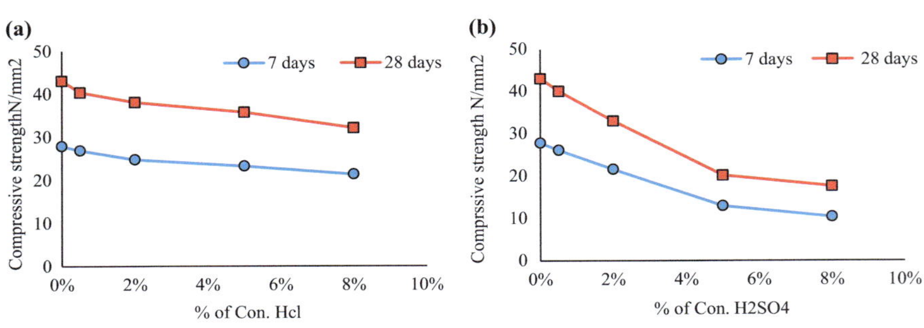

Fig. 6.8 Loss of compressive strength of 3D-printed solution in **a** HCl and **b** H_2SO_4 solution

influence of acidic exposure on top of layers of mortar that have been normally cast and 3D printed. In their work, samples had been immersed in two different sulfuric acid concentration (1 and 3%) till 56 days. The findings indicated that interfacial regions behaved as monolithic structures and did not facilitate the penetration of acid solutions. However, with increasing immersion time, material degradation intensified. By the 56th day, porous surfaces, greater corrosion, and damage to the cement paste were evident, leading to specimen size reduction. Based on this study [8], it was observed that printed specimens exhibited lower mass loss compared to cast ones under 3% sulfuric acid exposure. This difference was attributed to variations in pore size distribution and connectivity between printed and cast specimens. Moreover, the inherent differences in compaction between the two fabrication methods may also influence permeability and acid ingress, suggesting a need for further comparative research in this domain. A similar mass-loss trend was documented in Ref. [25], where sulfuric acid with a 0.5% concentration was employed to approximate real-world exposure while permitting gradual penetration without causing complete surface failure. The solution volume was maintained at four times the volume of the immersed specimen, following ASTM C1012/C1012M-18b guidelines [25].

Zhang et al. [11] further reported that 3D printed concrete (3DPC) exhibits enhanced durability in opposition to sulfate attack and carbonation as compared to traditional concrete cast, likely attributed for its developed content of additional cementitious materials. Their results indicated that 3D-printed specimens showed less deterioration than cast ones, even when tested along the x-axis, where greater pore connectivity would be expected. This was linked to a reduced presence of macropores and overall pore connectivity (aside from interfaces) in printed concrete compared to conventionally cast samples. Conversely, Ref. [26] emphasized that carbonation susceptibility in 3DPC is anisotropic, varying significantly depending on the testing axis and the pore structure within bonding between layers. Nevertheless, Zhang et al. [11] also highlighted that when the supplementary cementitious material content is low, 3DPC sections exhibit relatively poor resistance to damage under frost action and due to the penetration of chloride ions. Importantly, the interlayer pore network was identified as a critical factor influencing chloride permeability.

In another study, Moelich et al. [14] compared the conductivity of chloride penetration of mold-cast and 3D-printed samples under both laboratory and site conditions, with identical temperature (22.5 ± 0.5 °C) and humidity (60±3%), but both without and with air flow of speed 6.3±0.3 m/s. The outcomes showed that printed specimens demonstrated lower chloride conductivity, which was attributed to 3DPC's increased use of supplemental cementitious materials [14]. Furthermore, samples that went under water for curing exhibited considerably reduced chloride conductivity in comparison to those cured in ambient conditions, likely due to pore saturation reducing chloride absorption. The same study [14] also reported that restrained specimens showed higher chloride conductivity, explained by the presence of connected micro-cracks that allowed greater chemical ingress.

Additional insights were provided in Ref. [27], which explored the impact of printing time gap on porosity and chloride conductivity of cast versus 3DPC specimens. Results showed that an increase in the interlayer time from 0 to 30 min nearly doubled both chloride conductivity and porosity. This was attributed to weakened interlayer bonding and greater pore connectivity associated with longer printing intervals [27]. Malan et al. [27] further investigated carbonation resistance by exposing both printed and non-printed specimens to a carbonation chamber (23±2 °C, 65±5% RH, 2±0.1% CO_2). Four weeks later, the average weight decreased of cast and 3D-printed samples was 0.0296% and 0.0339% per week, respectively, indicating better resistance in printed specimens. However, carbonation depth was strongly influenced by layer pass time. Specimens printed with a 30-min interval reached a depth of ~23 mm compared to ~4 mm in those printed continuously, nearly six times greater. Poor interlayer fusion and increased pore connectivity brought on by longer intervals were associated with this impact. Interestingly, interlayer regions showed less carbonation depth in comparison to the printed layers themselves, a phenomenon attributed by Malan et al. [27] toward the larger exposed surface extent of layers, which facilitated greater carbonation penetration.

6.2.5 Shrinkage

In 3DPrinting Concrete, the lack of formwork and the possible bleeding from newly placed layers can considerably enhance the tendency of printed materials to shrink [28]. As reported in Ref. [28], the high surface area of mixture constituents, coupled with a low ratio of aggregate to binder and a higher binder content in additive manufacturing concrete, can lead to decrease in cracks appearing as early as 2 h after printing. These cracks may serve as pathways for chemical ingress, which in turn quickens steel corrosion and adversely affects the long-lasting durability in 3DPC [28]. Generally, shrinkage denotes the volumetric changes of concrete elements, furthermost often arising from ongoing chemical reactions or moisture exchange [22, 28]. For 3DPC, the main types of shrinkage include plastic, autogenous, and drying shrinkage. Plastic shrinkage occurs shortly after mixing, and in the absence of molds, rapid surface evaporation takes place. The water content in the cementitious system is expected to offset this loss; however, if inadequate, capillary pressure develops due to menisci formation in the pores. Consequently, capillary tensile stresses arise, which may cause cracking. As the interfaces of 3D-printed layers are directly exposed to the atmosphere, the risk of cracking and chemical infiltration pathways increases.

Le et al. [29] reported that shrinkage was greater for a 3DPC sample cured at 60% humidity and 20 °C, compared to water curing (175 μm). Moelich et al. [28] found higher amount of plastic shrinkage strains in printed specimen, attributed to the nonexistence of formwork, a lower value of sand-to-binder, and greater amount of fine compared to traditional concrete. Federowicz et al. [30] studied the impact of shrinkage-reducing admixtures on printed concrete. A dosage of 2% by cement weight decreased strains by 7%, while a 4% dosage achieved a 23% reduction. External curing resulted in an 80% strain reduction, proving more effective, though it limited continuous printing. Moelich et al. [28] further found that majority of shrinkage cracks formed within initial few hours after printing the specimen.

The introduction of shrinkage limiting reinforcement increased crack density, while differential shrinkage strains between layers caused slippage at the interfaces [28]. Due to the higher amount of binder content in printable mixtures, 3DPC is more vulnerable to cracks resulting from shrinkage. Shrinkage is also strongly influenced by curing conditions, underscoring the importance of adopting proper post-printing curing methods to reduce strains. To further control shrinkage, Ref. [31] suggested the combined use of admixtures, optimized water-to-cement and sand-to-binder content, along with various curing strategies.

- ***Linear Shrinkage and its Measurement***

Linear shrinkage, mainly driven by water evaporation, serves as a key parameter in evaluating the long-term durability of cementitious composites. Linear shrinkage of 3DPC specimens was measured in accordance with ASTM C531-18. Initially, the printed specimens were cut into blocks of length 160 mm and cross section 40 mm × 40 mm and cured for 24 h at 20 °C and 95% relative humidity. Subsequently, they were transferred to a curing chamber maintained at 20 °C and 50% humidity. An initial reading was taken before relocation, after which the specimens were measured every 6 h for the first 3 days and then once every 24 h. The linear shrinkage was calculated using Eq. (6.2).

$$S = \left[\frac{L_0 - L}{L_0}\right] X\,100\% \tag{6.2}$$

where L_0 denotes the length at the start of test and L represents the length after the curing period. An investigation [20] on slag quantity on the linear shrinkage of 3DPC revealed that it initially increases and then decreases with time. The rise in shrinkage occurs before 7 days, mainly due to the fast hydration of CSA cement during the early stage. Moreover, when the slag powder content is below 10%, shrinkage decreases considerably. This reduction is attributed to the smaller particle size of slag powder compared to cement particles, which enhances compactness and lowers the amount of free water. However, when the slag powder proportion exceeds 10%, linear shrinkage gradually rises again due to the excess slag powder. In this situation, slag powder negatively impacts the printability of the paste, causing more visible cracks in the material [22]. Furthermore, excessive slag powder weakens interlayer bonding and facilitates the entry of free water into the structure.

6.2.6 Freeze-Thaw Attack

Freeze-thaw deterioration refers to internal damage in cement-based materials caused by repeated freezing and thawing cycles [32]. This type of damage is common in regions where temperatures drop below zero. To promote the widespread use of 3DPC in such areas, it is essential to study and improve its freeze-thaw resistance. Unlike conventional concrete, the freeze-thaw performance of 3DPC differs due to its unique characteristics. The main cause of internal damage is the volume change of water in pores during freezing (water turning into ice), which generates fatigue stresses and eventually harms the structure.

Research on the mass loss and dynamic modulus of elasticity of 3D-printed CSACMs in various freeze-thaw cycles [20] shows a steady decline in both parameters with increasing cycles. In the test, 3D-printed specimens were cut into cubes (100 × 100 × 100 mm) and cured for three days at 20 °C and 95% humidity. Measurements were recorded after 0, 25, 50, 100, 200, 300, 400, 500, and 600 cycles. Compared with reference specimens, cracks in 3D-printed CSACMs significantly decreased as cycles progressed. This improvement is linked to the smaller particle size of slag powder compared to cement, reducing porosity and lowering the amount of free water available to freeze. However, when slag content exceeds 15%, the opposite effect is seen. This happens because the reduced cementitious material leads to fewer hydration products and lower compactness. Moreover, higher slag levels increase yield stress, causing poor extrusion and larger cracks. As a result, more free water penetrates the samples, leading to greater freeze-thaw damage.

In further studies, Assaad et al. [33] examined printable concrete under freeze-thaw exposure and tested the effects of adding air-entraining admixtures and styrene-butadiene rubber. Freeze-thaw cycles reduced strength under compression and flexure loading, but admixtures reduced the rate of damage. The interlayer bond strength was most affected. Air-entraining admixtures enhanced frost resistance by creating additional voids, while latex improved flexibility, compensating for frost-related damage. In contrast, [34] noted that pumping reduced void size and spacing, which are intentionally added to resist freezing and thawing. Similarly, Wang et al. [35] found that 3DPC had lower freeze-thaw resistance compared to conventional concrete, likely due to interlayer interfaces. On the other hand, Van Der Putten et al. [36] concluded there was no significant difference in freeze-thaw resistance between conventional and 3DPC specimens and highlighted that the interlayer interval time does not affect resistance. These conflicting results show that further research is needed to clarify freeze-thaw durability in 3DPC.

6.2.7 Fire Attack

Exposure of 3DPC to high temperatures decreases both mechanical and microstructural properties, like cast concrete. Cicione et al. [37] compared resistance under fire between printed and cast specimens by subjecting them to heat flux of 50–60 kW/m^2 until reaching 300 °C. Both performed similarly; however, cast concrete experienced spalling, while printed concrete showed layer separation at the interfaces. The improved resistance of 3DPC to spalling occurs due to its interconnected porosity and greater permeability. D'Hondt et al. [38] further tested printed concrete at temperatures of 120, 250, 400, and 600 °C, noting preservation of strengths under compression and flexure loading, stiffness. Kruger et al. [39] studied thermal behavior of 3D-printed elements and found delamination of layers in printed structures at high temperatures, unlike the spalling in cast high-strength concrete. Introducing steel fibers perpendicular to layer interfaces provided ductility after heating. Rapid heating can cause explosive spalling in 3DPC; however, adding polyethylene (PE) fibers mitigates this risk. PE melts at a temperature of 200 °C and evaporates at higher temperatures, forming micro-cracks that allow vapor to escape, thus lowering spalling risk [40].

6.3 Effect of Parameters on 3DPC Durability

Figure 6.9 highlights four primary factors influencing the durability of printed concrete: (1) mix design (employed materials), (2) printing process-based parameters, (3) transport properties of hardened concrete, and (4) the adjacent environment at the construction site.

- **Mix Design**: It includes type of binder, water-binder ratio, aggregate content, and dosages of admixtures (SP, retarders, viscosity modifiers) that ensure durability under aggressive conditions.
- **Transport Properties**: Pore size distribution, porosity, pore connectivity, and permeability determine the rate of ion penetration, which are strongly dependent on process parameters.
- **Process Parameters**: Nozzle height, time gap between layers, and printing speed influence microstructural characteristics.
- **Exposure Environment**: Conditions such as sulfate, chloride, or leaching govern degradation mechanisms and affect mechanical and microstructural properties.

Given the differences between 3DPC and conventional concrete, the influence of printing variables on durability requires deeper investigation, which is discussed in the following section.

The durability of 3DPC is directly linked to the process of printing and the properties of its state when it becomes hard [41]. The following influencing constraints include, but are not limited to, extrudability, construction time, rheological behavior, and buildability of the material that was placed. For example, Ref. [5] highlighted that pumping pressure and printing speed play a crucial role in determining the porosity of

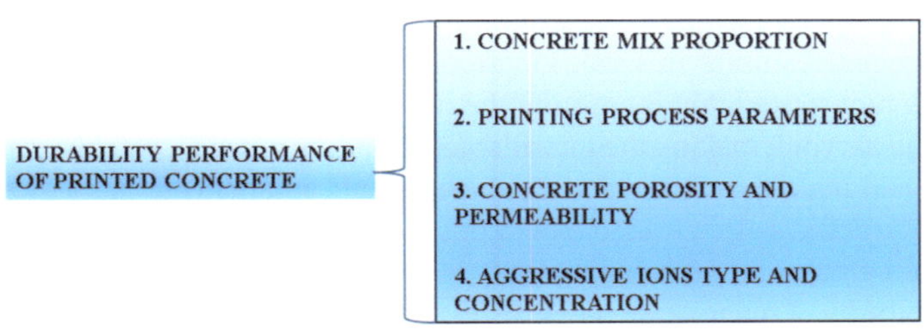

Fig. 6.9 Factors affecting 3D-printed concrete's durability performance [2]

deposited layers. On the other hand, Ref. [42] attributed both interlayer and internal porosity to the time of construction, specifically the time period between printing successive layers. Additionally, Ref. [13] pointed out that increased printing speed can significantly impact the hardened state characteristics, particularly the porosity of printed elements. Collectively, these findings suggest that the combined effect of printing parameters along with fresh and rheological properties largely governs the final quality and determines the success or failure of printed components.

6.3.1 Time of Construction

In 3DPC, time of construction encompasses print duration, elapsed time and delay time—essential aimed at ensuring that each printed mortar layer develops sufficient strength to support the mass of subsequent layers. This strength development is primarily influenced by factors such as printing speed, material workability (fresh state properties), and the waiting interval before the next layer is deposited [13, 42]. Broadly, construction time refers to the total duration required to fabricate a thing, element, or construction—from initiation to completion of the printing procedure—and can be described in Eq. (6.3) [43].

$$T_C = t_e - t_p + t_d \tag{6.3}$$

Another way to put this is as

$$\begin{aligned} T_c(s) &= (\text{required stop time for the duration of printing} + \text{Preparation time}) \\ &+ \text{Planned printing area}\left(\text{mm}^2\right)/\text{Rate of printing}\left(\text{mm}^2/\text{s}\right) \end{aligned}$$

Here, Tc denotes the whole construction time, t_e the overall forgotten time, t_d the postponement time, and t_p the necessary stoppage time, which refers to the waiting period required for each layer to achieve adequate strength to support the subsequent one—closely tied to the printing speed. Owing to a key basis of the quality and strength of printed structures, several studies have highlighted the impact of elapsed time on interlayer adhesion, cracking due to shrinkage and evolution of strength [6, 13]. For instance, Ref. [14] demonstrated that combining an extended print way with slow printing speed increases the printing interval, during which surface moisture loss may occur, resulting in weak long-term bonding between layers. Likewise, it has been demonstrated that increasing the interval between printed layers increases porosity and air void, making the interfilament bonding a crucial weakness in 3DPC's durability and making it easier for harsh chemicals and acids to enter [12]. Consequently, such defects often lead to reductions in mechanical properties due to elevated porosity and greater air void connectivity [6, 13].

In a study by Kruger et al. [12], (CT), i.e., micro-computed tomography was employed to analyze pore size, overall porosity, morphology, and dispersal within printed concrete layers. In printed samples, they found that tri-axial spheroidal air voids predominated and that there was a direct relationship between printing speed and void content [12]. Accordingly, extended construction time has been reported to elevate overall porosity, particularly at interlayer zones. Similarly, Refs. [6, 43] identified greater porosity and defects in interlayer regions when extended resting intervals were allowed between the deposition of successive layers. To address this issue, previous research suggests that incorporating microfibers (either as meshes or discrete reinforcements) and applying polymeric resins (such as epoxy) to bonding areas can significantly enhance interlayer adhesion in 3DPC by substantially lowering porosity [10].

6.3.2 Extrudability

The ability of a 3D printer to continually release concrete via its nozzle without obstruction is known as extrudability [44]. In essence, it symbolizes the mortar's equilibrium between flowability and workability, guaranteeing self-compaction and spray ability while preventing bleeding and segregation [44]. Segregation generally occurs when insufficient paste is present to lubricate aggregate or sand particles. Literature indicates that bleeding and segregation lead to uneven material distribution, which can severely compromise the long-term durability of 3DPC by weakening the bond between aggregates and cement paste [45]. Consequently, this increases the likelihood of cracking because of aggregate settlement and promotes shrinkage in areas that are rich in binder [45].

Admixtures are frequently used to enhance extrudability in order to produce a printable mix with sufficient working time and rheological characteristics. Nanomaterials, retarders, viscosity-modifying agents, and superplasticizers are frequently used. Kruger et al. [46] reported that nanoparticles such as nano-silica and nanoclay are especially effective in enhancing the thixotropy of 3D-printing mixes, primarily due to the high surface area to volume ratio. Even though nanoparticles can boost durability [47], ensuring uniform dispersion within the mix remains a challenge [46, 48]. A similar issue arises with microfibers, which, while beneficial to strength, increase friction inside the nozzle, reducing pumpability. Altogether, uneven dispersion of materials is a major source of anisotropy in 3DPC, significantly influencing its thermal durability.

6.3.3 Rheology

Fresh concrete can be characterized as dilatant, thixotropic, pseudo-plastic, or visco-plastic due to its non-Newtonian fluid behavior [46]. Non-Newtonian fluids are generally defined by viscosity that changes with flow history and shear rate [49]. In practical terms, this reflects the material's ability to resist gravitational stress deprived of flowing. In 3D printing, fresh mix viscosity is crucial for both flowability and shape retention of printed elements [49].

When solid particles interact with cement paste in the Bingham plastic model, thixotropy is commonly defined by yield stresses (both static and dynamic) and plastic viscosity [50]. According to this model, static yield stress corresponds to flow initiation, while dynamic yield stress represents continuous extrusion. Both must be balanced to achieve a printable, flowable mix [51]. Thixotropic behavior depends largely on cement composition, reaction time, and admixture dosage. The Bingham plastic model is expressed in Eq. (6.4):

$$\tau = \tau_y + \mu\gamma \tag{6.4}$$

where shear stress is denoted by τ, viscosity by μ, shear rate by γ, and yield stress by τ_y.

For 3DPC, rheological behavior is influenced by printing speed, aggregate size and shape, SCM incorporation, and resting intervals between layers. Zhang et al. [52] studied highly thixotropic mixes and found that the sand-to-cement ratio has a strong influence on viscosity, yield stress, and thixotropy. An increase in fine aggregates reduced thixotropy, negatively affecting dispersion [52]. To adjust rheology, nanoparticles are among the most effective additives. Dejaeghere et al. [53] showed that nanoclay in combination with coal fly ash altered rheological properties by reducing slump and bleeding while enhancing cohesion, hydration heat, yield stress, and viscosity. While rheology does not directly control durability, it governs successful, defect-free printing and uniform material distribution, both critical for long-term thermal durability of 3DPC.

6.3.4 Buildability

The mechanisms of strength development and crack formation in 3DPC differ from those in conventionally cast concrete [54]. The ability of each printed layer to stiffen and retain shape under self-weight and subsequent loads is termed buildability or stability. This property is strongly linked to self-compacting and shotcrete concrete, where rapid strength gain, flowability, and resistance to plastic shrinkage cracking are vital. Factors influencing buildability include rheology, printing speed, sand-to-cement ratios and binder-to-cement, and the surface characteristics (friction and roughness) of mixed materials [55].

6.4 Effect of SCMs on Durability of 3DPC

Supplementary cementitious materials (SCMs) are essential components in printable mixes, acting as shrinkage-reducing agents, rheology modifiers, [30], and most importantly, enhancers of both durability and mechanical properties of the resulting concretes. In 3D printing, silica fume and coal fly ash (class F) are among the most widely utilized SCMs. It has been demonstrated that their addition affects the concrete's permeability, acid resistance, and porosity while also reducing its chemical reactivity. However, Ref. [5] described the weak bonding at the interlayer zones of 3DPC units can increase the absorption of damaging substances, thereby raising the risk of corrosion and compromising durability. Such issues may arise due to bleeding [28], shrinkage [14], excessive use of low-reactivity SCMs, or other contributing factors.

For instance, Ref. [28] found that due to high reactivity, silica fume may exacerbate plastic shrinkage by raising capillary pressure and fine porosity. Additionally, Ref. [56] highlighted that silica fume accelerates the time of setting, necessitating higher dosages of retarding and superplasticizers admixtures to maintain printability. The greater surface area of silica fume in comparison to Portland cement is the cause of this need. Similarly, nanoclay rises the initial viscosity of mixes, thereby improving the shape stability of printed layers [57]. Reference [58] ascribed this to the fact that nanomaterials' greater surface area alters yield stress and plastic viscosity. Given that a fluid's resistance to flow is known as its plastic viscosity [59], increased internal friction from a larger surface area improves layer stability and form retention [60].

In contrast, (class F) coal fly ash behaves differently due to its coarser particle size and lower hydration rate, resulting from its chemical structure [61]. While higher viscosity supports buildability, it may also cause uneven material distribution, leading to greater porosity, permeability, and settlement risks [45]. Other SCMs, such as ground granulated blast furnace slag (GGBFS) and metakaolin, have also been explored. Their use has generally resulted in reduced slump values, implying higher viscosity [62]. Metakaolin, being porous, tends to absorb water, causing rapid structural build-up and flocculation. While this can improve mechanical properties by lowering the water-to-binder ratio, it also affects porosity, shrinkage, and printability, reducing working time and demanding more admixtures [62]. However, by improving compaction due to its high aluminum content, which actively participates in geopolymerization and the creation of Al–O–Al and Al–O–Si links, metakaolin can help 3D print alkali-activated and geopolymer mortars [23]. Thus, the chemical composition of SCMs plays a vital role in the overall porosity of 3DPC sections.

Similarly, GGBFS has been related with improved shape retention and buildability in both traditional concrete in addition to 3D-printed systems, including geopolymers. Its impact on fresh and rheological properties of 3DPC, however, depends largely on its dosage in the mix. At lower contents (e.g., 5%), it disperses efficiently, leading to better workability, whereas higher proportions reduce extrudability and reactivity, resulting in settlement issues in printed elements. Table 6.1 provides a

Table 6.1 Influence of commonly employed SCM on durability properties of 3DPC [63, 64]

Material	Major composition	Durability properties effect on
Silica fume	Si	Increases compaction and reduces voids
GGBFS	Ca–Si–Al	Acid resistance and higher sulfate
Fly ash	Si–Al/Si–Ca–Al	Higher compaction, acid and sulfate resistance
Metakaolin	Si–Al	A reduction of mechanical properties and increase in porosity
Rice husk ash	Si–C–K_2O	Often micropores and higher absorption

summarized comparison of commonly used SCMs, their effects on mix rheology, and corresponding durability outcomes.

6.5 Closure

Given that 3DPC elements are directly exposed to diverse environmental conditions such as temperature fluctuations, acid rain, carbonation, and drying, durability remains one of the most critical challenges for ensuring their long-term performance. To enhance the durability of 3DPC, strategies such as designing tailored geometries (e.g., hollow sections) to meet performance requirements, utilizing higher proportions of SCMs, and incorporating suitable admixtures can be highly effective [65]. Nevertheless, more comprehensive studies are still necessary to accurately assess the durability behavior of 3D-printed concrete under harsh environmental exposures.

References

1. Hou S, Duan Z, Xiao J, Ye J (2021) A review of 3D printed concrete: performance requirements, testing measurements and mix design. Constr Build Mater 273:121745
2. Rehman AU, Kim J-H (2021) 3D concrete printing: a systematic review of rheology, mix designs, mechanical, microstructural, and durability characteristics. Materials 14:3800
3. Vijay K, Murmu M, Deo SV (2017) Bacteria based self healing concrete—a review. Constr Build Mater 152:1008–1014
4. Nodehi M (2021) A comparative review on foam-based versus lightweight aggregate-based alkali-activated materials and geopolymer. Innov Infrastruct Solut 6:231
5. Baz B, Aouad G, Kleib J, Bulteel D, Remond S (2021) Durability assessment and microstructural analysis of 3D printed concrete exposed to sulfuric acid environments. Constr Build Mater 290:123220
6. Jo JH, Jo BW, Cho W, Kim J-H (2020) Development of a 3D printer for concrete structures: laboratory testing of cementitious materials. Int J Concr Struct Mater 14:1–11
7. Nerella VN, Hempel S, Mechtcherine V (2019) Effects of layer-interface properties on mechanical performance of concrete elements produced by extrusion-based 3D-printing. Constr Build Mater 205:586–601
8. Van Der Putten J, Azima M, Van den Heede P, Van Mullem T, Snoeck D, Carminati C, Hovind J, Trtik P, De Schutter G, Van Tittelboom K (2020) Neutron radiography to study the water

ingress via the interlayer of 3D printed cementitious materials for continuous layering. Constr Build Mater 258:119587
9. Van Der Putten J, De Volder M, Van den Heede P, De Schutter G, Van Tittelboom K (2020) 3D printing of concrete: the influence on chloride penetration. In: RILEM international conference on concrete and digital fabrication. Springer, pp 500–507
10. Wang L, Tian Z, Ma G, Zhang M (2020) Interlayer bonding improvement of 3D printed concrete with polymer modified mortar: experiments and molecular dynamics studies. Cem Concr Compos 110:103571
11. Zhang Y, Zhang Y, Yang L, Liu G, Chen Y, Yu S, Du H (2021) Hardened properties and durability of large-scale 3D printed cement-based materials. Mater Struct 54:1–14
12. Kruger J, du Plessis A, van Zijl G (2021) An investigation into the porosity of extrusion-based 3D printed concrete. Addit Manuf 37:101740
13. Van Der Putten J, De Schutter G, Van Tittelboom K (2018) The effect of print parameters on the (micro) structure of 3D printed cementitious materials. In: RILEM international conference on concrete and digital fabrication. Springer, pp 234–244
14. Moelich GM, Kruger PJ, Combrinck R (2021) The effect of restrained early age shrinkage on the interlayer bond and durability of 3D printed concrete. J Build Eng 43:102857
15. Yu S, Xia M, Sanjayan J, Yang L, Xiao J, Du H (2021) Microstructural characterization of 3D printed concrete. J Build Eng 44:102948
16. Build NT (1995) 443: Nordtest method, concrete, hardened: accelerated chloride penetration, Espoo, Finland
17. Blaakmeer J, Lobo B (2020) A robust mortar and printing system. In: Second RILEM international conference on concrete and digital fabrication: digital concrete, vol 2. Springer, pp 1091–1103
18. Build N (1999) Concrete, mortar and cement-based repair materials: chloride migration coefficient from non-steady-state migration experiments. Nordtest Method 492
19. Weger D, Kim H, Talke D, Henke K, Kränkel T, Gehlen C (2020) Lightweight concrete 3D printing by selective cement activation–investigation of thermal conductivity, strength and water distribution. In: Second RILEM international conference on concrete and digital fabrication: digital concrete, vol 2. Springer, pp 162–171
20. Chen M, Jin Y, Sun K, Wang S, Zhao P, Wang L, Liu J, Yue G, Li Q, Lu L (2023) Study on the durability of 3D printed calcium sulphoaluminate cement-based materials related to rheology control. J Mater Res Technol 26:2481–2494
21. Yang L, Zhao P, Liang C, Chen M, Niu L, Xu J, Sun D, Lu L (2021) Characterization and adaptability of layered double hydroxides in cement paste. Appl Clay Sci 211:106197
22. Moelich GM, Kruger PJ, Combrinck R (2022) Mitigating early age cracking in 3D printed concrete using fibres, superabsorbent polymers, shrinkage reducing admixtures B-CSA cement and curing measures. Cem Concr Res 159:106862
23. Nodehi M, Taghvaee VM (2022) Alkali-activated materials and geopolymer: a review of common precursors and activators addressing circular economy. Circ Econ Sustain 2:165–196
24. Gu L, Bennett T, Visintin P (2019) Sulphuric acid exposure of conventional concrete and alkali-activated concrete: assessment of test methodologies. Constr Build Mater 197:681–692
25. El Inaty F, Baz B, Aouad G (2023) Long-term durability assessment of 3D printed concrete. J Adhes Sci Technol 37:1921–1936
26. Sun X, Zhou J, Wang Q, Shi J, Wang H (2022) PVA fibre reinforced high-strength cementitious composite for 3D printing: mechanical properties and durability. Addit Manuf 49:102500
27. Malan JD, van Rooyen AS, van Zijl GPAG (2021) Chloride induced corrosion and carbonation in 3D printed concrete. Infrastructures (Basel) 7:1
28. Moelich GM, Kruger J, Combrinck R (2020) Plastic shrinkage cracking in 3D printed concrete. Compos B Eng 200:108313
29. Le TT, Austin SA, Lim S, Buswell RA, Law R, Gibb AGF, Thorpe T (2012) Hardened properties of high-performance printing concrete. Cem Concr Res 42:558–566
30. Federowicz K, Kaszyńska M, Zieliński A, Hoffmann M (2020) Effect of curing methods on shrinkage development in 3D-printed concrete. Materials 13:2590

31. Hoffmann M, Skibicki S, Pankratow P, Zieliński A, Pajor M, Techman M (2020) Automation in the construction of a 3D-printed concrete wall with the use of a lintel gripper. Materials 13:1800
32. Zhao Y-R, Wang L, Lei Z-K, Han X-F, Shi J-N (2018) Study on bending damage and failure of basalt fiber reinforced concrete under freeze-thaw cycles. Constr Build Mater 163:460–470
33. Assaad JJ, Hamzeh F, Hamad B (2020) Qualitative assessment of interfacial bonding in 3D printing concrete exposed to frost attack. Case Stud Constr Mater 13:e00357
34. Chatterji S (2003) Freezing of air-entrained cement-based materials and specific actions of air-entraining agents. Cem Concr Compos 25:759–765
35. Wang L, Xiao W, Wang Q, Jiang H, Ma G (2022) Freeze-thaw resistance of 3D-printed composites with desert sand. Cem Concr Compos 133:104693
36. Van Der Putten J, De Volder M, Van den Heede P, Deprez M, Cnudde V, De Schutter G, Van Tittelboom K (2022) Transport properties of 3D printed cementitious materials with prolonged time gap between successive layers. Cem Concr Res 155:106777
37. Cicione A, Kruger J, Walls RS, Van Zijl G (2021) An experimental study of the behavior of 3D printed concrete at elevated temperatures. Fire Saf J 120:103075
38. D'Hondt M, Rémond S, Leblond P, Iea B, Hynek E, Pinoteau N (2020) Fire behavior of a printed sample for building. In: RILEM international conference on concrete and digital fabrication. Springer, pp 428–438
39. Kruger J, Cicione A, Bester F, van den Heever M, Cho S, Walls R, van Zijl G (2020) Facilitating ductile failure of 3D printed concrete elements in fire. In: RILEM international conference on concrete and digital fabrication. Springer, pp 449–458
40. Xiao J, Han N, Zhang L, Zou S (2021) Mechanical and microstructural evolution of 3D printed concrete with polyethylene fiber and recycled sand at elevated temperatures. Constr Build Mater 293:123524
41. Jayathilakage R, Rajeev P, Sanjayan JG (2020) Yield stress criteria to assess the buildability of 3D concrete printing. Constr Build Mater 240:117989
42. Sanjayan JG, Nematollahi B, Xia M, Marchment T (2018) Effect of surface moisture on inter-layer strength of 3D printed concrete. Constr Build Mater 172:468–475
43. Diggs-McGee BN, Kreiger EL, Kreiger MA, Case MP (2019) Print time versus elapsed time: a temporal analysis of a continuous printing operation for additive constructed concrete. Addit Manuf 28:205–214
44. Liu G, Cheng W, Chen L, Pan G, Liu Z (2020) Rheological properties of fresh concrete and its application on shotcrete. Constr Build Mater 243:118180
45. Panesar DK, Shindman B (2012) The effect of segregation on transport and durability properties of self consolidating concrete. Cem Concr Res 42:252–264
46. Kruger J, Zeranka S, van Zijl G (2019) An ab initio approach for thixotropy characterisation of (nanoparticle-infused) 3D printable concrete. Constr Build Mater 224:372–386
47. Safiuddin M, Gonzalez M, Cao J, Tighe SL (2014) State-of-the-art report on use of nano-materials in concrete. Int J Pavement Eng 15:940–949
48. Bautista-Gutierrez KP, Herrera-May AL, Santamaría-López JM, Honorato-Moreno A, Zamora-Castro SA (2019) Recent progress in nanomaterials for modern concrete infrastructure: advantages and challenges. Materials 12:3548
49. Coussot P (2012) Introduction to the rheology of complex fluids. In: Understanding the rheology of concrete. Elsevier, pp 3–22
50. Weng Y, Li M, Tan MJ, Qian S (2018) Design 3D printing cementitious materials via Fuller Thompson theory and Marson-Percy model. Constr Build Mater 163:600–610
51. Tay YWD, Qian Y, Tan MJ (2019) Printability region for 3D concrete printing using slump and slump flow test. Compos B Eng 174:106968
52. Zhang Y, Zhang Y, She W, Yang L, Liu G, Yang Y (2019) Rheological and harden properties of the high-thixotropy 3D printing concrete. Constr Build Mater 201:278–285
53. Dejaeghere I, Sonebi M, De Schutter G (2019) Influence of nano-clay on rheology, fresh properties, heat of hydration and strength of cement-based mortars. Constr Build Mater 222:73–85

54. Ma G, Li Y, Wang L, Zhang J, Li Z (2020) Real-time quantification of fresh and hardened mechanical property for 3D printing material by intellectualization with piezoelectric transducers. Constr Build Mater 241:117982
55. Al-Qutaifi S, Nazari A, Bagheri A (2018) Mechanical properties of layered geopolymer structures applicable in concrete 3D-printing. Constr Build Mater 176:690–699
56. Manikandan K, Wi K, Zhang X, Wang K, Qin H (2020) Characterizing cement mixtures for concrete 3D printing. Manuf Lett 24:33–37
57. Kazemian A, Yuan X, Cochran E, Khoshnevis B (2017) Cementitious materials for construction-scale 3D printing: Laboratory testing of fresh printing mixture. Constr Build Mater 145:639–647
58. Ma G, Wang L (2018) A critical review of preparation design and workability measurement of concrete material for largescale 3D printing. Front Struct Civ Eng 12:382–400 (n.d.)
59. Mahmoodzadeh F, Chidiac SE (2013) Rheological models for predicting plastic viscosity and yield stress of fresh concrete. Cem Concr Res 49:1–9
60. Jeong H, Han S-J, Choi S-H, Lee YJ, Yi ST, Kim KS (2019) Rheological property criteria for buildable 3D printing concrete. Materials 12:657
61. Langan BW, Weng K, Ward MA (2002) Effect of silica fume and fly ash on heat of hydration of Portland cement. Cem Concr Res 32:1045–1051
62. Chen M, Yang L, Zheng Y, Huang Y, Li L, Zhao P, Wang S, Lu L, Cheng X (2020) Yield stress and thixotropy control of 3D-printed calcium sulfoaluminate cement composites with metakaolin related to structural build-up. Constr Build Mater 252:119090
63. Ganesan K, Rajagopal K, Thangavel K (2008) Rice husk ash blended cement: assessment of optimal level of replacement for strength and permeability properties of concrete. Constr Build Mater 22:1675–1683
64. Paiva H, Velosa A, Cachim P, Ferreira VM (2012) Effect of metakaolin dispersion on the fresh and hardened state properties of concrete. Cem Concr Res 42:607–612
65. Nodehi M, Aguayo F, Nodehi SE, Gholampour A, Ozbakkaloglu T, Gencel O (2022) Durability properties of 3D printed concrete (3DPC). Autom Constr 142:104479

Chapter 7
Fire Behavior of 3D Printed Concrete

7.1 Introduction

With the growing use of 3D concrete printing (3DCP) in modern construction, understanding the behavior under fire of 3D-printed structures has become critical for ensuring structural safety and code compliance. Unlike conventional cast concrete, 3D-printed concrete (3DPC) exhibits unique geometric and microstructural features, such as interlayer interfaces, anisotropic mechanical properties, and altered porosity distributions, all of which can influence its thermal and fire resistance.

This chapter explores the fundamental principles governing the fire behavior of both conventional concrete and 3DPC, reviews recent studies, the influence of various material constituents on the thermo-mechanical behavior, identifies challenges and gaps, and outlines strategies to enhance fire performance through material and process optimization.

7.2 Fire Exposure and Thermal Degradation in Cementitious Materials

When exposed to elevated temperatures, different components in concrete undergo several physicochemical transformations. For instance, under elevated temperature exposure, such as in fire conditions, the constituents of concrete respond differently. Aggregates generally exhibit thermal expansion, the extent of which depends on their mineralogical composition. Conversely, hardened cement paste undergoes only a minor initial expansion, followed by shrinkage due to dehydration, as chemically bound water is progressively released with increasing temperature. This incompatibility in thermal response induces internal stresses and microcracking, adversely affecting the mechanical performance of concrete [1]. Due to the differential thermal

A. Chourasia et al., *Advances and Challenges in 3D Printed Concrete (3DPC)*, Materials Horizons: From Nature to Nanomaterials, https://doi.org/10.1007/978-981-95-8557-1_7

expansion and contraction of aggregates and paste, the boundary between the aggregate surface and the surrounding cement paste turns to be the most vulnerable region in concrete. This region, referred to as the Interfacial Transition Zone (ITZ), serves as the initiation site for crack formation under elevated temperature exposure. The incompatibility in thermal response between the two phases accelerates microcracking in the ITZ, leading to a more pronounced deterioration of elasticity and strength in the mortar than in the bulk paste. Fu et al. [2] examined the deterioration of strength under compression and the elastic modulus of both paste and mortar under elevated temperatures. The results indicated that mortar exhibits a higher rate of deterioration in both elasticity and strength under compression compared to hardened cement paste. This accelerated degradation in mortar is primarily attributed to the formation of micro-cracks arising from the thermal mismatch between sand particles and the surrounding paste.

Analysis of the individual constituents of concrete, namely paste and aggregates, under elevated temperatures reveals the occurrence of various reversible and irreversible physical and chemical transformations within each component. These transformations induce contrasting volumetric changes in the constituents, which in turn significantly influence the overall mechanical behavior of concrete [1, 2]. For instance, C–S–H gels begin to dehydrate at approximately 105 °C, and this process continues up to about 800 °C, while the decomposition of CSH gel occurs at a temperature between 375 and 550 °C. Such dehydration causes the breakdown of chemical bonds and the release of chemical water from the hydration products. Consequently, the volume fraction of pores increases. In addition, the released chemical water develops pore pressure inside the cement paste, resulting in concrete failure at high temperatures. These physicochemical transformations in the paste influence mechanical behavior at elevated temperatures and also the residual performance of the concrete structure after cooling [3, 4]. For example, Portlandite in cement decomposes at around 400 °C, yielding calcium oxide (CaO). During the cooling phase, this CaO rehydrates and generates expansive products, leading to deterioration in OPC paste specimens that can break down after cooling [1].

Further, after cooling, some amount of the disintegrated hydrated products rehydrate, and CSH products are formed, resulting in recovering the strength of a post-cooled paste after fire exposure [3, 4]. Based on the experiment, it was reported in [4] that up to 60% of the total cracks of a concrete sample (following exposure to a temperature of 800 °C) could be healed after cooling with 30 days of curing. Thus, the strength in compression of concrete (which immediately after cooling was 65% of the strength before fire) recuperates up to 86% of the pre-heated strength after a curing period of 30 days. Studies carried out in [3] also showed similar strength recovery of the cement and concrete specimens subjected to a heating–cooling cycle, followed by curing for a specified duration. Concrete manufactured using firebrick or firebrick slag, after exposure to 500 °C temperature and 90 days of curing, can recover up to 90–100% of its pre-heating strength value [3]. A comparable recovery in the strength value of concrete specimens was also reported in [5].

On the other hand, siliceous aggregates go through a phase change process at 570 °C that results in considerable remaining strains after cooling [1]. Also, fissures that occur The damage induced in aggregates by this phase transformation is irreversible, unlike cement paste. Calcareous aggregates, on the other hand, undergo disintegration at around 750 °C and, upon cooling, absorb moisture to form calcium hydroxide. This reaction is accompanied by a volumetric expansion of up to 44%, which induces severe cracking in the aggregates and consequently contributes to the deterioration of concrete strength. [1, 6]. This phenomenon is referred to as post-cooling spalling phenomena or thermo-chemical spalling [7].

Owing to the degraded mechanical properties of concrete ingredients, part of the concrete specimens randomly spall out at different times during high temperatures, commonly known as spalling of concrete. Various thermo-mechanical or hygral stresses inside the concrete pores at high temperature, when reaches the strength value, spalling occurs. Such fire-induced spalling is considered as one of the potential factors that affect the post-fire residual capacity of a concrete structure [8]. In literatures, spalling of different types was reported, from an explosive nature to a less violent type. Similarly, the mechanism behind such spalling is also found to be conflicting in nature [8, 9]. Multiple influencing factors and their interdependency lead to contradicting observations on the nature of spalling and its mechanism.

However, in literatures, two broad theories are thought to be major reasons behind the spalling phenomenon [8, 9]: Accumulation of pore pressure and restrained thermal dilation occur during heating. Water released from chemical decomposition, along with the moisture previously present in the pores of concrete, generates vapor pressure. When this pressure cannot dissipate due to the low permeability of concrete, it leads to internal stress build-up (as more often observed in high-strength concrete), which starts increasing in magnitude. When the vapor pressure is high, it surpasses the tensile strength of concrete, which decreases with rising temperature, and parts of the concrete specimen break off sequentially. Such concrete failure is known as "pore pressure induced spalling," which is explosive in nature. Another theory behind concrete spalling assumes that when concrete structures' thermal expansion is restrained, compressive stress is developed parallel and near the heated surface. Owing to such higher compressive stress, compressive fracture or spalling can occur in concrete. However, as stated in [8], due to the interdependency of the above two mechanisms, they complement each other and might lead to the same prediction of spalling. In fact, Dwaikat and Kodur [8] reported that the mechanism, governed by pore pressure, is mainly responsible for concrete spalling, where restrained thermal dilation implicitly affects the pore pressure development. The occurrence of spalling leads to a decrease in the cross section of the concrete member and consequently reduces the load-resisting capacity of a structural member. Hence, a realistic forecast of concrete spalling is of utmost interest to assess the structural protection of a concrete member subjected to great temperature. These mechanisms are well documented in conventional concrete but may behave differently in 3DPC due to its unique layered deposition and typically higher paste content, which requires extensive research on the same.

7.3 Role of Distinct Characteristics on the Fire Performance of 3DPC

7.3.1 Layer Interfaces and Anisotropy

3DPC is printed layer by layer without formwork, creating cold joints or interfaces that are potential weak zones in the structural elements. These interlayer regions exhibit different thermal conductivity and permeability and act as preferential crack initiation sites under fire. Thus, the interlayer joints reduce the overall integrity of the printed structural elements when heated unevenly.

7.3.2 Porosity and Microstructure

The extrusion-based printing process often leads to higher porosity in comparison to conventional mold-cast concrete. These altered capillary structure which can affect moisture migration and vapor pressure build-up during fire exposure.

7.3.3 Influence of Mix Design

The presence of type and amount of different primary constituents (e.g., OPC, FA, and GGBFS) and admixtures (e.g., SP and VMA) in the concrete mix has a stronger effect on the microstructure and porosity of printed concrete. These compositional changes influence the thermal performance such as thermal diffusivity and thermal conductivity of the printed concrete, like mold-cast concrete. Any change in the thermal properties affects the spalling performance of 3DPC structures.

7.4 Experimental Studies on Thermal Performance of 3DPC

In the literature, limited research was found that focuses on the fire behavior of 3DPC structural elements. In general, the International Standard ISO 834 was employed to replicate real fire conditions, in which the temperature rises exponentially with heating duration. However, different researchers [10] employed various heating rates in evaluating the thermos-mechanical performance of concrete. For instance, Liang et al. [11] observed that the heating rate does not influence the likelihood of explosive spalling in high-strength concrete; however, higher heating rates intensify the severity of spalling when it occurs. A heating rate of 4 °C/min was considered appropriate for evaluating the performance of high-strength concrete after exposure to elevated temperatures. Hence, taking into account the high compactness and low permeability of high-strength concrete, the heating rate for the test was set at 4 °C/min in accordance with ASTM E831. After heating up to the target temperature, the constant target temperature was maintained for 2 h to achieve a uniform temperature distribution within the test specimens. Heating was applied and then stopped. For ensuring temperature homogeneity, a thermocouple was positioned at the center of each specimen prior to testing, followed by a preliminary heating trial. In the test, cylinder specimens of various sizes, such as 42 mm × 82 mm and 68 mm × 135 mm cores, are extracted from the printed beam. It was observed that the temperatures at the center of the specimens closely matched the target values, indicating that a uniform temperature distribution had been achieved throughout the specimens. Here, it is required to mention that due to slower thermal propagation inside concrete, reaching a uniform temperature from surface to core of the specimen takes a variable amount of time depending on the targeted temperature. During the fire exposure, color changes and formation of cracks of the specimen with an increase in exposed temperature were observed.

7.4.1 Measurement of Thermal Conductivity

Thermal conductivity represents the rate of heat transfer through a material per unit thickness and unit area, measured perpendicular to the surface, under a unit gradient of temperature [12]. The thermal conductivity of 3D-printed concrete (3DCP) is measured using standardized laboratory methods such as ASTM C177 or ASTM C518. Figure 7.1 exhibits the fundamental schematic of different methodologies to determine thermal conductivity at lab, various experimental setups. However, in general, thermal conductivity is typically determined through "steady-state" or "transient" test methods. First, 3DCP specimens are prepared, typically in slab or cube form with smooth and flat surfaces, and are properly cured for at least 28 days to ensure consistency in results. The specimens are conditioned to a specified moisture level, either dry or saturated, depending on the test requirement. In the test setup, the sample is placed between two plates—one heated and the other cooled—to create a steady temperature gradient across the specimen. The equipment then measures the rate of heat movement and the difference in temperature across the thickness of the sample. Figure 7.2 presents the one-dimensional flow of heat and, accordingly measurement of thermal conductivity. Once the system reaches steady-state conditions (no further variation in temperature readings), for calculating the thermal conductivity (k) formula is used:

$$q = -k\frac{\Delta T}{\Delta L} \tag{7.1}$$

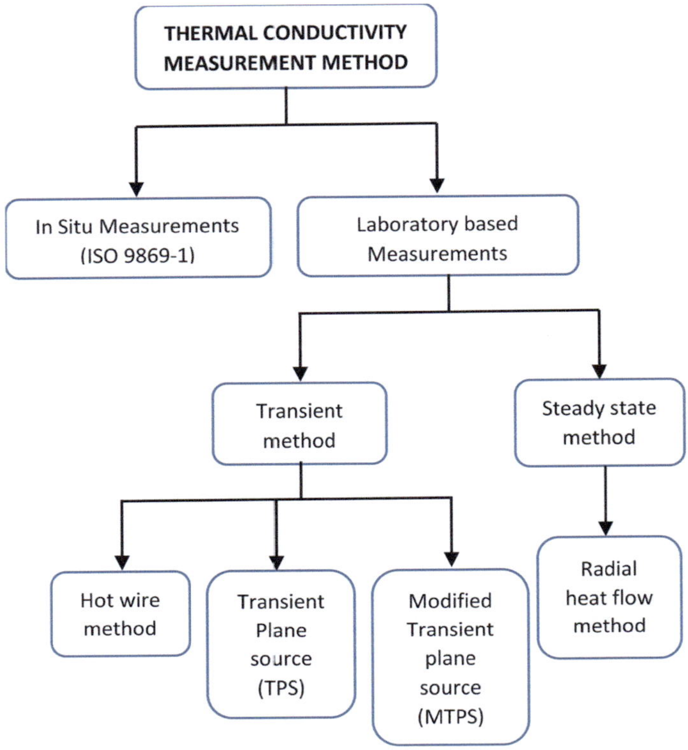

Fig. 7.1 Thermal behavior evaluation techniques [13]

where q is the heat movement per unit area, Δ L is the sample thickness, and ΔT is the temperature difference. The minus (−) sign comes because heat transfer happens in the direction of decreasing temperature. The final value of thermal conductivity at different temperatures for 3DPC and mold-cast concrete is shown in Fig. 7.3. Thermal conductivity for plain concrete is found to be 0.98 to 1.8 W/m-°K, and for agro-industrial waste-based 3DPC is found to be 0.68 to 1.29 W/m-°K.

Such a drop in thermal conductivity in the case of the 3DPC specimen could be due to the utilization of agro-industrial waste along with the higher amount of larger-sized pores than mold-cast concrete. Optimizing the heat transfer properties of construction materials is considered a smarter strategy to attain higher environmental benefits owing to the fact that "*K*" plays a vital role in designing insulation networks compliant with building codes [12].

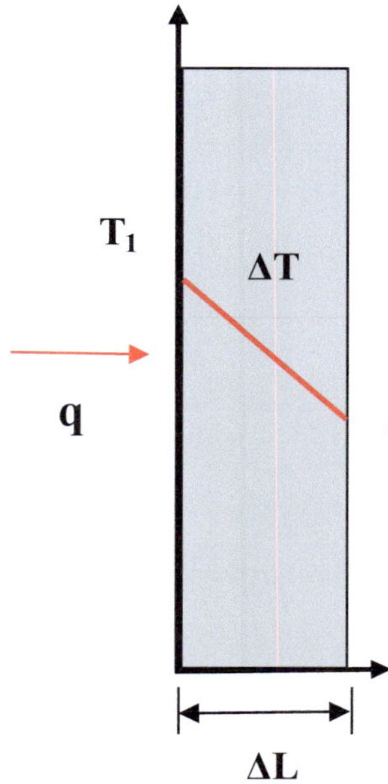

Fig. 7.2 One-dimensional diffusion of heat energy

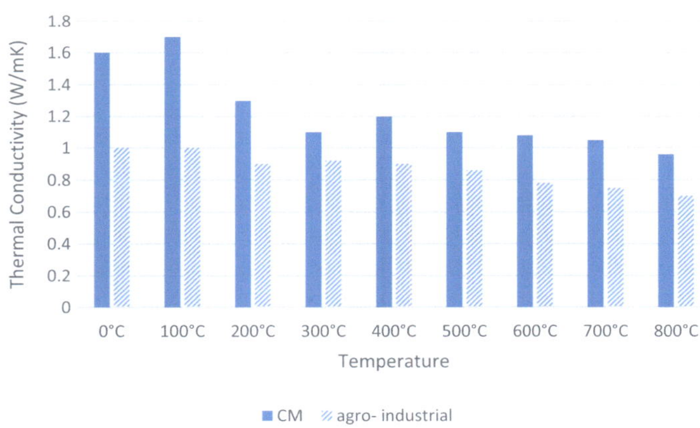

Fig. 7.3 Thermal conductivity of plain concrete and agro-industrial waste 3DCP

7.4.2 Specific Heat Capacity

Specific heat capacity of 3D-printed concrete (3DPC) represents the amount of heat energy needed to increase the temperature of one kilogram of the material by 1° Celsius (or Kelvin). It is a key thermal property that affects how 3DPC structures

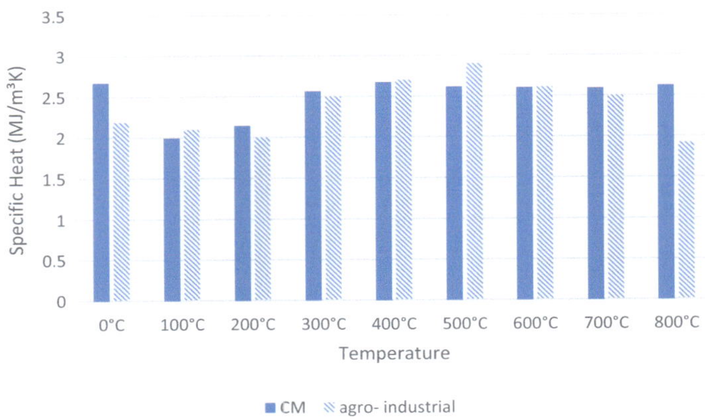

Fig. 7.4 Specific heat capacity of plain concrete and Agro-Industrial waste 3DCP

respond to temperature changes, including environmental heating and fire exposure. The thermal capacity, representing the product of the specific heat of concrete and concrete density, is a common way to express specific heat. Concrete's specific heat depends on physical as well as chemical processes that occur at higher temperatures. The obtained specific heat capacity of 3DPC is shown in Fig. 7.4. It is observed that the specific heat capacity of plain concrete and 3DPC are found to be in a similar range for different temperatures. However, at room temperature, the specific heat capacity for the sustainable 3DPC mix is higher than plain concrete, which implies the positive effect of thermal comfort in the case of using agro-industrial byproducts-based 3D-printed building.

In literatures, other methods were used to determine the specific heat capacity of concrete. For instance, Islam et al. [14] used a foam icebox as a calorimeter, with a cylindrical specimen replacing the beam sample. Afterward, the box was constructed from thermally insulating material. Researchers recorded the masses of both the water and the asphalt concrete (AC) specimen. The temperature at the surface of the specimen was measured, after which it was immersed in water maintained at 50.5 °C inside the box. After 42 min, thermal equilibrium was reached, and the heat capacity was determined using the following Eq. (7.3).

$$C = \frac{m_w c_w(\theta_w - \theta_f)}{m_{AC}(\theta_f - \theta_s)} \tag{7.2}$$

where m_w, C_w, and θ_w are the mass, specific heat capacity, and initial temperature of water, m_{AC}, θ_s are the mass and original temperature of the specimen, respectively, and h_f is the final temperature of the mix.

Hansen et al. [15] utilized a two-stage process of temperature measurement to determine the C-value of the paste specimen. The rise in temperature was first recorded without the specimen and then while the sample was placed inside the calorimeter. C-value of the paste was subsequently evaluated using the given Eq. (7.3).

$$C = \frac{1}{m}\left(\frac{E_2}{\Delta T_2} - \frac{E_1}{\Delta T_1}\right) \tag{7.3}$$

where C represents the specific heat of the paste (KJ/Kg °K), m stands for mass of the material (Kg), E_2 and E_1 are the power with and without the sample (KW), ΔT_2 and ΔT_1 are the variation in temperature value with and without the paste (°K).

7.4.3 Thermal Diffusivity

Thermal diffusivity of 3D-printed concrete (3DPC) is a critical parameter that influences its thermal performance, especially in building applications. It defines how quickly heat spreads through the material and is mathematically defined as the ratio of thermal conductivity to the product of specific heat capacity and density. Due to the layer-by-layer deposition process, 3DPC often exhibits anisotropic thermal properties—meaning the thermal diffusivity may vary in the horizontal and vertical directions. Thermal diffusivity is measured using the laser flash analysis (LFA) method, which is commonly used, wherein a small disc sample is subjected to a small laser pulse on one side, and the temperature response is recorded on the opposite surface. Using the characteristic time-dependent rise of temperature, thermal diffusivity is determined using the given relation. Equation (7.4) [16, 17]

$$\alpha = \frac{138x^2}{\pi^2 t_{1/2}} \tag{7.4}$$

where α represents thermal diffusivity (m^2 /s), x stands for thickness (m), and $t_{1/2}$ is the time required for the back face to reach half of the extreme temperature growth.

Alternatively, the transient plane source (TPS) method is employed for larger or anisotropic samples. Experimental results typically show that the thermal diffusivity of 3DPC is influenced by factors such as mix composition, porosity, interlayer bonding, and printing orientation. Studies have reported values ranging from 0.2 to 0.6 mm^2/s, depending on the material constituents and printing parameters. Enhanced interlayer bonding and reduced porosity generally lead to higher thermal diffusivity, indicating improved heat transfer characteristics. Howlader et al. [18] measured thermal diffusivity using a cylindrical specimen with a 9.5 mm axial hole, into which a thermocouple was inserted. The sample was first immersed in a hot water bath till its center reached the target temperature, and then it was transferred to a constant-temperature cold-water bath. Temperature readings were taken at an interval of one-minute, and a semi-logarithmic plot of time versus change in temperature was prepared. Thermal diffusivity was then determined using the corresponding equation.

$$\alpha = \frac{60\,In\left(\frac{T_1}{T_2}\right)}{(t_2 - t_1)\left(\frac{5.783}{r^2} + \frac{\pi^2}{l^2}\right)} \tag{7.5}$$

Here, α represents the thermal diffusivity (m^2/s), T_1 and T_2 represents temperature at time t_1 and t_2 in °C, r is the radius of the specimen (m), and l is its length (m).

7.5 Experimental Studies on Mechanical Response of 3DPC

Experimental studies on the mechanical response of 3D-printed concrete (3DPC) exposed to fire are crucial for assessing its structural integrity and fire resistance in real-world applications. The unique layered construction of 3DPC introduces potential weaknesses at the interlayer interfaces, which may behave differently under elevated temperatures compared to monolithic, cast concrete. Fire exposure can lead to degradation of the cement matrix, thermal cracking, spalling, and a significant reduction in terms of mechanical properties like compressive and flexural strength. Experimental investigations typically involve subjecting 3DPC specimens to standard fire curves or controlled furnace conditions, followed by post-fire mechanical testing. These studies aim to evaluate the influence of printing orientation, layer bonding, and material composition on residual strength and failure mechanisms after fire exposure. The results are essential for understanding the fire behavior of 3DPC and for developing fire-safe design and construction practices for printed concrete structures.

In the context of evaluating the mechanical behavior of fire-exposed 3D-printed concrete, very few research articles were found. Yaqub et al. [18] evaluated the mechanical response and fire resistance characteristics of cementitious mixtures fabricated using 3D-printed concrete (3DPC) technology with the integration of recycled plastic waste, specifically polyethylene terephthalate (PET), serving as a substitute for a portion of ground granulated blast furnace slag (GGBFS). The researchers prepared five different concrete mix designs incorporating 0%, 5%, 10%, 15%, and 20% PET plastic waste. Details mix composition can be found in [18]. Each mix was evaluated through two construction techniques: the traditional mold casting method and advanced 3D printing. These two fabrication strategies were compared to understand the influence of printing direction, plastic content, and temperature on material behavior under both normal and elevated thermal conditions.

The testing protocol involved analyzing the strength in compressive and flexural, modulus of elasticity (MOE), and flexural toughness index of the specimens at ambient and high temperatures of 100, 300, and 600 °C.

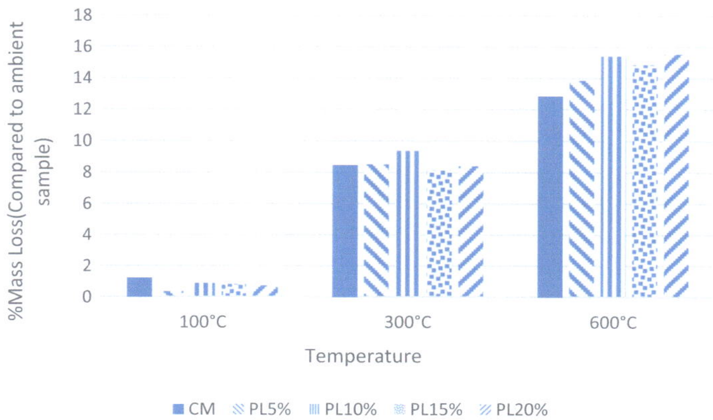

Fig. 7.5 Percentage mass loss in different mixes (CM-Control Mix, PL-PET plastic waste) [18]

7.5.1 Evaluation of Mass Loss

When it came to the effect of elevated temperatures, the fire resistance of the materials varied considerably based on plastic content and fabrication techniques. Figure 7.5 presents the mass loss of different 3D-printed concrete samples made with various amounts of plastic fibers along with control mold-cast sample. It is observed that (Fig. 7.5), at 100 °C, the samples exhibited minimal mass loss (less than 1.25%), primarily due to the evaporation of free moisture. However, exposure to 300 °C led to a noticeable increase in mass loss (8.16% to 9.38%), and at 600 °C, mass loss peaked between 12.88% and 15.57%, especially in plastic-containing samples as shown in Fig. 7.5. These losses were accompanied by thermal degradation of the plastic, leading to the formation of voids, surface discoloration, internal cracking, and in severe cases, visible spalling.

7.5.2 Evaluation of Mechanical Properties

7.5.2.1 Evaluation of Compressive Strength

Figure 7.6 shows the variation in compressive strength of cast mortar (CM) and plastic fiber incorporated composite at different temperatures. At ambient temperature, the CM shows a strength of around 55 N/mm^2, while the strength gradually decreases as the plastic content increases, with PL 5% at approximately 50 N/mm^2, PL 10% around 45 N/mm^2, PL 15% at 40 N/mm^2, and PL 20% down to 35 N/mm^2. When the temperature is raised to 100 °C, the compressive strength of CM significantly increases to nearly 67 N/mm^2, possibly due to the continued hydration

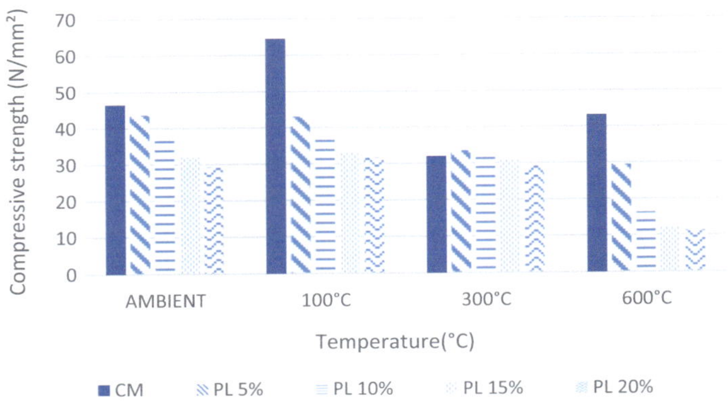

Fig. 7.6 Effect of raised temperature and plastics addition on the strength under compression loading [18]

and densification of the matrix. The plastic-modified mixes also show slight gains in strength at this temperature, although still lower than CM.

At 300 °C, the compressive strength begins to decline in all mixes, with CM falling to approximately 37 N/mm^2, and the plastic-containing samples showing slightly lower but closely grouped values, indicating the onset of thermal degradation. The most drastic drop in strength is observed at 600 °C, where thermal damage becomes more severe. While CM still retains a relatively higher strength of around 43 N/mm^2, all PL-modified concretes suffer significant losses. PL 5% records about 28 N/mm^2, PL 10% drops to 20 N/mm^2, PL 15% reaches only 16 N/mm^2, and PL 20% decreases sharply to just 13 N/mm^2. Overall, the results indicate that while small amounts of plastic may slightly influence the strength, higher plastic contents substantially compromise the thermal resistance and structural integrity of concrete under elevated temperatures, as shown in Fig. 7.6.

7.5.2.2 Evaluation of Flexural Strength

The bar graph presents the variation in flexural strength (in N/mm^2) of concrete mixtures containing different percentages of plastic (PL 5%, PL 10%, PL 15%, and PL 20%) in comparison with a control mix (CM), across various temperatures: ambient, 100 °C, 300 °C, and 600 °C. At ambient temperature, the flexural strength of all mixes is relatively close, with CM and PL-modified concretes showing values around 7–8 N/mm^2, indicating minimal impact of plastic content under normal conditions. However, at 100 °C, a slight decrease is observed in the flexural strength of CM, while PL-containing mixes (especially PL 5% and PL 10%) exhibit marginally higher or comparable strengths, suggesting that a small percentage of plastic may contribute positively to flexural performance at mild heat exposure.

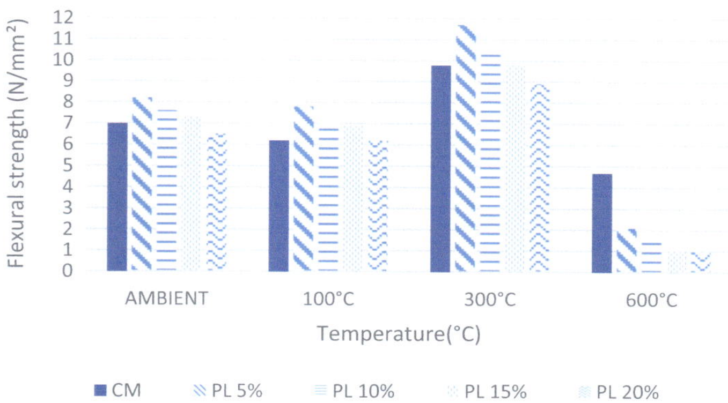

Fig. 7.7 Effect of raised temperature and inclusion of plastic on flexural strength [18]

At 300 °C, a notable increase in flexural strength is observed in all mixes, with PL 5% achieving the highest value (~11 N/mm^2), followed by PL 10%, PL 15%, and CM. This suggests that moderate heating may enhance matrix cohesion or reduce microcracking temporarily, especially in plastic-modified concretes. However, at 600 °C, the flexural strength drops sharply for all specimens. CM retains only around 5 N/mm^2, while plastic-modified concretes exhibit even lower values—ranging from approximately 1 to 3 N/mm^2—indicating severe thermal degradation. Among them, PL 20% performs the worst, suggesting that higher plastic content significantly compromises the thermal stability of concrete in terms of flexural performance. Overall, the data shows that while plastic additions may slightly enhance flexural strength at certain temperatures (notably 300 °C), they negatively impact the durability of concrete at higher temperatures such as 600 °C as shown in Fig. 7.7.

7.5.2.3 Evaluation of Modulus of Elasticity

The graph (Fig. 7.8) illustrates the variation in the elasticity modulus (in GPa) for different concrete mixes—namely the control mix (CM) and those incorporating polymeric additives (PL) at 5%, 10%, 15%, and 20%—when subjected to temperatures of ambient, 100 °C, 300 °C, and 600 °C. At ambient temperature, the control mix shows a relatively high modulus of elasticity of around 28 GPa, while all PL mixes exhibit lower values, indicating a reduction in stiffness due to the inclusion of polymers [18]. As the temperature increases to 100 °C, the elasticity modulus of the control mix increases to its peak value of approximately 33 GPa, suggesting some densification or internal stress relief in the matrix. The PL mixes also show a slight increase in modulus at this temperature, with PL 10% and PL 15% showing comparatively better performance than PL 5% and PL 20%. However, at 300 °C, a noticeable drop is observed in all mixes. The control mix reduces to about 28 GPa, while the PL mixes drop significantly, showing values between 10 and 15 GPa, highlighting a loss

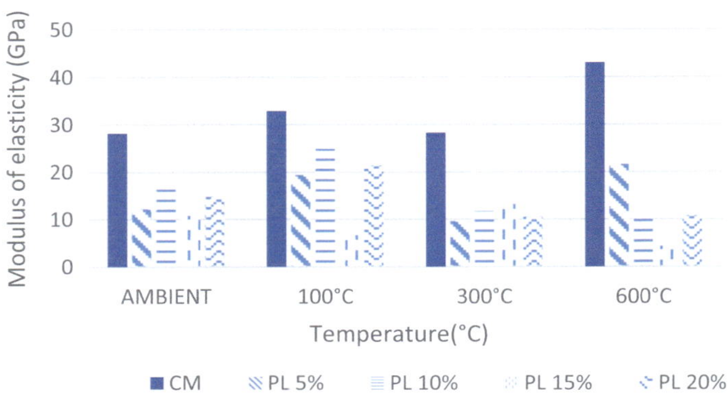

Fig. 7.8 Effect of raised temperature and plastic addition on modulus of elasticity [18]

in stiffness with increasing temperature. At 600 °C, an unusual trend is seen where the modulus of elasticity of the control mix sharply rises to over 40 GPa, which may be attributed to anomalous thermal effects or experimental variation, whereas all PL mixes show a significant decrease, with modulus values dropping to nearly 5–10 GPa. This overall trend suggests that while polymer-modified concretes may enhance post-crack performance, as seen in earlier graphs, they tend to compromise stiffness, especially under high-temperature exposure. The control mix, though more brittle, retains or even improves its stiffness at elevated temperatures, contrasting with the softening behavior of polymer-enhanced mixes as shown in Fig. 7.8.

7.5.2.4 Evaluation of Fracture Properties

Figure 7.9 presents the post-crack strength (PCS) of various concrete mixes exposed to different temperatures—ambient, 100 °C, 300 °C, and 600 °C. The control mix (CM) consistently exhibits the lowest PCS values across all temperature conditions, indicating limited resistance to cracking under thermal exposure [18]. At ambient temperature, the PCS values are generally low for all mixes, ranging around 1 MPa for the control and slightly higher for the PL (polymeric additive) mixes, with PL 15% showing a modest improvement. As the temperature increases to 100 °C, a slight enhancement in PCS is observed, particularly in the PL 15% mix, which reaches approximately 2.2 MPa, suggesting improved thermal resistance with fiber incorporation. A significant improvement is evident at 300 °C, where PL 15% achieves the highest PCS of around 8.5 MPa, followed closely by PL 20%. This indicates that the inclusion of polymeric fibers greatly enhances the post-crack behavior of concrete at elevated temperatures, likely due to improved ductility and crack-bridging mechanisms activated by heat. However, at 600 °C, while the PCS values decline from the peak at 300 °C, they remain higher than those at ambient and 100 °C. Notably, PL 20% performs best at this temperature, suggesting it offers better long-term heat

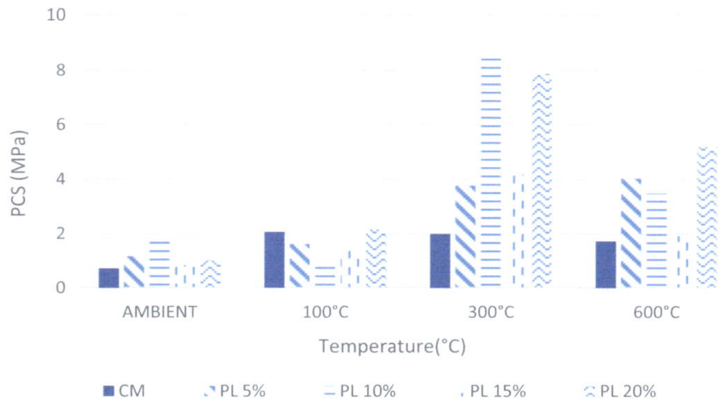

Fig. 7.9 Effect of raised temperature and plastic addition on PCS results [18]

resistance. Overall, the results highlight that incorporating 15–20% PL content significantly enhances the thermal and mechanical performance of concrete, especially at higher temperatures, while the control mix shows severe degradation in strength as shown in Fig. 7.9.

Thus, summarizing the results related to the evaluation of various mechanical properties, it is observed that while the control mix (0% plastic) showed only a 7.3% reduction strength in compressive after exposure to 600 °C, the plastic-modified samples suffered a drastic decline, with compressive strength decreasing by an average of 43.5% and flexural strength dropping by as much as 81.3%. This confirmed that while plastic-enhanced 3DCP offers certain mechanical benefits, it significantly compromises the thermal integrity and fire resistance of concrete, particularly under prolonged exposure to high temperatures.

Another key observation was the behavior of the modulus of elasticity (MOE) and flexural toughness across the different mixes and temperatures. MOE values for the control mix remained consistently higher across all temperatures compared to plastic-based mixes, indicating harder and less elastic behavior. Conversely, plastic-containing samples demonstrated increased ductility and higher flexural stiffness, particularly at moderate temperatures like 100 °C and 300 °C, due to the bridging effect of aligned plastic particles. The mix with 5% PET (PL5%) yielded the best flexural performance among the plastic-based groups, showing favorable energy absorption and post-peak load-carrying capacity.

In comparing 3D-printed and mold-cast samples, it was evident that mold-cast concrete remains superior for applications demanding high compressive strength and thermal durability. However, 3D-printed samples showed potential for lightweight, flexible, and sustainable construction elements where flexural performance and material conservation are prioritized. Despite its limitations under thermal stress, 3DCP with plastic waste offers a promising avenue for sustainable construction practices, especially when optimized with appropriate print orientation and plastic content. The study recommends future research to explore the use of plastic in fiber form, which could potentially improve bond characteristics and structural integrity. It also suggests further investigation into the effects of plastic content on shrinkage, printability, workability, and long-term durability of 3D-printed components.

7.6 Numerical Modeling to Predict Fire Behavior of 3DPC

Computational modeling is increasingly used to predict fire behavior in 3DPC. With regards to numerical modeling of fire behavior of 3D-printed material, one could employ thermo-mechanical or an advanced thermo-hygro-mechanical model [19]. Moreover, in this analysis scheme, various governing phenomena are coupled in many aspects. Choice of a particular analysis method depends on the precision required, but that would come at the expense of an increase in computational effort.

7.6.1 Thermal Analysis

7.6.1.1 Governing Equation

Thermo-mechanical analysis of any material/system is composed of the governing equations of thermal and mechanical analysis that are coupled in the governing behavior, i.e., the solution of one equation depends on the solution of the other. In general, thermal analysis inside a domain Ω can be written as

$$\rho C_p \frac{\partial T}{\partial t} - \nabla.(\lambda \nabla T) + Q = 0 \tag{7.6}$$

In the above equation, ρC_p and λ presents the specific heat and thermal conductivity of the material,T is the temperature, and Q is a scalar demonstrating an internal heat volumetric source.

7.6.1.2 Boundary Conditions

To solve the partial differential Eq. (7.6) describing the thermal equilibrium of the solid domain Ω, it is essential to specify boundary conditions. These conditions define how Ω interacts with its external surroundings. Such interaction is typically represented on the boundary of Ω either through a prescribed flux density ϕ · n or through a specified temperature T. On the boundary portion ∂Ωq, where the flux density q is given, the condition is usually expressed in the following form:

$$\vec{\phi}.\vec{n} = \left(-\underline{\lambda} \cdot \overrightarrow{grad}(T)\right)\vec{n} = -q(T) \tag{7.7}$$

Here, $\vec{n}$ denotes the outward unit normal at the considered point, and q represents the flux density entering the domain Ω at that location. This flux may depend on the local temperature T. Such a boundary condition is typically used to model processes like convective heat transfer, expressed as:

$$q(T) = h(T_{ext} - T)$$

where h is the local heat transfer coefficient (W/m^2·K), and T_{ext} is the temperature of the surrounding environment. Whereas, heat transfer through radiation in an infinite environment is

$$q(T) = \sigma\varepsilon\left(T_{\propto}^4 - T4\right)q(T) \tag{7.8}$$

where σ is the Stefan constant (σ = 5.67·10 − 8 W/m^2/K^4), ε is the emissivity of the surface, T is the temperature at the point considered, and T_∞ is the temperature of the infinite environment.

When the surface flux density q is independent of temperature, the boundary condition is referred to as a Neumann condition. This condition is often used to represent situations such as adiabatic boundaries (q = 0) or symmetry boundaries (q = 0 along a plane of symmetry). On the other hand, when the temperature is prescribed on a portion ∂ΩT of the boundary, it is expressed as T = Td at every point of ∂ΩT. Such a boundary condition is known as a Dirichlet condition.

7.6.2 Mechanical Analysis

Mechanical behavior of a material can be described by balancing the linear momentum as follows [20]

$$\nabla.\sigma + \rho g = 0 \tag{7.9}$$

where σ is the stress tensor, ρ is the density, and g represents gravitational acceleration. Cementitious composites are quasi-brittle materials that exhibit distinct stress–strain behavior in compression and tension. To realistically model their mechanical response, which includes strain-softening, a non-linear approach is required. One such method is the isotropic damage-based model proposed by Mazars [22], which effectively characterizes the stress–strain relationship as

$$\sigma = (1 - d)D : \varepsilon_m \tag{7.10}$$

Here, D denotes the initial stiffness tensor, while d represents the overall scalar damage variable. When a concrete structure is subjected to elevated temperatures, its mechanical properties degrade as a result of both applied mechanical loads and various thermo-chemical-physical processes occurring within the individual constituents of the concrete.

Such damages can be calculated by employing the damage due to mechanical load and thermo-chemical degradation [19]. ε_m is the mechanical strain component that is related to the total strain component (ε_t), through the following relation

$$\delta\varepsilon_t = \delta\varepsilon_m + \delta\varepsilon_{th} + \delta\varepsilon_{tr} + \delta\varepsilon_0 \tag{7.11}$$

In the above equation, ε_{th} represents the thermal dilation strain and is defined as $\delta\varepsilon_{th} = \alpha(T)dTI$ where $\alpha(T)$ stands for the thermal dilation coefficient at any arbitrary temperature T, $\boldsymbol{\varepsilon}_{tr}$ is the transitory creep strain which occurs when a mechanically loaded material specimen is heated for the first time, $\boldsymbol{\varepsilon}_o$ stands for other strain components, such as hygral or thermo-chemical strain [19]. Thus, in a coupled thermo-mechanical analysis, it is the thermal dilation strain ε_{th} that considers the effect of temperature change through the heat transfer equation defined by Eq. (7.6) and then used in the mechanical Eqs. (7.9) through (7.10 and 7.11).

7.6.3 Numerical Simulation of Heat Transfer Analysis

The current section present heat transfer investigation of a 3D-printed concrete beam for a given heat source. For the analysis, the 3D-printed beam as used in Chap. 8 (Sect. 8.3) is used, i.e., all the geometric details and material properties are as mentioned in Sect. 8.3. Thermal properties of concrete, such as conductivity of thermal and specific heat, are taken as 1.8 W/(m–K) and 880 J/(kg-K). In general, such thermal properties have a strong dependency on the temperature, which can be incorporated through Eq. (7.6). However, for illustrative purposes, the same have been considered as constant here. Figure 7.10 shows a 3D-printed beam of size 120 mm × 120 mm × 580 mm with each extruded layer of size 30 mm × 12 mm. In the heat transfer analysis, two different time-variant furnace temperatures are considered, in which the heating rates are 5 °C/min and 15 °C/min. Such furnace temperature is then transferred from the heat source to the concrete surface through convection and radiation. The convective heat transfer coefficient, h, and emissivity ϵ for the concrete surface are taken as 0.015 W/(m^2-K) and 0.8, respectively. In the analysis, it is also assumed that only the bottom surface of the specimen is subjected to the heating load, whereas other faces are exposed to ambient conditions. The analysis is conducted in finite element (FE) software Abaqus®.

Figure 7.11a presents the evolution of temperature across the beam depth (at bottom, mid, and top surface) along with applied furnace temperature for the heating rate of 5 °C/min. Figure 7.11b shows the temperature profile across the section when furnace temperature is 800 °C for the heating rate of 5 °C/min. In the simulated surface plot of temperature (Fig. 7.11b), the variable NT11 represents the nodal temperature. Similarly, Fig. 7.12a presents the temperature evolution across the surface when the heating rate is 15 °C/min and Fig. 7.12b shows the temperature profile of the cross

Fig. 7.10 3D-printed beam of size 120 mm × 120 mm × 580 mm with each extruded layer of size 30 mm × 12 mm

section at a furnace temperature of 800 °C. A comparative analysis of the temperature evolution for two different heating rates (Figs. 7.11a and 7.12a) implies that the temperature inside the specimen lags behind the furnace temperature, with a lower time lag found for the slower heating rate of 5 °C/min till the furnace temperature reaches 800 °C, whereas a larger time lag is found for higher heating rate 15 °C/min at the same furnace temperature. Such time lag from furnace temperature to the surface temperature at the fire-exposed faces strongly depends on the convective and radiative heat transfer parameters, which are dependent on the outside environment, characteristics of surfaces, etc. Since the degradation of concrete under fire is strongly dependent on the maximum temperature reached during fire, such analysis would be helpful in predicting the potential damage under different heating rates for varying specimen shape and size, avoiding time consuming and costly experiments.

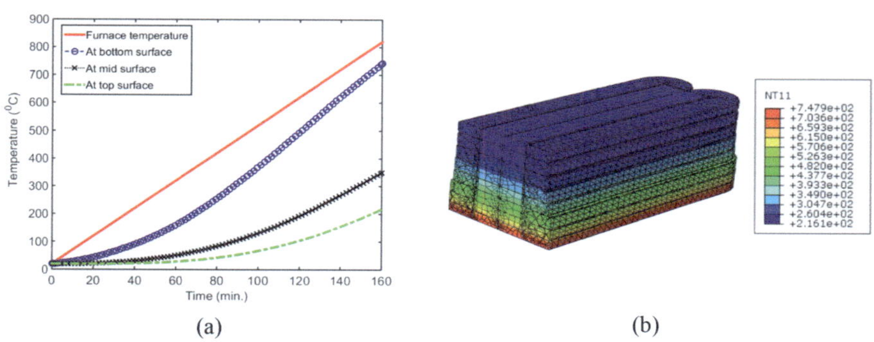

Fig. 7.11 **a** Evolution of temperature across the beam depth along with applied furnace temperature for the heating rate of 5 °C/min, **b** temperature profile across the section when furnace temperature is 800 °C for the heating rate of 5 °C/min

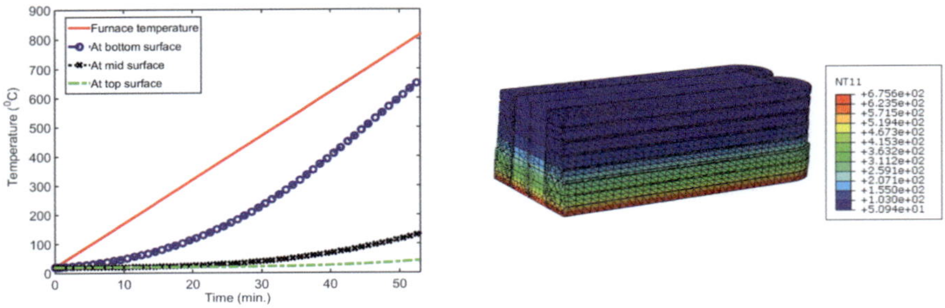

(a) (b)

Fig. 7.12 **a** Evolution of temperature across the beam depth along with applied furnace temperature for the heating rate of 15 °C/min, **b** temperature profile across the section when furnace temperature is 800 °C for the heating rate of 15 °C/min

7.7 Strategies for Enhancing Fire Resistance in 3D Printed Concrete

Fire resistance in 3D-printed concrete (3DPC) can be significantly improved through both material-level and process-level interventions. These strategies aim to mitigate spalling, reduce thermal conductivity, and preserve structural integrity under elevated temperatures.

7.7.1 Material-Level Modifications

- **Fiber Reinforcement**: Incorporating fibers such as polypropylene (PP), basalt, or steel can enhance thermal stability and reduce explosive spalling by improving crack bridging and facilitating vapor pressure relief. PP fibers, in particular, melt at low temperatures, creating micro-channels for steam escape.
- **Supplementary Cementitious Materials (SCMs)**: Blending ordinary Portland cement (OPC) with SCMs such as fly ash, metakaolin, or ground granulated blast furnace slag (GGBFS) improves high-temperature performance through denser microstructure formation and reduced calcium hydroxide content, thereby minimizing thermal degradation.
- **Lightweight Aggregates**: Using expanded clay, pumice, or other lightweight aggregates lowers thermal conductivity, slowing heat penetration and reducing thermal gradients within the printed elements.

7.7.2 *Process Improvements*

- **Optimized Print Parameters**: Adjusting layer height, extrusion rate, and print speed can minimize interlayer porosity and interfacial weakness, which are potential pathways for heat-induced cracking.
- **Tailored Curing Regimes**: Controlled curing (e.g., humidity retention, stepwise temperature increase) enhances microstructural homogeneity and reduces internal stresses during thermal exposure.
- **Infill Pattern and Geometry Design**: Strategic internal geometry, such as cellular or lattice infill, can control heat flow paths and dissipate thermal energy more effectively, reducing the risk of localized failure.

7.8 Challenges and Future Research Directions

Despite notable advancements in understanding the fire behavior of 3D-printed concrete (3DPC), several critical gaps hinder its widespread adoption in fire-sensitive applications:

- **Absence of Standardized Testing Frameworks**: Current fire testing protocols are primarily designed for conventionally cast concrete and do not account for the anisotropic nature, interlayer interfaces, or unique geometries inherent to 3DPC.
- **Limited Insight into Long-Term Fire Exposure Effects**: Most studies focus on short-duration or single-event fire scenarios, with insufficient data on repeated thermal cycling, post-fire residual strength, and durability over extended service life.
- **Scarcity of Real-Scale Performance Data**: Experimental research is often confined to small-scale specimens, leaving uncertainties in predicting full-scale structural response under realistic fire loads and boundary conditions.
- **Complexity in Modeling Interlayer Behavior**: Thermal stresses at layer interfaces introduce unique failure modes in 3DPC, which remain difficult to simulate accurately due to the coupled effects of temperature-dependent material properties and microstructural heterogeneity.

Thus, future research on the fire behavior of 3DPC should prioritize:

- Large-scale and full-building fire testing to capture realistic heat transfer, structural deformation, and failure patterns.
- Hybrid reinforcement strategies combining steel, basalt, polymeric, or smart fibers with fire-protective coatings to enhance thermal stability and structural resilience.
- Development of next-generation fire-resistant 3DPC mixes through engineered binders, nano-additives, and optimized aggregate selection.
- Advanced computational models incorporating thermo-mechanical coupling, moisture migration, and interface-specific behavior to improve fire performance prediction.

7.9 Closure

As 3D concrete printing (3DCP) transitions from experimental applications to mainstream construction, fire safety will be a critical determinant of regulatory compliance, structural reliability, and public acceptance. The inherent characteristics of 3D-printed concrete (3DPC)—including its directional anisotropy, interlayer porosity, and interface-dependent mechanical behavior—necessitate tailored fire assessment methodologies that go beyond conventional concrete design frameworks.

A holistic approach, integrating large-scale experimental testing, advanced numerical modeling, and innovative material engineering, offers the most effective pathway to ensuring that 3DPC structures achieve the necessary fire resistance while retaining the architectural freedom and construction efficiency that 3D printing enables. By proactively addressing these challenges, the construction industry can fully leverage the transformative potential of 3DCP without compromising safety, durability, or occupant protection in fire scenarios.

References

1. Hager I (2003) Behaviour of cement concrete at high temperature. Bull Pol Acad Sci Techn Sci 61:145–154
2. Fu YF, Wong YL, Poon CS, Tang CA, Lin P (2004) Experimental study of micro/macro crack development and stress-strain relations of cement-based composite materials at elevated temperatures. Cem Concr Res 34:789–797. https://doi.org/10.1016/j.cemconres.2003.08.029
3. Sarshar R, Khoury GA (1993) Material and environmental factors influencing the compressive strength of unsealed cement paste and concrete at high temperatures. Mag Concr Res 45:51–61
4. Li L, Jia P, Dong J, Shi L, Zhang G, Wang Q (2017) Effects of cement dosage and cooling regimes on the compressive strength of concrete after post-fire-curing from 800 C. Constr Build Mater 142:208–220
5. Poon C-S, Azhar S, Anson M, Wong Y-L (2001) Strength and durability recovery of fire-damaged concrete after post-fire-curing. Cem Concr Res 31:1307–1318
6. Xing Z, Beaucour AL, Hebert R, Noumowe A, Ledesert B (2011) Influence of the nature of aggregates on the behaviour of concrete subjected to elevated temperature. Cem Concr Res 41:392–402. https://doi.org/10.1016/j.cemconres.2011.01.005
7. Liu JC, Tan KH, Yao Y (2018) A new perspective on nature of fire-induced spalling in concrete. Constr Build Mater 184:581–590. https://doi.org/10.1016/j.conbuildmat.2018.06.204
8. Dwaikat MB, Kodur VKR (2009) Hydrothermal model for predicting fire-induced spalling in concrete structural systems. Fire Saf J 44:425–434. https://doi.org/10.1016/J.FIRESAF.2008.09.001
9. Kodur VKR (2004) Spalling in high strength concrete exposed to fire: concerns, causes, critical parameters and cures. In: Structures congress 2000: advanced technology in structural engineering, vol 103, pp 1–9. https://doi.org/10.1061/40492(2000)180
10. Abed F, Khalaf S, Alhoubi Y, Moustafa MA, Al Jamal M (2024) Effect of fiber types on fire-induced spalling and thermal performance of UHPC circular columns. Dev Built Environ 19:100523
11. Liang X, Wu C, Su Y, Chen Z, Li Z (2018) Development of ultra-high performance concrete with high fire resistance. Constr Build Mater 179:400–412
12. Latif E, Bevan R, Woolley T (2019) Thermal insulation materials for building applications. Institute of Civil Engineers

13. Bajwa AUR, Samarasinghe DAS, Flemmer CL, Bao DW (2025) A systematic literature review on the thermal behaviour of building elements constructed through 3D concrete printing (3DCP). In: CIB conferences, p 172
14. Islam MR, Tarefder RA (2014) Determining thermal properties of asphalt concrete using field data and laboratory testing. Constr Build Mater 67:297–306
15. Hansen PF, JH H, Kjaer U, EJ P (1982) Thermal properties of hardening cement paste. In: Proceedings of RILEM international conference on concrete at early ages. Paris
16. Parker WJ, Jenkins RJ, Butler CP, Abbott GL (1961) Flash method of determining thermal diffusivity, heat capacity, and thermal conductivity. J Appl Phys 32:1679–1684
17. Gaal PS, Thermitus M-A, Stroe DE (2004) Thermal conductivity measurements using the flash method. J Therm Anal Calorim 78:185–189
18. Yaqub W, Aslani F (2024) Mechanical properties and fire resistance of 3D-printed cementitious composites with plastic waste. Int J Concr Struct Mater 18:84
19. Pal B, Ramaswamy A (2024) A Meso-scale study to predict high-temperature behavior of concrete structures in a hygral-thermal-chemical-mechanical framework. Constr Build Mater 421:135671
20. Pal B, Ramaswamy A (2023) A multi-physics-based approach to predict mechanical behavior of concrete element in a multi-scale framework. Mech Mater 176:104509. https://doi.org/10.1016/J.MECHMAT.2022.104509

Chapter 8
Reinforcements in 3D Printed Concrete (3DPC)

8.1 Overview

3D Concrete Printing is revolutionizing the construction industry by making it possible to create complex structures with unmatched efficiency and flexibility. However, maintaining the integrity and durability aspects of printed concrete structures still poses a significant challenge. Reinforcement techniques—particularly the application of reinforcement such as steel—play a critical role in improving the mechanical properties and performance of 3D-printed concrete, much like in conventional concreting. This chapter provides an in-depth look into these reinforcement strategies, covering their types, working mechanisms, practical implementation, and the challenges they bring.

8.2 Different Reinforcements Required in 3DPC

Just like conventional concrete structures, reinforcement in 3D-printed concrete (3DPC) can mainly be categorized into two types: longitudinal reinforcement and transverse (or shear) reinforcement. Longitudinal reinforcement, aligned with the principal stress direction or the printing path, primarily resists tensile and bending forces. It plays a significant role in enhancing the flexural strength and crack resistance of structural elements such as beams and columns. On the other hand, transverse reinforcement, positioned perpendicular to the longitudinal axis, is intended to improve shear resistance, provide confinement, and control crack propagation between layers—a function that is especially crucial due to the inherently weak interlayer bonding of 3DPC. In addition, vertical reinforcement can be incorporated to counteract out-of-plane forces, particularly in slender elements such as walls. The successful integration of these reinforcement types—whether through conventional steel bars, fiber additions, or automated placement techniques—demands careful

A. Chourasia et al., *Advances and Challenges in 3D Printed Concrete (3DPC)*, Materials Horizons: From Nature to Nanomaterials, https://doi.org/10.1007/978-981-95-8557-1_8

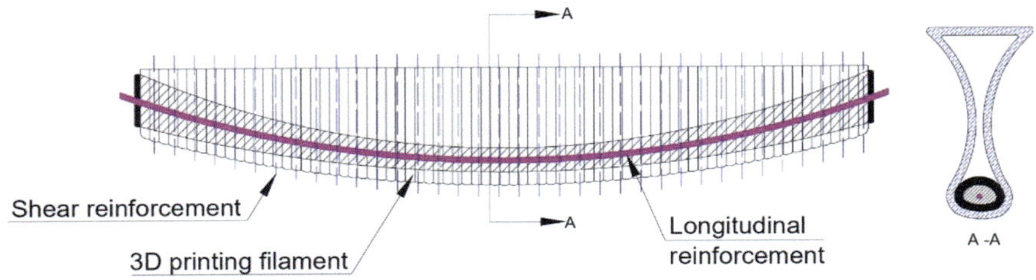

Fig. 8.1 Different types of reinforcement in a beam of complex geometry

synchronization with the layer-by-layer deposition process to maintain structural continuity and ensure performance.

8.2.1 Longitudinal Reinforcement

Longitudinal reinforcement serves a fundamental purpose in carrying the primary structural loads. As a result, reinforcement strategies with proven effectiveness in conventional cast-in-place concrete are generally favored. Reinforcement must be continuous, ductile, and compliant with structural design codes, which are generally grounded in the lower-bound theorem of plasticity. In 3DPC, a practical approach to achieving such continuity is to embed reinforcement—such as conventional passive steel bars, prestressed bars, or tendons—into pre-formed voids within the printed element. For elements with simple geometries, straight and rigid bars can be used effectively. Conversely, for more complex geometries, curved tendons can be employed to follow non-linear paths; however, the deviation forces resulting from such curvature must be accounted for in the structural design (see Fig. 8.1).

After the reinforcement is positioned, the voids can either be left unfilled or grouted to create a bond between the steel bar and the concrete around it. For example, Gebhard et al. [1] examined the structural behavior of beams reinforced with straight longitudinal elements. In their study, one group utilized unbonded post-tensioning techniques, while another employed bonded passive steel bars. While similar reinforcement approaches have been explored by other researchers, detailed documentation of the resulting structural behavior remains scarce.

8.2.2 Shear Reinforcement

The structural requirement for shear reinforcement in concrete elements is generally less than that for reinforcement placed in the longitudinal direction. The distinction is reflected in design codes, which usually specify much lesser minimum reinforcement for shear components like stirrups compared to those needed to resist axial or bending loads. Importantly, even shear reinforcement that shows strain-softening behavior can still contribute to a strain-hardening response in the overall structural member, as long as the longitudinal reinforcement can accommodate the additional stress [2].

This characteristic creates opportunities for innovative design methods in 3D-printed concrete (3DPC), where shear reinforcement can be incorporated during the printing process to form structurally active components rather than just acting as sacrificial formwork. Figure 8.1 also shows a schematic diagram of shear reinforcement placement within a 3DPC structural element. To handle complex geometries and maintain printing efficiency, these reinforcement strategies must provide high geometric flexibility and be smoothly integrated into the printing process—either by mixing fibers directly into the printable matrix (i.e., fiber-reinforced concrete) or by simultaneously embedding reinforcement elements during printing. These placement methods are explained in more detail in Sect. 8.3.3.

One reported approach for interlayer shear reinforcement [3] involves embedding a steel cable during the printing process. This technique has demonstrated effectiveness for the automated placement of continuous reinforcement [3]. Nevertheless, challenges related to anchoring high-strength, smooth cables [4] have prompted the exploration of alternative strategies. Consequently, an alternative approach using aligned interlayer steel fibers as a substitute for cables was investigated. This method addresses several problems associated with directly mixing fibers into 3DPC materials—such as changes in rheological properties and limitations in pumping long fibers. A detailed discussion on fiber integration strategies in 3D-printed concrete structures is available in Sect. 8.3.2.

8.3 Reinforcement Placement Strategies in 3DPC

The integration of reinforcement into 3D-printed concrete (3DPC) structural elements—an essential feature of any reinforced concrete system—continues to be a major challenge in developing structurally sound 3D-printed components. However, significant research advancements have been achieved in recent years to facilitate the inclusion of reinforcement, whether as solid steel bars or fibers. This section presents the two main reinforcement strategies used in 3DPC: steel reinforcement and fiber reinforcement.

8.3.1 Steel Reinforcement

The use of steel reinforcement in 3DPC is increasingly being studied at both laboratory and construction scales to satisfy structural performance requirements and regulatory standards [5]. Several steel reinforcement techniques have been proposed, including:

- Prepositioned reinforcement integrated with 3DPC processes.
- Horizontal reinforcement is placed directly onto printed walls.
- Reinforcement used to connect parallel printed wall sections.
- Reinforcement embedded within printed formworks.

These strategies are illustrated in Fig. 8.2. Figure 8.2a shows a large-scale 3DPC technique that uses pre-positioned reinforcement, where steel bars are arranged first, followed by an inverted U-shaped print head depositing concrete on both sides of the reinforcement. This technique has been practically implemented on construction sites by HuaShang Tenda [9].

The placement of horizontal reinforcement in printed walls is the most commonly used technique in both small- and large-scale applications [6]. In this approach, a layer of printed filament is first deposited, after which horizontal steel reinforcement is placed. Then, additional filament layers are printed on top, and this cycle is repeated (Fig. 8.2b).

In an advanced method, Bos et al. [3] developed a 3DPC system that can simultaneously print concrete filaments and place metal cables. Their results showed that this technique improves post-crack resistance, although the bond strength between the printed concrete and metal cable was observed to be lower than that of conventional ribbed rebar used in cast concrete. Nonetheless, the study presented a promising

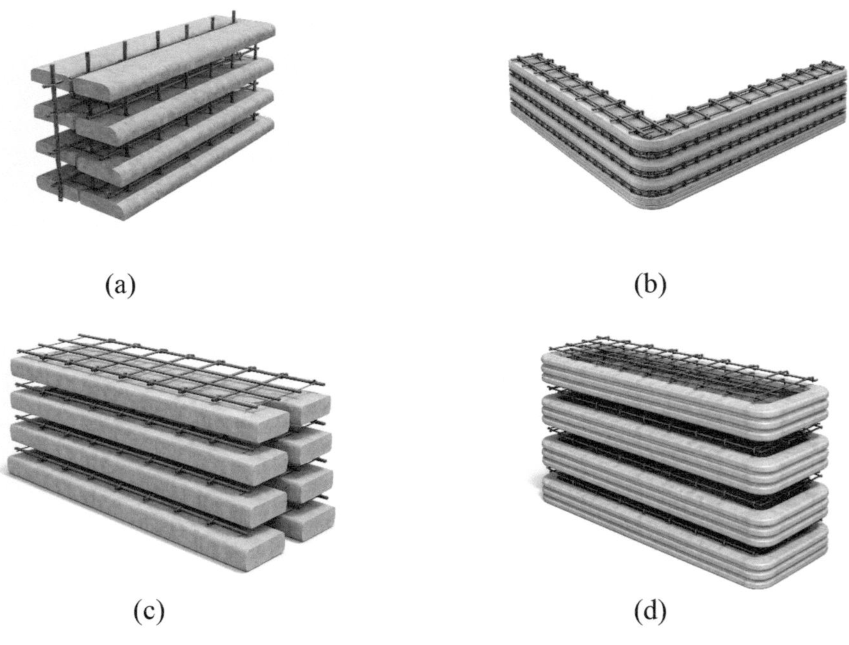

Fig. 8.2 Reinforcement constructions: **a** 3DPC with pre-positioned reinforcement, **b** horizontal reinforcements placed on printed walls, **c** horizontal reinforcements connecting printed walls, **d** reinforcement placing in printed formworks

automation method that could be scaled up for large-scale 3D printing applications. Another technique involves placing horizontal reinforcements between two parallel printed walls to strengthen the overall structural integrity (Fig. 8.2c) [7].

In some studies, reinforcements are placed inside printed formworks, after which concrete is cast in place to create fully reinforced concrete (RC) elements. This hybrid approach has been applied both in factory-controlled settings and directly at construction sites [8], as depicted in Fig. 8.2d. Further examples and detailed discussions about this method are available in Sect. 6.2.

8.3.2 Existing Approaches for Steel Bar Integration in 3DPC Structures

One of the key technological challenges in 3D concrete printing (3DCP) is developing an effective reinforcement technique that can be applied during the printing process. For 3DCP to be a practical method for building structurally sound, load-bearing elements, a reinforcement strategy that is both reliable and compatible is crucial. Several approaches have been investigated to achieve this, including embedding continuous cables or fibers within the printed layers [10, 11], as well as mixing chopped fibers into the concrete [12, 13]. However, these techniques mainly reinforce the structure in the horizontal (layer-aligned) direction and provide limited support for interlayer (vertical) reinforcement.

At present, most large-scale 3DCP projects focus on printing vertical wall elements, which are subsequently reinforced using post-processing methods, often involving manual placement. A widely used approach is to employ the 3D-printed outer shell as permanent formwork, into which conventional concrete is poured after steel reinforcements are installed. This hybrid method has been adopted by commercial 3DCP firms like Apis Cor [14] and WinSun [15], and was successfully demonstrated by U.S. military researchers for constructing in-situ barracks [16].

Other post-processing methods include printing voids or channels to accommodate reinforcement, which are later filled and grouted after printing [17]. Additional techniques involve externally attaching reinforcements or integrating them with hardened truss systems [18]. A notable example is the pedestrian bridge project by Eindhoven University of Technology (TU/e), where reinforcement was installed and post-tensioned after the printing process [19]. While effective, these approaches rely extensively on manual labor, which limits the full automation and efficiency potential of 3DCP.

To address the absence of vertical reinforcement in 3DPC, experimental methods such as deploying vertical steel meshes from spools [20] and using robotic welding for reinforcement [21, 22] have been investigated. However, large-scale implementation of these methods is still limited due to the challenges of integrating them with the printing process. Most existing solutions rely on labor-intensive pre- or post-processing techniques, including:

- Printing shells and post-tensioning.
- Inserting reinforcement into pre-formed voids and then grouting or casting.
- Printing over pre-positioned reinforcements [23, 24].

While these approaches offer some degree of effectiveness, they do not fully utilize the inherent automation potential of 3DCP. To further develop extrusion-based 3DCP technology, an in-process solution for vertical reinforcement is essential. One promising approach to address this need is the bar penetration technique.

8.3.3 Bar Penetration Technique

The bar penetration method entails the controlled insertion of pre-cut conventional reinforcement bars into the freshly deposited concrete layers. This process is carried out sequentially to form vertically overlapping bars, resulting in continuous vertical reinforcement (Fig. 8.3a). One of the key advantages of this technique is its potential for automation and smooth integration with the printing process.

Similar approaches, such as inserting short elements like nails [25], staples [26], and fibers [27], have also been studied. While these are easier to embed, their penetration depth is limited to about 1–5 layers, which restricts their overall contribution to structural strength and ductility. In contrast, the bar penetration method employs standard-sized reinforcing bars capable of penetrating 30 to 50 layers,

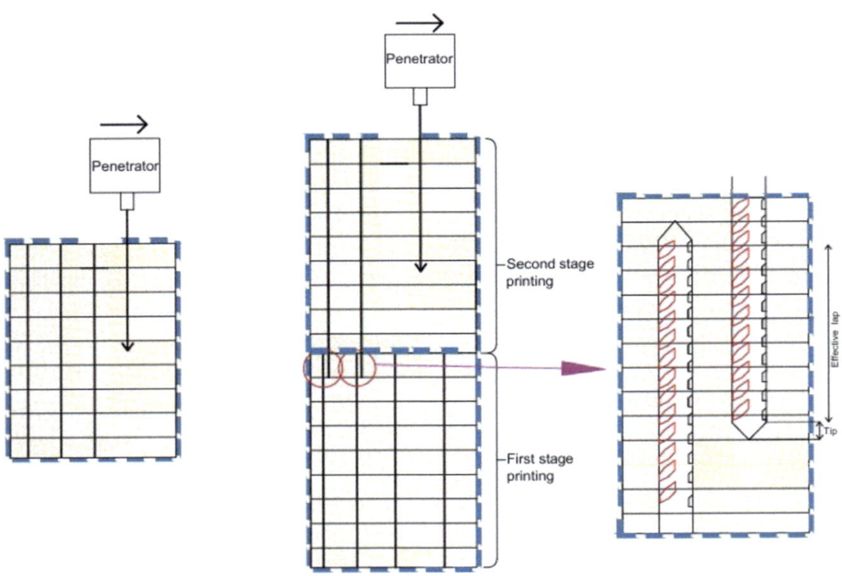

Fig. 8.3 Illustration of an automated bar penetrator system and lap joint for reinforcement

thereby reducing the need for frequent overlaps and improving both load-bearing capacity and ductility.

Experiments have been carried out using different reinforcement types, including deformed bars and helical bars, both of which have shown promising results [28, 29]. While helical bars require a slower twisting mechanism for insertion, deformed bars can be placed more quickly using a pushing system. However, a key challenge with both methods is the formation of voids around the reinforcement, which reduces bond strength and may lead to slippage [4, 30].

Recent studies [31] examining the bond mechanism of penetrated bars revealed that voids form along 70–80% of the bar length, with the upper 20% showing the weakest bond due to greater material disturbance during penetration. In contrast, no macro-voids were observed in the lower 20%, likely because of bar-induced compaction. In this lower region, the bond strength was found to be comparable to that of conventionally cast concrete (Fig. 8.4, left).

To address this challenge, a new method has been developed that involves simultaneously coating the reinforcement bars with a cementitious paste during their insertion. The underlying hypothesis is that the paste will travel downward along with the bar, filling voids and thereby enhancing the bond with the surrounding concrete (Fig. 8.4, right). This approach aims to deliver consistent bonding performance without compromising the simplicity or automation potential of the process.

A related method—pre-forming holes and then grouting the reinforcement—was previously studied by Freund et al. [28], but the present technique differs in important ways. In particular, the paste coating encourages in-situ compaction and can be seamlessly integrated into the reinforcement insertion system through a spray mechanism.

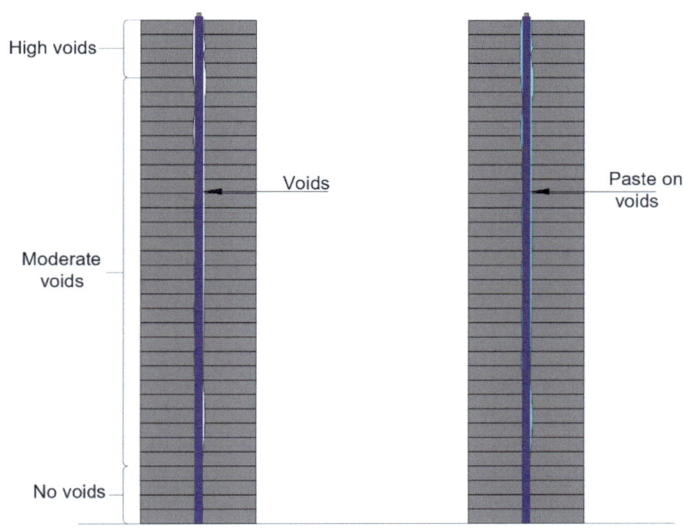

Fig. 8.4 Schematic representation of the void during bar penetrations technique [31] (left) and application of paste for void (right)

A conceptual automated setup is illustrated in Fig. 8.4, in which coating and insertion take place simultaneously. This can be achieved using either a printer-mounted penetrator or a secondary robotic arm.

In the proof-of-concept study, a single bar was inserted manually and then coated within a 350 mm-tall 3D-printed wall. Bars coated with paste and those without coating were compared to conventionally cast samples using three-point bending tests. To visually evaluate the distribution and coverage efficiency, a red oxide pigment was mixed with the binder.

The parameters in the study were confined to small-scale, single-bar penetrations tested under controlled laboratory conditions using semi-manual insertion, which may not entirely reflect real-world applications. However, the findings provide strong evidence that paste-coated bar penetrations perform comparably to conventional cast-in-place reinforcement, validating the technique's potential for integration into automated 3D printing systems. To efficiently reinforce larger structural elements, multiple overlapping steel bars will be necessary to maintain continuity. A limitation observed was the lack of shear or transverse reinforcement, resulting in shear failures in some specimens. To overcome this, adding short fibers to the print mix could improve both ductility and horizontal (layerwise) reinforcement. Furthermore, combining bar penetration with cable or fiber entrainment methods may yield additional structural advantages. Another benefit of the paste coating seems to be the reduction of voids around the bars, potentially enhancing durability and corrosion resistance—an important aspect that requires further research due to its relevance to long-term structural integrity.

8.3.4 Pre-installed Reinforcement Strategies

The Chinese construction company HuaShang Tengda Ltd. has developed a large-scale technique for integrating both vertical and horizontal reinforcement by pre-installing steel reinforcements prior to the start of the concrete printing process. In this method, the reinforcements are manually fabricated and positioned, after which concrete is extruded layerwise with two nozzles—each operating on opposite sides of the pre-fixed reinforcement. The procedure is exemplified in Fig. 8.5a, b.

This approach is particularly well-suited for building vertical structural elements such as walls and columns, where conventional steel mesh can be accommodated with ease. However, its use is inherently limited to vertical and straight geometries, as the nozzle system moves along a vertical axis and requires access to both sides of the reinforcement. As a result, design flexibility is constrained by the need for manual placement and stabilization of the reinforcement during the printing process. Despite these constraints, the method has demonstrated practical feasibility for large-scale projects, including structures engineered to withstand seismic events [32].

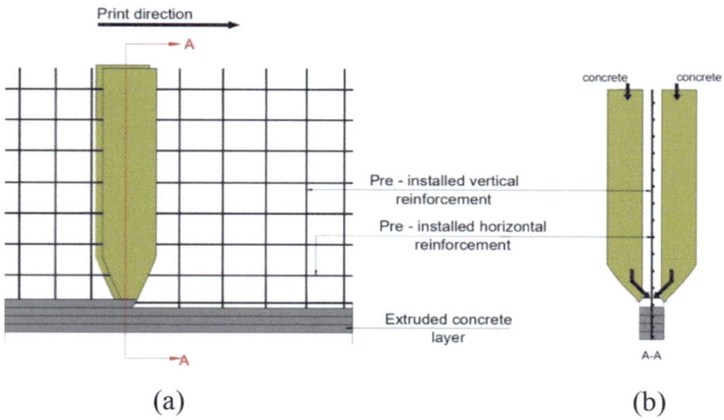

Fig. 8.5 **a** Schematic of the pre-installed and **b** a photo of the pre-installed reinforcement method in 3DPC [32]

8.3.5 Reinforcement Welding Method

A recent study conducted by researchers at TU Dresden investigated a novel method for addressing the challenge of cross-layer reinforcement in 3D-printed concrete, while preserving design flexibility, by simultaneously printing reinforcement and concrete [21]. The research centerd on the robotic fabrication of steel reinforcement using free-form gas-based metal arc welding—a process similar to that used by MX3D in the Netherlands to produce a 3D-printed steel bridge [33]. Although the simultaneous printing of concrete was not implemented in this study, the team assessed the structural performance of short, 3D-printed steel bars against that of conventional steel reinforcement.

Although promising, this welding-based method is currently resource-intensive, requiring substantial welding time and cost, which limits its immediate applicability. At present, 3DCP reinforcement technologies lack a fully integrated, time- and cost-efficient system capable of reinforcing in both the layer (horizontal) and interlayer (vertical) directions, while also providing a high degree of geometric.

To overcome these limitations and minimize dependence on post-processing, a new technique was developed that allows mesh reinforcement to be integrated directly during the printing process. This approach introduces a key innovation—*lapping* the mesh reinforcement—to ensure structural continuity across printed layers. In this method, the mesh is embedded simultaneously with the concrete filament, using a specially designed nozzle that positions the mesh as the extrusion takes place.

This technique is entirely novel and has not been attempted before. It is designed to address several challenges associated with earlier reinforcement methods. The two main innovations are:

1. The lapping of mesh provides continuous interlayer reinforcement.

2. A custom nozzle design capable of embedding mesh during extrusion, enabling strong bonding between the reinforcement and the printed material.

A lab-scale prototype was constructed to validate the concept. Tests were conducted to evaluate the performance of the lapped mesh in achieving continuity and to assess the effectiveness of the nozzle design in ensuring strong bonding between the mesh and the concrete.

8.3.6 *In-Layer Direction Reinforcement*

Several researchers have explored the simultaneous integration of a continuous steel-reinforced cable inside the extruded concrete layer. The concept was first introduced by Khoshnevis in 2006, involving the use of a steel coil; however, it did not advance beyond the conceptual stage [34]. Subsequently, researchers at TU/e refined the approach by employing high-tensile steel cables in place of coils [11]. This method was later applied in conjunction with post-tensioned reinforcement in the previously mentioned pedestrian bridge project, primarily to enhance confinement. With suitable print head modifications, the technique has also been successfully implemented using single or multi-strand cables, as well as chains [35, 36]. Compared to fiber reinforcement, this method provides significantly greater strength and continuity. Nonetheless, a key limitation persists: reinforcement is embedded only along the layer direction, with no provision for reinforcement across layers—typically in the vertical direction.

8.4 Fibers in 3DPC

Fiber-reinforced concrete has been extensively researched and widely implemented in conventional construction, and its adaptation to 3D-printed concrete (3DPC) has generated considerable research interest. In traditional applications, fibers are incorporated during the mixing stage, where they help transfer tensile stresses within the binder matrix, thereby improving tensile strength and allowing limited flexural cracking. This principle translates effectively to 3DPC, as the reinforcement can be integrated directly during the printing process, streamlining construction by eliminating the need for separate reinforcement placement.

Fiber-reinforced 3DPC offers several mechanical advantages, including enhanced tensile strength, improved crack resistance, increased durability, and reduced shrinkage and deformation during the plastic phase. These benefits are especially valuable in 3D printing, where incorporating conventional steel reinforcement is often impractical. In extrusion-based 3D printing, fibers can be introduced either prior to or during material deposition. However, despite these advantages, fiber reinforcement has inherent limitations: the fibers are typically short and discontinuous, restricting

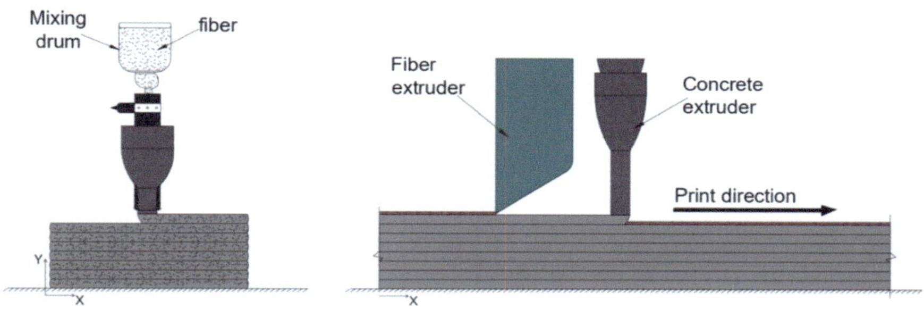

(a) (b)

Fig. 8.6 Two different processes of fiber addition during 3D printing of concrete **a** before and **b** during printing

their ability to ensure structural continuity and ductility. Moreover, reinforcement is generally aligned with the material deposition direction, providing minimal to no reinforcement across interlayer (usually vertical) planes. Various fiber integration methods employed in 3DPC are illustrated in Fig. 8.6 and discussed in detail in the following sections [37].

8.4.1 Strategies of Fiber Inclusion During Mixing

In this method, fibers are added during the preparation of the printable concrete mix, specifically during the mixing of the constituents. As a result, a uniform three-dimensional distribution of fibers throughout the material occurs. When the concrete is subjected to loading and subsequent cracking, fibers oriented perpendicular to the crack plane help resist and control crack propagation. Similar to conventional steel reinforcement, fibers enhance the toughness and tensile performance of concrete elements, although their primary effectiveness is observed after the formation of cracks [38]. Incorporating fiber as reinforcement in 3DPC sample is relatively simple, since the extrusion of short fibers from the print nozzle does not cause clogging issues [39]. When smaller fibers are blended into fresh concrete and printed, they generally align with the direction of material deposition, resulting in improved uniaxial strength properties [40].

This section outlines the use of different types of short fibers in 3D printable concrete structures. When fibers are incorporated before the printing process, three primary methods are commonly employed:

- ***Pre-blended Fibers***: In this step, fibers are evenly blended into the concrete mixture prior to the start of printing.
- ***Fiber Treatment***: Fibers are pre-treated or coated with a binder (e.g., resin) before being added to the mixture.

- ***Fiber Dispersion***: Fibers are dispersed in a liquid medium and subsequently added to the concrete to be printed.

8.4.2 Fiber Integration During 3D Printing

In one method, fibers are introduced separately from the concrete extrusion process. After each layer of concrete is deposited, a machine-based fiber placer positions the fibers in the required orientation and quantity over the freshly printed layer. When the nozzle resumes printing the next layer, these fibers become embedded within the structure. This approach eliminates the constraints imposed by the concrete matrix on fiber length, allowing the use of longer fibers—restricted only by the geometry of the printed component. It also facilitates fiber alignment in the anticipated tensile strain direction and enables customized fiber distribution in accordance with the structural requirements.

The mechanical behavior of such interlayer reinforcement by using 60 mm hooked steel fibers at two dosages, 0.7% and 1.4%, was assessed employing bending tests [37]. The results indicated that aligned and locally concentrated fibers substantially enhanced structural performance. However, at higher dosages, a sharp post-peak load drop was observed due to interlayer delamination, suggesting that this reinforcement approach is not suitable as the primary longitudinal reinforcement. Rather, it is more appropriate for application as transverse (shear) reinforcement when used in conjunction with an alternative form of longitudinal reinforcement, as discussed further in the following chapter.

For incorporating fibers during the 3D printing process, various methods can be utilized (see the section on interlayer shear reinforcement for additional details):

- **Aligned Fiber Dosing**: Fibers are introduced directly into the nozzle during the printing operation.
- **Layer-by-Layer Fiber Placement**: Fibers are deposited sequentially on each freshly printed layer.
- **Fiber Spraying**: Specialized equipment sprays fibers onto newly printed layers.
- **Fiber Insertion**: Fibers are inserted—either manually or robotically—into designated regions.
- **Fiber Interlacing**: Fibers are cross-linked with the structure as it is printed.

8.4.3 Types of Fibers Employed in 3DPC Structures

Researchers have investigated a variety of short fibers to improve the performance of 3D-printed concrete (3DPC). These fibers include carbon and basalt fibers, green tow flax, glass fibers, polyethylene (PE) fibers, polypropylene (PP) fibers, hooked-end steel fibers, polyvinyl alcohol (PVA) fibers, steel fibers, and wollastonite microfibres. Most of these fibers are generally added to the concrete mix prior to the printing process. However, only a select few, such as carbon and steel fibers, have been effectively incorporated during the printing itself. The following section describes the types of fibers used in 3DPC along with their specific roles.

8.4.3.1 Polypropylene Fibers

Polypropylene (PP) fibers are synthetic linear polymers produced through the polymerization of propylene. They provide several benefits, including low density, high strength, excellent toughness, corrosion resistance, and thermally insulated [41]. In a work conducted by Shakor et al. [42], the addition of 1% PP fiber by volume resulted in a 30% increase in compressive strength. Flexural strength was assessed using three-point bending tests on samples with different numbers of printed layers. The results demonstrated that PP fiber steadily enhanced flexural strength, although the rate of improvement varied depending on the layer number.

Nematollahi et al. [43] reported that while PP fibers improved flexural strength, they caused a slight drop in interlayer bond. The results indicate that fibers placed parallel to the printing direction mainly enhance compressive strength in that specific direction. According to Heever et al. [44], the ductility of 3D-printed components is greatly improved when PP microfibers are directionally aligned. Their study observed a 6.5 times rise in ultimate tensile crack mouth opening displacement (CMOD) and a 15 times surge in fracture energy along the longitudinal direction. However, Al-Qutaifi et al. [45] observed that adding PP fibers may adversely affect interlayer bond strength. Furthermore, van den Heever et al. [46] found that 3D-printed samples showed higher porosity compared to cast specimens—reaching up to 10.8% in the perpendicular direction and 10.3% in the longitudinal direction.

8.4.3.2 Carbon Fibers

Carbon fibers, with diameters between 5 and 10 μm and composed entirely of carbon atoms, possess a variety of advantageous properties. These include exceptional stiffness, tensile strength, an excellent strength-to-weight ratio, superior chemical resistance, high thermal stability, and low thermal expansion. Their use as reinforcement in concrete provides a promising alternative to traditional steel fibers. Unlike steel fiber-reinforced concrete (SFRC) elements, carbon fiber-made concrete can withstand greater strains beyond the initial linear-elastic phase, categorizing it as a cementitious composite with strain-hardening properties [40].

Hambach and Volkmer [47] demonstrated that the addition of carbon fibers increased the flexural strength from 10 to 30 MPa. Remarkably, carbon fiber-reinforced samples showed no cracking, unlike those reinforced with glass fibers. Incorporation of 0.5% carbon fiber did not adversely affect extrudability, although it increased the superplasticizer demand to 2.5%. The stiffness of the reinforcement material clearly influenced shear performance. Due to the flexible characteristics of carbon fibers, shear strength improved by 12%, 25%, and 43% at longitudinal reinforcement ratios of 0.3%, 0.45%, and 0.6%, respectively. Moreover, the presence of carbon fibers greatly enhanced form retention, leading to a more stable and efficient extrusion process. The fibers combine excellent tensile strength with high stiffness, making them particularly suitable for demanding, high-performance applications. Their corrosion resistance is an additional benefit in harsh environmental conditions.

8.4.3.3 Basalt Fibers

Basalt fiber is obtained from naturally produced basalt rock, which predominantly contains plagioclase and olivine minerals. Although like glass, basalt fiber provides superior physio-mechanical properties and is considerably more cost-effective than carbon fiber. Two critical factors in developing 3D printable concrete mixtures—extrudability and buildability—are vital for determining the optimal fiber content [47]. A cementitious material containing 0.5% basalt fibers by weight was found to be optimum that achieve both good printability and mechanical performance suitable for extrusion-based 3D printing [47].

The study by Ma et al. [47] revealed that filaments oriented perpendicular to the direction of applied load showed superior mechanical strength (43 MPa) compared to those aligned parallel (31.65 MPa). Under split tensile and bending conditions, fibers oriented perpendicular to the load direction consistently performed better than those aligned parallel. Additionally, Liu et al. [48] reported that increasing the number of reinforcement layers resulted in a gradual improvement in the splitting tensile strength of printed specimens. Their results also showed that reinforcement techniques applied in 3D-printed specimens increased flexural strength by 167% (up to 6.61 MPa) and maximum tensile strength by 128% (up to 3.25 MPa) at a fiber content of 0.67% and 0.97%.

8.4.3.4 Polyethylene (PE) Fibers

Polyethylene (PE) fiber, composed of polymerized polyethylene units, is commonly available as monofilaments, continuous filament, or staple fibers. The mechanical performance of 3D-printed PE fiber-reinforced concrete was evaluated under ambient and heated conditions, focusing on compressive, flexural, and interlayer bond strengths [49]. Exposure to elevated temperatures caused a 56% reduction in compressive strength due to water evaporation, while bond shear strength decreased by 65%, 60%, and 48% at 30 s, 2.5 h, and 4 h, respectively. While the mold-cast specimen exhibited a flexural strength of 6.36 MPa, printed prisms reinforced with 1.4% PE fibers (12 mm) achieved an 18% higher strength. Ding et al. [50] reported that PE fiber addition mitigated interface weakness and significantly enhanced flexural strength in all directions. Five mixtures with PE fiber ratios of 0–1.4% by weight were tested, revealing that post-peak behavior was highly fiber-length dependent. At low dosages (0.25%), longer fibers were ineffective, whereas at 1.4% dosage,

12 mm fibers doubled fracture energy compared to 6 mm fibers. Thus, 12 mm fibers were found optimal, balancing post-peak performance, extrusion smoothness, and avoiding clogging during printing. Longer fibers may further enhance flexural behavior, but demand improved mixing and nozzle designs. In a subsequent study, Ding et al. [51] assessed flexural properties of PE fiber-reinforced 3D-printed concrete with recycled sand using a four-point bending experiment. Although recycled sand initially reduced flexural strength, PE fiber addition compensated for this loss, yielding significant improvements. Fiber length had a limited influence on peak flexural strength but strongly affected post-peak behavior, with 12 mm fibers achieving two to three times the fracture energy of 6 mm fibers.

8.4.3.5 Polyvinyl Alcohol (PVA) Fibers

Polyvinyl alcohol (PVA) fiber is manufactured using polyvinyl alcohol as the base material through several processes, including dissolution, spinning, cutting, etc. This produces a fiber with enhanced strength and higher elastic modulus. In concrete, PVA fibers are distributed as monofilaments inside the matrix, creating a multidirectional fibrous system. Such a fiber network plays a crucial role in reducing shrinkage, improving abrasion resistance, adhesion [53], and providing thermal stability by reducing the contraction and expansion effects [52].

Soltan and Li [53] developed a 3D printable mix incorporating PVA fibers, considering key factors such as extrudability and buildability. Their printability assessment involved manual extrusion using a caulk gun. Yu et al. [54] studied the stress-strain behavior of PVA-based strain-hardening cementitious composites (SHCCs), noting strain-hardening responses especially in directions perpendicular to the print path. Larger-scale experimental studies involving both high-density polyethylene and PVA fibers were carried out by Ogura et al. [55] and Chaves Figueiredo et al. [56].

PVA fibers significantly affect both fresh and hardened concrete properties, comprising slump flow, strength under compression, and flexure. At a 1.5% dosage, compressive strength decreased by 36% at 7 days and 22% at 28 days, mainly due to increased porosity caused by fiber entanglement within the paste matrix. Flexural strength showed varied results: while 1.5% PVA resulted in a modest 16% improvement at 7 days, a lower 0.5% dosage caused a substantial 43% increase. The improved crack resistance in PVA-reinforced printed concrete is attributed to the collective effect of fiber bridging and aggregate interlock [57].

8.4.3.6 Wollastonite Microfibers

Wollastonite, a white calcium metasilicate mineral with a needle-like morphology, has been shown to improve both the mechanical performance and durability of cementitious composite formed by conventional mold casting [58]. As a natural and inorganic microfiber, wollastonite offers a more economical and sustainable alternative to short synthetic or metallic fibers.

A study by Bong et al. [59] investigated the mechanical performance of 3D-printed concrete (3DCP) reinforced solely with wollastonite fibers. The research assessed fresh properties, e.g., extrudability, buildability, and rheology, and hardened properties, including strength under compression and flexural loading. Two important conclusions emerged from this study: First, natural wollastonite microfibers exhibit strong potential as an effective and eco-friendly reinforcement for 3DCP, presenting a cost-efficient substitute for synthetic fibers. Second, employing a sustainable binder further improves the environmental footprint of the printed concrete.

Figure 8.7 presents the effect of wollastonite fiber on the compressive and flexural strength of 3DPC. The addition of 10% wollastonite led to improved flexural strength in the Y- and Z-directions of the printed specimens. In mold-cast samples, increasing the wollastonite content resulted in flexural strength gains ranging from 20 to 54%, depending on the dosage. However, this improvement in flexural strength was accompanied by a slight decrease in compressive strength, which dropped by 3% to 10%. A mixture with 10% wollastonite achieved the highest improvement in flexural strength by 54%; however, its compressive strength was 4% lower than that of the control specimen.

8.4.3.7 Glass Fibers

Glass fibers are widely recognized for their capacity to enhance the tensile and flexural strength of concrete while also reducing shrinkage cracking, thereby improving the durability of concrete structures. They exhibit good corrosion resistance and provide thermal insulation; however, their resistance to alkalis and abrasion is relatively low [60]. Panda et al. [61] investigated the influence of glass fibers of 3 mm, 6 mm, and 8 mm length and volume fractions (from 0.25 to 1%) on the mechanical properties of 3D-printed specimens. The results showed that although glass fibers had a limited impact on compressive strength, they notably increased flexural and tensile strength.

In another study, Shakor et al. [62] examined the incorporation of 1% glass fibers (by weight of cementitious material) into 3D printable concrete. Their findings revealed significant improvements: compressive strength rose by approximately 108%, and flexural strength increased by about 68%. Further testing of six-layer printed specimens demonstrated that adding glass fibers led to a 46.98% enhancement in overall mechanical strength compared to non-reinforced samples.

Specifically, aligned glass fibers have demonstrated a strong effect on flexural performance. Compared to unreinforced concrete paste, flexural strength improvements of up to 600% were observed with the addition of 1% fiber (by volume). Raising the aligned fiber content to 2% can further increase the ultimate flexural strength to approximately 12.05 MPa [63]. However, exceeding this 2% limit can significantly raise the mixture's viscosity, resulting in extrusion nozzle clogging during printing. This may adversely affect the quality and behavior of the printed sample [63].

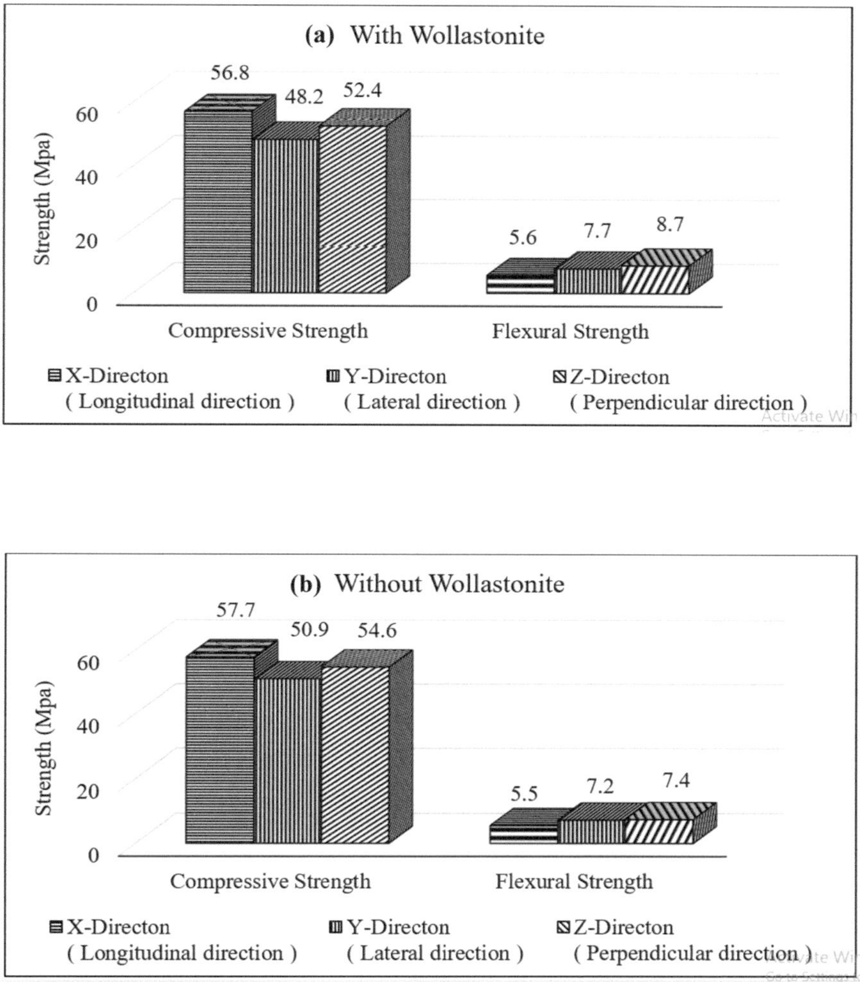

Fig. 8.7 Compressive and flexural strength of additively manufactured concrete samples **a** with and **b** without fiber [59]

8.4.3.8 Steel Fibers

Steel fibers are highly effective in enhancing the toughness, impact resistance, and overall structural strength of concrete, making them particularly valuable in applications that demand high load-bearing capacity. In 3D-printed concrete (3DPC), steel fibers play a vital part in improving fracture toughness, fatigue resistance, impact performance, bending strength, tenacity, and long-term durability. These fibers come in five main categories, classified according to their production process and shape, e.g., cold-drawn wire and modified wire [64].

Incorporating steel fibers can result in significant improvements in flexural strength—often up to 500% [65]. However, due to challenges such as pump blockages, fiber segregation, and uneven distribution, the fiber content in 3DPC is commonly restricted to between 1 and 3%. A study by Nguyen et al. [66] demonstrated that adding 2% steel fibers to ultra-high-performance 3D printable concrete allowed the fabrication of slender structural elements while substantially reducing or even removing the necessity for conventional steel reinforcement.

Pham et al. [69] conducted comprehensive research on the effects of steel fiber lengths (3 mm and 6 mm) and various printing patterns on the mechanical properties of 3DPC samples. Their results showed that, with fiber volume fractions of 0.25%, 0.5%, 0.75%, and 1%, the compressive strength of printed specimens ranged from 70 to 111 MPa, while flexural strength varied from 6 to 15 MPa. By comparison, cast control samples exhibited compressive strengths between 90 and 113 MPa and flexural strengths between 11 and 14 MPa [67]. Additional studies by Nair et al. [68] highlighted that reducing layer height can enhance flexural and fracture properties, although extremely thinner layer heights in comparison to fiber length may cause misalignment, fiber protrusion, or nozzle drag, resulting in surface defects.

The concept of hybrid fiber reinforcement, where two or more types of fibers are combined, has also demonstrated promising outcomes in 3DPC. Examples include combinations of PVA and steel fibers [69], as well as carbon and steel fibers [70]. These combinations have been used to improve both compressive and tensile performance. For instance, integrating steel with carbon fibers enhanced flexural strength and ductility while partially decreasing porosity [70].

Cao et al. [69] further assessed impact resistance through a four-point test. Concrete beams reinforced with a hybrid mix of steel fibers (1.5%) and PVA fibers (0.5%) absorbed approximately 15 times more impact energy at failure compared to plain concrete beams. Although drop-weight impact performance has been extensively studied in conventionally cast fiber-reinforced concrete [71], research specifically focused on the impact behavior of 3D-printed fiber-reinforced concrete remains limited.

8.4.3.9 Natural Fibers

Natural fibers originate from renewable sources such as plants, animals, or geological processes. These fibers often display superior characteristic properties compared to synthetic fibers, primarily due to natural variations in their composition influenced by environmental conditions during growth. A key advantage of natural fibers over conventional reinforcements like steel is their considerably lower cost.

However, the performance of natural fiber-reinforced cement composites is affected by multiple factors. Research [72] has shown that the moisture content in fiber-reinforced samples can influence both the mechanical strength and the failure modes of cementitious materials. Regarding workability, Mansur and Aziz [73] reported that increasing the length and content of jute fibers reduced the workability of the cementitious binder. Similarly, Savastano [74] noted a reduction in workability when using combinations of eucalyptus pulp, coir, or sisal fibers, attributing this effect to the hydrophilic nature of these fibers, which absorb mixing water. The primary factors contributing to workability loss in cement composites reinforced with natural fibers are the fiber's aspect ratio and volume fraction.

Varshoee et al. [75] observed that the inclusion of sugarcane bagasse fibers in cement composites delayed the setting time and reduced the heat of hydration. This effect was attributed to the presence of water-soluble sugars released through lignin hydrolysis and the partial dissolution of hemicellulose within the fibers. Similarly, Mansur and Aziz [73] and Savastano [74] reported shorter setting times in bamboo fiber-reinforced Portland cement composites, which they associated with the high sugar content of the fibers. These sugars react with cement to form calcium compounds that lower the hydration temperature and delay the development of hydration products. A similar delay was also noted by Sedan [76] in hemp fiber-reinforced cement composites, attributed to the presence of pectins that inhibit the development of calcium-silicate-hydrate (C–S–H).

In terms of mechanical performance, Ramakrishna and Sundararajan [77] reported a substantial increase in impact strength—ranging from 3 to 18 times—in plant fiber-reinforced mortar plates compared to unreinforced ones. Despite these benefits, natural plant-based fibers can also present certain drawbacks. Toledo Filho [78] found that the addition of short sisal and coconut fibers (by volume) resulted in a 2–3% increase in drying shrinkage in cement mortar. The shrinkage effect was more pronounced in sisal fiber composites due to their higher water absorption and rougher surface texture compared to coconut fibers. A follow-up study by the same authors [79] confirmed that increased porosity caused by the fibers contributed to this behavior. Therefore, the drying shrinkage of natural fiber-reinforced cement composites largely depends on fiber characteristics, fiber content, and their influence on the pore structure of the cement matrix. Figure 8.8 shows the different printing paths employed in [80] for flexural (path A and B) and compressive strength (path C and D) evaluation to study the influence of different types of fibers. Figure 8.9 presents the influence of different fibers, carbon, glass, and basalt on the flexural (Fig. 8.9a) and compressive strength (Fig. 8.9b). It is observed that, along with the printing path and direction of loading, fibers also play a crucial role in enhancing the mechanical performance of 3D-printed composites. In this regard, it should be noted that further studies are required to investigate the effect of detailed fiber characteristics, such as amount, variety of fibers, its surface characteristics, on the printing and mechanical performance of the 3DPC sample.

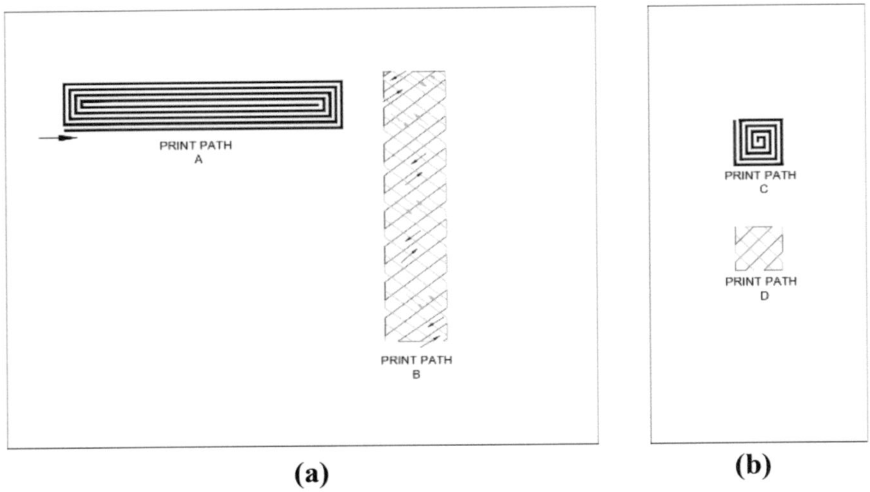

Fig. 8.8 Plan view of different printing paths [80]

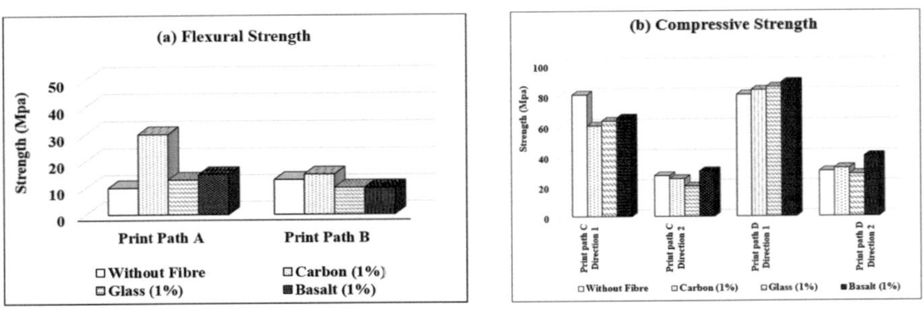

Fig. 8.9 Flexural and compressive strength of 3D-printed concrete with different fibers (for details of printing path see Fig. 8.8)

8.5 Mechanisms of Fiber Reinforcement

Fiber reinforcement in 3D Concrete Printing (3DCP) plays an essential role in improving the mechanical properties and overall performance of printed concrete structures. The ways in which fibers strengthen concrete in 3DCP can be explained through several key mechanisms.

8.5.1 Crack Bridging and Control

- **Crack Arresting**: Fibers intercept and bridge micro-cracks that develop within the concrete matrix, preventing their propagation into larger fractures and thereby enhancing the material's durability and toughness.
- **Crack Width Reduction**: By bridging cracks, fibers help limit crack width, reducing water penetration and lowering the risk of corrosion in any embedded steel reinforcements.

8.5.2 Stress Distribution

- **Load Transfer**: Fibers aid in distributing applied loads more uniformly across the concrete matrix. This load transfer mechanism minimizes stress concentrations and enhances the overall load-bearing capacity of the structure.
- **Ductility Enhancement**: The inclusion of fibers increases the ductility of concrete, enabling it to undergo greater deformations before failure. This characteristic is especially important for structures exposed to dynamic or impact loads.

8.5.3 Improved Tensile and Flexural Strength

- **Tensile Reinforcement**: Since concrete is naturally weak in tension, fibers improve its tensile strength by carrying tensile loads and lowering the risk of tensile cracking.
- **Flexural Strength**: Fibers enhance the flexural strength of concrete, improving its ability to resist bending stresses. This is particularly advantageous for structural components such as beams and slabs.

8.5.4 Improved Toughness and Energy Absorption

- **Energy Dissipation**: Fibers increase the toughness of concrete by dissipating energy during crack initiation and propagation. This leads to a more damage-tolerant material capable of absorbing and dissipating greater amounts of energy before failure.
- **Impact Resistance**: The added toughness also results in better impact resistance, making fiber-reinforced concrete well suited for applications where impact loads are a critical concern.

8.5.5 Enhanced Durability

- **Fatigue Resistance**: Fibers enhance the fatigue resistance of concrete, enabling it to better withstand cyclic loading conditions.
- **Durability Against Environmental Factors**: The inclusion of fibers can improve resistance to environmental effects such as freeze-thaw cycles, chemical attacks, and abrasion, thereby extending the structure's service life.

8.5.6 Layer Adhesion in 3DCP

- **Inter-Layer Bonding**: In 3D concrete printing, fibers improve the adhesion between successive layers, reducing delamination and ensuring a more uniform and integrated structure.
- **Shape Stability**: Fibers help preserve the shape and stability of printed layers during the printing process, minimizing the chances of deformation or sagging.

8.5.7 Reduction of Shrinkage and Creep

- **Shrinkage Control**: Fibers help limit both plastic and drying shrinkage by providing internal restraint, thus reducing the risk of shrinkage cracks.
- **Creep Reduction**: The presence of fibers can also decrease creep deformation over time, helping maintain the long-term structural integrity of printed elements.

8.5.8 Mechanisms Specific to Different Fiber Types

- **Steel Fibers**: Deliver high-tensile strength and stiffness, making them ideal for structural applications that require significant load-bearing capacity.
- **Synthetic Fibers**: Provide good ductility and toughness, well-suited for situations where flexibility and impact resistance are essential.
- **Natural Fibers**: Promote sustainability and offer good thermal and acoustic insulation, though they may need treatments to mitigate issues such as moisture absorption and biodegradability.

8.6 Closure

To realize the full potential of 3D concrete printing (3DCP) as a practical and competitive alternative to conventional cast-in-place construction, developing effective reinforcement strategies remains a crucial challenge. Although significant progress has been made in exploring a range of reinforcement methods—including fiber reinforcement, cable embedding, post-tensioning, and hybrid solutions—each approach presents specific limitations, especially regarding structural continuity and long-term performance.

One of the most persistent and technically important challenges lies in reinforcing the interlayer direction, which represents the weakest plane in 3D-printed elements. Most current techniques focus primarily on in-plane reinforcement along the print path, leaving vertical bonding between layers insufficiently supported. This absence of effective through-thickness reinforcement can result in reduced ductility, poor shear resistance, and a higher risk of delamination under mechanical loads or environmental effects.

Overcoming this critical limitation demands not only the use of innovative materials and integration methods but also a holistic understanding of the relationship between fresh-state behavior, printing process parameters, and structural requirements. Future research should prioritize the development of reinforcement systems capable of bridging layers without compromising printability, while also ensuring mechanical strength, automation compatibility, and design adaptability. Ultimately, the successful use of 3DCP in structural applications will depend on addressing this reinforcement challenge. Advancing reinforcement strategies—particularly those that can guarantee structural integrity in all directions—will be essential for enabling the wider adoption of 3D-printed concrete in mainstream construction.

References

1. Gebhard L, Mata-Falcón J, Anton A, Dillenburger B, Kaufmann W (2021) Structural behaviour of 3D printed concrete beams with various reinforcement strategies. Eng Struct 240:112380
2. Kaufmann W, Mata Falcón J, Amin A (2019) Compression field analysis of fiber-reinforced concrete based on cracked membrane model. ACI Struct J 116:213–224
3. Bos FP, Ahmed ZY, Jutinov ER, Salet TAM (2017) Experimental exploration of metal cable as reinforcement in 3D printed concrete. Materials 10:1314
4. Bos F, Dezaire S, Ahmed Z, Hoekstra A, Salet T (2020) Bond of reinforcement cable in 3D printed concrete. In: Second RILEM international conference on concrete and digital fabrication: digital concrete 2020, vol 2. Springer, pp 584–600
5. Xiao J, Ji G, Zhang Y, Ma G, Mechtcherine V, Pan J, Wang L, Ding T, Duan Z, Du S (2021) Large-scale 3D printing concrete technology: Current status and future opportunities. Cem Concr Compos 122:104115
6. Liu M, Zhang Q, Tan Z, Wang L, Li Z, Ma G (2021) Investigation of steel wire mesh reinforcement method for 3D concrete printing. Arch Civ Mech Eng 21:1–18
7. Ding T, Xiao J, Duan Z, Ji G, Du S, Li J (2019) A 3D printed ready-mixed concrete power distribution substation: materials and construction technology. Materials (1996–1944):12
8. Classen M, Ungermann J, Sharma R (2020) Additive manufacturing of reinforced concrete—development of a 3D printing technology for cementitious composites with metallic reinforcement. Appl Sci 10:3791
9. Scott C (2016) Chinese construction company 3D prints an entire two-story house onsite in 45 days. https://3dprint.com/138664/huashang-tengda-3d-print-house/

10. Mechtcherine V, Michel A, Liebscher M, Schmeier T (2020) Extrusion-based additive manufacturing with carbon reinforced concrete: concept and feasibility study. Materials 13:2568
11. Bos FP, Ahmed ZY, Wolfs RJM, Salet TAM (2018) 3D printing concrete with reinforcement. In: High tech concrete: where technology and engineering meet: proceedings of the 2017 fib symposium, held in Maastricht, The Netherlands, 12–14 June 2017. Springer, pp 2484–2493
12. Bong SH, Nematollahi B, Arunothayan AR, Xia M, Sanjayan J (2020) Effect of wollastonite micro-fiber addition on properties of 3D-printable 'just-add-water' geopolymers. In: Second RILEM international conference on concrete and digital fabrication: digital concrete 2020, vol 2. Springer, pp 23–31
13. Arunothayan AR, Nematollahi B, Ranade R, Bong SH, Sanjayan J (2020) Development of 3D-printable ultra-high performance fiber-reinforced concrete for digital construction. Constr Build Mater 257:119546
14. Apis Cor (n.d.) Accessed 1 May 2019. https://www.apis-cor.com/
15. Wu P, Wang J, Wang X (2016) A critical review of the use of 3-D printing in the construction industry. Autom Constr 68:21–31
16. Kreiger EL, Kreiger MA, Case MP (2019) Development of the construction processes for reinforced additively constructed concrete. Addit Manuf 28:39–49
17. Lim S, Buswell RA, Le TT, Austin SA, Gibb AGF, Thorpe T (2012) Developments in construction-scale additive manufacturing processes. Autom Constr 21:262–268
18. Asprone D, Auricchio F, Menna C, Mercuri V (2018) 3D printing of reinforced concrete elements: technology and design approach. Constr Build Mater 165:218–231
19. Salet TAM, Ahmed ZY, Bos FP, Laagland HLM (2018) Design of a 3D printed concrete bridge by testing. Virtual Phys Prototyp 13:222–236
20. Marchment T, Sanjayan J (2020) Mesh reinforcing method for 3D concrete printing. Autom Constr 109:102992
21. Mechtcherine V, Grafe J, Nerella VN, Spaniol E, Hertel M, Füssel U (2018) 3D-printed steel reinforcement for digital concrete construction–manufacture, mechanical properties and bond behaviour. Constr Build Mater 179:125–137
22. Müller J, Grabowski M, Müller C, Hensel J, Unglaub J, Thiele K, Kloft H, Dilger K (2019) Design and parameter identification of wire and arc additively manufactured (WAAM) steel bars for use in construction. Metals (Basel) 9:725
23. Baz B, Aouad G, Leblond P, Al-Mansouri O, D'hondt M, Rémond S (2020) Mechanical assessment of concrete–Steel bonding in 3D printed elements. Constr Build Mater 256:119457
24. Vantyghem G, De Corte W, Shakour E, Amir O (2020) 3D printing of a post-tensioned concrete girder designed by topology optimization. Autom Constr 112:103084
25. Perrot A, Jacquet Y, Rangeard D, Courteille E, Sonebi M (2020) Nailing of layers: a promising way to reinforce concrete 3D printing structures. Materials 13:1518
26. Geneidy O, Kumarji S, Dubor A, Sollazzo A (2020) Simultaneous reinforcement of concrete while 3D printing. In: Second RILEM international conference on concrete and digital fabrication: digital concrete 2020, vol 2. Springer, pp 895–905
27. van Overmeir A (2020) Designing an interlayer reinforcement solution for printable strain-hardening cement-based composites: practical research on various bond improvement concepts. TU Delft Repositories, Delft, The Netherlands
28. Freund N, Dressler I, Lowke D (2020) Studying the bond properties of vertical integrated short reinforcement in the shotcrete 3D printing process. In: Second RILEM international conference on concrete and digital fabrication: digital concrete, vol 2. Springer, pp 612–621
29. Marchment T, Sanjayan J (2020) Penetration reinforcing method for 3D concrete printing. In: Second RILEM international conference on concrete and digital fabrication: digital concrete 2020, vol 2. Springer, pp 680–690
30. Baz B, Aouad G, Remond S (2020) Effect of the printing method and mortar's workability on pull-out strength of 3D printed elements. Constr Build Mater 230:117002
31. Marchment T, Sanjayan J (2020) Bond properties of reinforcing bar penetrations in 3D concrete printing. Autom Constr 120:103394

32. New China TV (2016) World's first 3D-printed house that can withstand 8.0-magnitude quake. YouTube Video. https://www.youtube.com/watch
33. MX3D (n.d.) MX3D Bridge. Accessed 6 July 2025. https://mx3d.com/projects/bridge-2/
34. Khoshnevis B, Hwang D, Yao K-T, Yeh Z (2006) Mega-scale fabrication by contour crafting. Int J Ind Syst Eng 1:301–320
35. Ma G, Li Z, Wang L, Bai G (2019) Micro-cable reinforced geopolymer composite for extrusion-based 3D printing. Mater Lett 235:144–147
36. Lim JH, Panda B, Pham Q-C (2018) Improving flexural characteristics of 3D printed geopolymer composites with in-process steel cable reinforcement. Constr Build Mater 178:32–41
37. Warsi SBF, Panda B, Biswas P (2023) Exploring fibre addition methods and mechanical properties of fibre-reinforced 3D printed concrete: a review. Dev Built Environ 16:100295
38. Wu Z, Memari AM, Duarte JP (2022) State of the art review of reinforcement strategies and technologies for 3D printing of concrete. Energies (Basel) 15:360
39. Baduge SK, Navaratnam S, Abu-Zidan Y, McCormack T, Nguyen K, Mendis P, Zhang G, Aye L (2021) Improving performance of additive manufactured (3D printed) concrete: a review on material mix design, processing, interlayer bonding, and reinforcing methods. In: Structures. Elsevier, pp 1597–1609
40. Lauff P, Pugacheva P, Rutzen M, Weiß U, Fischer O, Volkmer D, Peter MA, Grosse CU (2021) Evaluation of the behavior of carbon short fiber reinforced concrete (CSFRC) based on a multi-sensory experimental investigation and a numerical multiscale approach. Materials 14:7005
41. Marais H, Christen H, Cho S, De Villiers W, Van Zijl G (2021) Computational assessment of thermal performance of 3D printed concrete wall structures with cavities. J Build Eng 41:102431
42. Shakor P, Nejadi S, Paul G (2019) A study into the effect of different nozzles shapes and fibre-reinforcement in 3D printed mortar. Materials 12:1708
43. Nematollahi B, Vijay P, Sanjayan J, Nazari A, Xia M, Naidu Nerella V, Mechtcherine V (2018) Effect of polypropylene fibre addition on properties of geopolymers made by 3D printing for digital construction. Materials 11:2352
44. van den Heever M, Bester F, Kruger J, van Zijl G (2021) Mechanical characterisation for numerical simulation of extrusion-based 3D concrete printing. J Build Eng 44:102944
45. Al-Qutaifi S, Nazari A, Bagheri A (2018) Mechanical properties of layered geopolymer structures applicable in concrete 3D-printing. Constr Build Mater 176:690–699
46. van den Heever M, du Plessis A, Kruger J, van Zijl G (2022) Evaluating the effects of porosity on the mechanical properties of extrusion-based 3D printed concrete. Cem Concr Res 153. https://doi.org/10.1016/j.cemconres.2021.106695
47. Ma G, Li Z, Wang L, Wang F, Sanjayan J (2019) Mechanical anisotropy of aligned fiber reinforced composite for extrusion-based 3D printing. Constr Build Mater 202:770–783
48. Liu M, Huang Y, Wang F, Sun J, Ma G (2021) Tensile and flexural properties of 3D-printed jackets-reinforced mortar. Constr Build Mater 296:123639
49. Alchaar AS, Al-Tamimi AK (2021) Mechanical properties of 3D printed concrete in hot temperatures. Constr Build Mater 266:120991
50. Ding T, Xiao J, Zou S, Zhou X (2020) Anisotropic behavior in bending of 3D printed concrete reinforced with fibers. Compos Struct 254:112808
51. Ding T, Xiao J, Zou S, Yu J (2021) Flexural properties of 3D printed fibre-reinforced concrete with recycled sand. Constr Build Mater 288:123077
52. Ayegba BO, Egbe KJI, Matin Nazar A, Huang M, Hariri-Ardebili MA (2022) Resource efficiency and thermal comfort of 3D printable concrete building envelopes optimized by performance enhancing insulation: a numerical study. Energies (Basel) 15:1069
53. Soltan DG, Li VC (2018) A self-reinforced cementitious composite for building-scale 3D printing. Cem Concr Compos 90:1–13
54. Yu J, Leung CKY (2018) Impact of 3D printing direction on mechanical performance of strain-hardening cementitious composite (SHCC). In: RILEM international conference on concrete and digital fabrication. Springer, pp 255–265

55. Ogura H, Nerella VN, Mechtcherine V (2018) Developing and testing of strain-hardening cement-based composites (SHCC) in the context of 3D-printing. Materials 11:1375
56. Figueiredo SC, Rodríguez CR, Ahmed ZY, Bos DH, Xu Y, Salet TM, Çopuroğlu O, Schlangen E, Bos FP (2019) An approach to develop printable strain hardening cementitious composites. Mater Des 169:107651
57. Yang S, Lan T, Sun Z, Xu M, Wang M, Feng Y (2022) A predictive model to determine tensile strength and fracture toughness of 3D printed fiber reinforced concrete loaded in different directions. Theoret Appl Fract Mech 119:103309
58. Bong SH, Nematollahi B, Xia M, Nazari A, Sanjayan J (2020) Properties of one-part geopolymer incorporating wollastonite as partial replacement of geopolymer precursor or sand. Mater Lett 263:127236
59. Bong SH, Nematollahi B, Xia M, Ghaffar SH, Pan J, Dai JG (2022) Properties of additively manufactured geopolymer incorporating mineral wollastonite microfibers. Constr Build Mater 331. https://doi.org/10.1016/j.conbuildmat.2022.127282
60. Mohajerani A, Hui SQ, Mirzababaei M, Arulrajah A, Horpibulsuk S, Abdul Kadir A, Rahman MT, Maghool F (2019) Amazing types, properties, and applications of fibres in construction materials. Materials 12:2513
61. Panda B, Paul SC, Tan MJ (2017) Anisotropic mechanical performance of 3D printed fiber reinforced sustainable construction material. Mater Lett 209:146–149
62. Shakor P, Nejadi S, Sutjipto S, Paul G, Gowripalan N (2020) Effects of deposition velocity in the presence/absence of E6-glass fibre on extrusion-based 3D printed mortar. Addit Manuf 32:101069
63. Zhang P, Lu X (2021) Study on bending strength of cementitious composites based on fiber alignment. J Phys Conf Ser, IOP Publishing, p 12033
64. Han B, Yu X, Ou J (2014) Self-sensing concrete in smart structures. Butterworth-Heinemann
65. Bos FP, Bosco E, Salet TAM (2019) Ductility of 3D printed concrete reinforced with short straight steel fibers. Virtual Phys Prototyp 14:160–174
66. Nguyen NT, Bui T-T, Bui Q-B (2022) Fiber reinforced concrete for slabs without steel rebar reinforcement: assessing the feasibility for 3D-printed individual houses. Case Stud Constr Mater 16:e00950
67. Pham L, Tran P, Sanjayan J (2020) Steel fibres reinforced 3D printed concrete: influence of fibre sizes on mechanical performance. Constr Build Mater 250:118785
68. Nair SAO, Tripathi A, Neithalath N (2021) Examining layer height effects on the flexural and fracture response of plain and fiber-reinforced 3D-printed beams. Cem Concr Compos 124:104254
69. Cao M, Li L, Zhang C, Feng J (2018) Behaviour and damage assessment of a new hybrid-fibre-reinforced mortar under impact load. Mag Concr Res 70:905–918
70. Liu J, Li S, Fox K, Tran P (2022) 3D concrete printing of bioinspired Bouligand structure: a study on impact resistance. Addit Manuf 50:102544
71. Jia PC, Wu H, Wang R, Fang Q (2021) Dynamic responses of reinforced ultra-high performance concrete members under low-velocity lateral impact. Int J Impact Eng 150:103818
72. El-Ashkar NH, Nanko H, Kurtis KE (2007) Effect of moisture state on mechanical behavior and microstructure of pulp fiber-cement mortars. J Mater Civ Eng 19:691–699
73. Mansur MA, Aziz MA (2015) Jute fiber reinforced concrete materials for building construction. Int J Cem Compos Lightweight Concrete 4:116–123
74. Savastano H Jr, Agopyan V, Nolasco AM, Pimentel L (1999) Plant fibre reinforced cement components for roofing. Constr Build Mater 13:433–438
75. Varshoee A, Nasiri H, Kargarfard A, Hoseinzadeh A, Ezati E (2012) Investigating cement-bagasse fiber composite properties. World Appl Sci J 16:744–747
76. Sedan D, Pagnoux C, Smith A, Chotard T (2008) Mechanical properties of hemp fibre reinforced cement: influence of the fibre/matrix interaction. J Eur Ceram Soc 28:183–192
77. Ramakrishna G, Sundararajan T (2005) Impact strength of a few natural fibre reinforced cement mortar slabs: a comparative study. Cem Concr Compos 27:547–553

78. Toledo Filho RD, Ghavami K, Sanjuán MA, England GL (2005) Free, restrained and drying shrinkage of cement mortar composites reinforced with vegetable fibres. Cem Concr Compos 27:537–546
79. de Andrade Silva F, Toledo Filho RD, de Almeida Melo Filho J, Fairbairn EMR (2010) Physical and mechanical properties of durable sisal fiber–cement composites. Constr Build Mater 24:777–785
80. Hambach M, Rutzen M, Volkmer D (2019) Properties of 3D-printed fiber-reinforced Portland cement paste. In: 3D concrete printing technology, Elsevier, pp 73–113

Chapter 9
Sustainability in 3DPC

9.1 Overview

The amount of cement used in a 3D printed mix is comparatively higher than the conventional concrete mix [1]. Such higher amount of cement is generally employed to accommodate the eliminated coarse aggregate content in order to maintain the desired printability and rheological properties of 3D printed concrete mix. Consequently, from a concrete mix-design perspective, 3D printing technology could result in a higher carbon footprint owing to the utilization of higher amount of ordinary Portland cement (OPC). In general, manufacturing of 1 kg of OPC produces around 0.9 kg of CO_2 and the cement industry is responsible for more than 7% of the global CO_2 emissions [2]. The process involves the calcination of limestone (calcium carbonate), which releases carbon dioxide. Additionally, the high energy consumption during the production process further exacerbates its environmental impact. Further, utilization of OPC puts a thread on the depletion of natural resources. The above-mentioned issues associated with OPC could sideline the potential advantages of 3D printing in construction.

The most effective strategy to overcome the problem of lower CO_2 emissions and depletion of natural resources such as limestone or natural aggregates is to search for alternate materials. Utilizing waste materials as potential replacements for cement or for natural aggregate is a promising strategy for promoting sustainability in the construction industry. This approach not only addresses the environmental concerns associated with cement production but also provides a solution for waste management. Moreover, such an approach could help in preserving the natural resources that are used in cement production or as natural aggregates. Hence, in recent times, one of the focal points of research in the civil engineering field is to develop greener building materials utilizing agro-industrial wastes. Over the years, several studies have been conducted where various industrial wastes, such as fly ash (FA) [3], ground granulated blast furnace slag (GGBFS) [4], ladle furnace slag (LF) [5], lead–zinc tailings [6], etc., have been employed in preparing waste-based cementitious binder. Researchers

A. Chourasia et al., *Advances and Challenges in 3D Printed Concrete (3DPC)*, Materials Horizons: From Nature to Nanomaterials, https://doi.org/10.1007/978-981-95-8557-1_9

also focused on utilizing agricultural wastes such as bagasse ash (BA) [7], rice husk ash (RHA) [8], hazelnut shell ash [9] etc., in fabricating the cementitious binder. In the above-mentioned studies, OPC is partially replaced with each of the agro-industrial waste and the resultant waste-based cementitious binder was developed. Further, the integration of waste materials as aggregates in 3D printing mortars offers a sustainable solution, addressing both the environmental impact of construction waste and the demand for innovative construction materials. This chapter explores the various waste materials either as a replacement of cement or as aggregates in 3D printing mortars, supported by contemporary research findings from the existing research articles as well as from the study conducted at CSIR-CBRI.

9.2 Different Waste Materials

9.2.1 Fly Ash

Fly ash, a byproduct of coal combustion in thermal power plants, has garnered significant attention as a sustainable alternative to Portland cement in concrete production. Its utilization not only mitigates the environmental impact of cement production but also addresses the disposal challenges associated with fly ash.

9.2.1.1 Chemical and Physical Properties

Fly ash primarily consists of silicon dioxide (SiO_2), aluminum oxide (Al_2O_3), and calcium oxide (CaO). These components impart pozzolanic properties to fly ash, enabling it to react with calcium hydroxide in the presence of water to form cementitious compounds. The physical characteristics, such as particle size and specific surface area, play a crucial role in influencing the performance of fly ash in concrete.

9.2.1.2 Benefits in Concrete

i. **Strength and Durability**: Concrete with fly ash often exhibits improved long-term strength and durability. The pozzolanic reaction enhances the microstructure of the concrete, resulting in a denser matrix that reduces permeability and increases resistance to chemical attack [10]. Studies have shown that replacing up to 30% of cement with fly ash can achieve comparable or superior compressive strength in concrete [11].
ii. **Workability**: Fly ash improves the workability of fresh concrete due to its spherical particle shape, which acts as a lubricant, reducing water demand. This leads to better pumpability and flowability of concrete mixes [12].

iii. **Reduced CO_2 Emissions and Waste Management**: By substituting cement with fly ash, the CO2 emissions associated with cement production can be significantly reduced. Further, utilizing fly ash in concrete helps address the disposal issues of this industrial byproduct, diverting it from landfills and promoting a circular economy.

9.2.1.3 Challenges and Considerations

i. **Quality Variability**: The properties of fly ash can vary depending on the source and combustion process, which necessitates stringent quality control measures to ensure consistent performance in concrete.
ii. **Curing Practices**: Proper curing practices are essential to maximize the benefits of fly ash in concrete, as inadequate curing can adversely affect the pozzolanic reaction and the resultant properties of the concrete.

9.2.2 Bagasse Ash

Bagasse ash, a byproduct from the sugarcane industry, emerges as a promising material for partial or full replacement of cement in concrete. Utilization of this waste not only provides an eco-friendly solution to waste disposal issues but also contributes to the reduction of CO2 emissions associated with cement production. This document explores the properties, benefits, challenges, and applications of bagasse ash in cement replacement, supported by contemporary research.

9.2.2.1 Chemical and Physical Properties

Bagasse ash primarily comprises silicon dioxide (SiO_2), aluminum oxide (Al_2O_3), and calcium oxide (CaO), similar to fly ash and other pozzolanic materials. The high silica content, often in the amorphous form, makes bagasse ash reactive in the presence of calcium hydroxide and water, leading to the formation of additional cementitious compounds. The physical properties, such as particle size and specific surface area, also significantly affect its reactivity and performance in concrete [13].

9.2.2.2 Benefits in Concrete

i. **Mechanical Properties**: Research indicates that incorporating bagasse ash in concrete can enhance its mechanical properties. Studies have demonstrated that concrete mixes with up to 20% bagasse ash replacement can exhibit improved compressive strength, tensile strength, and flexural strength compared to conventional concrete [14].

ii. **Durability**: Bagasse ash contributes to the durability of concrete by refining its microstructure. Cordeiro et al. [15] conducted microstructural analysis of bagasse ash concrete and observed a denser matrix with reduced porosity. This was attributed to the filling effect of the fine ash particles and the pozzolanic reaction, which improved the overall quality of the concrete. As a result, the permeability of concrete reduces, thus enhancing resistance to chloride ion penetration, sulfate attack, and alkali-silica reaction (ASR). This improvement in durability extends the lifespan of concrete structures exposed to aggressive environments.
iii. **Workability and Setting Time**: The incorporation of bagasse ash generally improves the workability of concrete due to its finer particles and spherical shape, which act as a lubricant in the mix. However, the setting time can be affected depending on the replacement level and the specific characteristics of the ash. Proper optimization of mix design is necessary to balance these effects.
iv. **Environmental Benefits**: Using bagasse ash as a cement replacement reduces the environmental impact of concrete production. Cement manufacturing is a significant source of CO2 emissions; substituting a portion of cement with bagasse ash lowers the carbon footprint of concrete. Additionally, this practice helps manage the disposal of bagasse ash, promoting sustainable waste management.

9.2.2.3 Challenges and Considerations

i. **Quality Variability**: The chemical composition and physical properties of bagasse ash can vary significantly depending on factors such as the type of sugarcane, harvesting practices, and combustion conditions. Ensuring consistent quality and performance requires stringent quality control measures and proper characterization of the ash.
ii. **Optimization of Mix Design**: The optimal replacement level of bagasse ash in concrete depends on various factors, including the intended application, desired properties, and specific characteristics of the ash. Various recent studies have focused on optimizing the mix design for bagasse ash concrete. A collection of various existing studies on identifying the optimum proportion of bagasse ash in the mix in different curing conditions can be found in [13]. However, extensive experimental investigations are necessary to determine the appropriate mix proportions that maximize the benefits while maintaining performance.
iii. **Curing Practices**: Adequate curing is essential to realize the full potential of bagasse ash in concrete. The pozzolanic reaction is slower than the hydration of cement, requiring extended curing periods to achieve optimal strength and durability. Proper curing practices must be followed to ensure the long-term performance of bagasse ash concrete.

Hence, in summary, it can be stated that bagasse ash presents a viable and sustainable alternative to Portland cement in concrete production. Its pozzolanic properties, combined with its environmental benefits, make it an attractive option for enhancing the performance and sustainability of concrete. While challenges related to quality variability and optimization of mix design exist, ongoing research and proper characterization of bagasse ash can address these issues. The incorporation of bagasse ash in structural and non-structural concrete applications, as well as blended cements, offers significant potential for reducing the environmental impact of construction and promoting sustainable development.

9.2.3 Ground Granulated Blast Furnace Slag (GGBFS)

Ground Granulated Blast Furnace Slag (GGBFS) is an industrial byproduct derived from the steel manufacturing process, specifically from the quenching of molten iron slag in water or steam to produce a glassy, granular product that is then dried and ground into a fine powder. GGBFS has gained prominence as a sustainable cement replacement material due to its beneficial properties and its contribution to reducing the environmental impact of concrete production.

9.2.3.1 Chemical and Physical Properties

GGBFS primarily comprises silica (SiO_2), alumina (Al_2O_3), and calcium oxide (CaO), with smaller amounts of magnesium oxide (MgO) and other minor components. The glassy nature of GGBFS and its chemical composition make it highly reactive in alkaline environments, promoting the formation of cementitious compounds when mixed with water and calcium hydroxide. The physical properties, such as fineness and specific surface area, significantly influence the reactivity and performance of GGBFS in concrete.

9.2.3.2 Benefits in Concrete

i. **Strength and Durability**: GGBFS is known to enhance the strength and durability of concrete. Its slow hydration process leads to a denser microstructure, which improves compressive strength over time. Such long-term strength development is attributed to the continued pozzolanic reaction and densification of the microstructure. [4]
ii. **Resistance to Chemical Attack**: The use of GGBFS improves the resistance of concrete to various chemical attacks, including sulfate and chloride ingress. The reduced permeability of GGBFS concrete limits the penetration of aggressive agents, thereby enhancing durability in harsh environments [16].
iii. **Workability and Heat of Hydration**: GGBFS can improve the workability of concrete due to its smooth, glassy particles, which act as a lubricant in the mix. Consequently, GGBFS-based concrete required less water to achieve the same slump than OPC concrete, indicating better workability and ease of placement.

Additionally, the heat of hydration is significantly lower in GGBFS concrete, reducing the risk of thermal cracking in mass concrete applications [17].

iv. **Sustainability and Environmental Impact**: Incorporating GGBFS as a cement replacement material significantly reduces the carbon footprint of concrete. The production of GGBFS generates considerably lower CO2 emissions compared to OPC. By diverting industrial byproducts from landfills and utilizing them in construction, GGBFS promotes sustainable waste management.

v. **Marine and Aggressive Environments**: Due to its superior resistance to chemical attacks, GGBFS concrete is particularly beneficial in marine environments and other aggressive conditions where exposure to sulfates and chlorides is a concern. This makes it a preferred choice for coastal structures, wastewater treatment plants, and chemical processing facilities [16].

9.2.3.3 Challenges and Considerations

i. **Activation Requirements**: GGBFS requires activation to achieve its cementitious properties. This activation can be achieved through the presence of calcium hydroxide (from OPC) or by using alkaline activators. The need for activation can complicate the mix design process and may require additional materials or adjustments.

ii. **Variability in Quality**: The quality of GGBFS can vary depending on the source and the manufacturing process. Variations in chemical composition and physical properties can affect the performance of GGBFS in concrete. Therefore, consistent quality control and characterization are essential to ensure reliable results [18].

iii. **Initial Setting Time**: The initial setting time of GGBFS concrete can be longer compared to OPC concrete, which may impact construction schedules. Adequate curing practices and proper mix design adjustments are necessary to manage this aspect.

Ground Granulated Blast Furnace Slag (GGBFS) is an effective and sustainable cement replacement material with numerous benefits, including enhanced strength, durability, and resistance to chemical attacks. Its use in concrete contributes to reduced carbon emissions and promotes sustainable waste management by utilizing industrial byproducts. Despite challenges such as activation requirements and quality variability, ongoing research and advancements in mix design optimization continue to improve the performance and applicability of GGBFS concrete. As a result, GGBFS plays a crucial role in the development of eco-friendly and durable construction materials for 3D printing applications.

9.2.4 *Rice Husk Ash (RHA)*

Rice Husk Ash (RHA) is a byproduct of rice milling and a promising alternative to traditional cement in concrete production. This material is produced by burning rice husks, resulting in an ash rich in amorphous silica. The use of RHA in concrete has garnered significant attention due to its potential to enhance concrete properties and reduce environmental impact.

9.2.4.1 Chemical and Physical Properties

RHA is primarily composed of silicon dioxide (SiO_2), which can constitute up to 90% of its composition. The silica in RHA is mostly in an amorphous form, which is highly reactive and contributes to its pozzolanic activity. Other components include small amounts of alumina (Al_2O_3) and iron oxide (Fe_2O_3). The physical properties of RHA, such as particle size and surface area, significantly influence its reactivity. RHA typically has a high specific surface area due to its fine particles, which enhances its pozzolanic reaction when mixed with lime or Portland cement [19].

9.2.4.2 Benefits in Concrete

i. **Enhanced Strength and Durability**: RHA improves the compressive strength of concrete, particularly at later ages. This is due to the pozzolanic reaction between the amorphous silica in RHA and the calcium hydroxide released during cement hydration, forming additional calcium silicate hydrate (C-S–H) gel. Research indicates that replacing 10–20% of cement with RHA can significantly enhance the compressive strength of concrete [20]. Similar to compressive strength, it was reported [20] that inclusion of RHA in concrete showed significant improvement in resistance to chloride penetration and sulfate attack, thereby improving the durability of concrete structures in aggressive environments.
ii. **Improved Workability and Reduced Water Demand**: The fine particles of RHA contribute to better workability and reduced water demand in concrete mixes. This is because RHA can fill voids between cement particles, leading to a denser and more cohesive mix. Studies have shown that RHA can reduce the water-cement ratio required to achieve the desired workability, thus enhancing the overall quality of the concrete [21].
iii. **Reduced Permeability**: Concrete containing RHA exhibits lower permeability compared to conventional concrete. The pozzolanic reaction helps refine the pore structure of the concrete, reducing the size and connectivity of pores. This leads to improved resistance to water and chemical ingress, thereby enhancing the durability of concrete structures exposed to aggressive environments [20].
iv. **Environmental Benefits**: The use of RHA as a cement replacement material contributes to sustainability in the construction industry. By utilizing agricultural

waste, RHA reduces the demand for Portland cement, which is associated with high CO2 emissions. Furthermore, the incorporation of RHA in concrete helps manage rice husk disposal, mitigating environmental pollution [15].

9.2.4.3 Challenges and Considerations

i. **Quality and Consistency**: The properties of RHA can vary depending on the burning conditions and the quality of rice husks. Factors such as combustion temperature and duration affect the amorphous silica content and the overall reactivity of RHA. Ensuring consistent quality requires controlled burning processes and proper characterization of the ash [22].
ii. **Optimization of Mix Design**: The optimal replacement level of RHA in concrete depends on various factors, including the specific characteristics of the ash and the desired properties of the concrete. Extensive experimental studies are necessary to determine the appropriate mix proportions that maximize the benefits of RHA while maintaining the required performance standards.
iii. **Curing Requirements**: Proper curing practices are essential to achieve the full potential of RHA in concrete. The pozzolanic reaction of RHA is relatively slow, requiring adequate curing time to ensure the development of strength and durability. Inadequate curing can negatively affect the performance of RHA concrete [21].
iv. **Workability and Water Demand**: Research in [23] demonstrated that the addition of RHA improved the workability of concrete mixes and reduced the water-cement ratio required for achieving the desired slump. This finding supports the use of RHA in producing high-quality, workable concrete.

Rice Husk Ash (RHA) presents a viable and sustainable alternative to traditional Portland cement in concrete production. Its pozzolanic properties, coupled with its environmental benefits, make it an attractive option for enhancing the performance and sustainability of concrete. While challenges such as quality variability and optimization of mix design exist, ongoing research and advancements in material processing continue to improve the applicability and reliability of RHA concrete. The incorporation of RHA in high-performance concrete, sustainable construction practices, and blended cements offers significant potential for reducing the environmental impact of construction and promoting the utilization of renewable resources.

9.2.5 Recycled Aggregates

Recycled aggregates (RAs) are derived from processing construction and demolition waste (CDW). They are increasingly being used as a substitute for natural aggregate (NA) in concrete production due to growing environmental concerns and the need for sustainable construction practices. This document explores the properties, benefits, challenges, and applications of recycled aggregates, supported by contemporary research.

9.2.5.1 Properties of Recycled Aggregates

Recycled aggregates (RA) often consist of crushed concrete, brick, and asphalt. They can be classified into recycled concrete aggregates (RCA) and mixed recycled aggregates (MRA), depending on their source material. Key properties influencing their use in concrete include:

i. **Particle Size and Shape**: RAs typically have a more irregular shape and rougher texture compared to natural aggregate (NA), affecting the workability and mechanical properties of concrete [24].
ii. **Density and Porosity**: RAs generally have lower density and higher porosity than NAs, due to the presence of adhered mortar. This can lead to increased water absorption and reduced concrete strength [25].
iii. **Contaminants**: The presence of impurities such as wood, plastic, and gypsum can affect the quality and performance of RAs. Proper processing and quality control are essential to minimize these contaminants.

9.2.5.2 Benefits of Using Recycled Aggregates

i. **Environmental Impact**: Utilizing RAs reduces the need for natural resource extraction and minimizes landfill waste. It also decreases the environmental footprint of concrete production by lowering CO_2 emissions and energy consumption associated with quarrying and transportation of NAs.
ii. **Structural aspects**: Although RAs are primarily employed in non-structural applications such as pavements, sidewalks, and road base materials, however, with proper quality control and mix design, RAs can also be used in structural concrete [26].
iii. **Economic Advantages**: The use of RAs can lead to cost savings in material acquisition, transportation, and landfill fees. This is particularly significant in urban areas where disposal of construction waste is expensive [27].
iv. **Sustainability**: Incorporating RAs in concrete supports sustainable construction practices and aligns with green building certifications and regulations. It promotes the circular economy by reusing waste materials in new construction.

9.2.5.3 Challenges and Considerations

i. **Quality and Performance**: The variable quality of RAs can affect the consistency and performance of concrete. This necessitates thorough testing and quality control measures to ensure reliable outcomes. The mechanical properties of concrete made with RAs, such as compressive strength and durability, can be inferior to those of conventional concrete [26].

ii. **Mix Design Adjustments**: Concrete mixes incorporating RAs often require adjustments to the mix design to compensate for the higher water absorption and lower density of RAs. This may involve increasing the cement content or using supplementary cementitious materials (SCMs) to achieve the desired properties.

iii. **Regulatory and Standards Compliance**: The use of RAs in structural applications is subject to building codes and standards, which may vary by region. Ensuring compliance with these regulations is crucial for the safe and effective use of RAs in construction.

9.2.5.4 Recent Research Findings

i. **Mechanical and durability properties**: Recent studies have shown that concrete with up to 30% RCA replacement can achieve comparable strength and durability to conventional concrete [28]. The study emphasized the importance of proper mix design and curing practices to achieve optimal results. Further, research by Su et al. [29] demonstrated that the durability performance improved with the use of supplementary cementitious materials like fly ash and slag.

ii. **Workability and Mix Design**: Study on the workability of RAC revealed that the use of superplasticizers and appropriate water-cement ratios can mitigate the adverse effects of the higher water absorption of RAC [24]. The study provided guidelines for optimizing mix design to achieve desired workability and mechanical properties.

iii. **Cost–Benefit Analysis**: Kou and Poon [27] performed a cost–benefit analysis of using RAs in concrete and concluded that the economic savings from reduced material costs and landfill fees outweigh the potential increase in production costs due to additional processing and quality control measures.

Hence, in overall, recycled aggregates (RAs) offer a viable and sustainable alternative to natural aggregates (NAs) in concrete production. Their use contributes to environmental conservation by reducing the need for natural resource extraction and minimizing construction waste. While challenges such as variability in terms of quality and adjustments of mix design exist, ongoing research and advancements in material processing continue to improve the performance and reliability of RAs in concrete. The incorporation of RAs in non-structural and structural applications, as well as in recycled aggregate concrete (RAC), demonstrates significant potential for reducing the environmental impact of construction and promoting sustainable practices.

9.2.6 LF Slag

Ladle Furnace (LF) slag, a byproduct of steel production, is gaining attention in the field of 3D concrete printing (3DCP) as a potential eco-friendly alternative to traditional materials. Incorporating LF slag into 3DCP can improve the mechanical properties of printed structures, enhance durability, and reduce environmental impact.

9.2.6.1 Physical Properties

Ladle Furnace (LF) slag, also known as secondary refining slag, has distinct physical properties that can vary based on its chemical composition and the processes used in steel production. When used in applications like 3D concrete printing, the physical properties of LF slag are critical for ensuring compatibility with the printing process and the desired performance of the final product. Here are some general physical properties of LF slag:

i. **Particle Size**: The particle size of LF slag can range from fine powders (micrometers) to coarse granules (millimeters). Consequently, depending on the size distribution, it is often ground into fine particles for use in concrete applications to improve workability and reactivity.
ii. **Density**: The density of LF slag can vary but typically ranges from 2.8 to 3.4 g/cm^3. The specific gravity is influenced by the slag's mineralogical composition and porosity.
iii. **Porosity**: LF slag may have a porous structure, with porosity depending on the cooling process and treatment.
iv. **Surface Area**: The specific surface area of the LF slag can vary widely, generally ranging from 300 to 500 m^2/kg. A higher specific surface area can enhance the reactivity of the slag when used in concrete mixtures.
v. **Thermal Properties**: LF slag has a relatively high melting point, typically around 1400–1500 °C. Its thermal properties make it stable under high-temperature conditions, which is beneficial in certain construction applications.

9.2.6.2 Chemical Properties:

i. **Mineralogical Composition**: Like physical properties, chemical composition of LF slags can vary based on the type of steel produced and the refining processes used, but it generally consists of a mixture of oxides, primarily calcium oxide (CaO) (40–60%), silicon dioxide (SiO_2) (10–20%), aluminum oxide (Al_2O_3) (10–25%), magnesium oxide (MgO) (5–15%). Additionally, some other oxides in minor amount, such as iron oxide (Fe_2O_3/FeO) (1–5%), manganese oxide (MnO) (1–3%), sulfur compounds and phosphorus pentoxide (P_2O_5) are also found in LF slag.
ii. **Hydraulic Properties**: The presence of free lime (CaO) and other reactive phases can give LF slag some hydraulic properties, making it capable of reacting

with water to form cementitious compounds. However, these properties are usually not as pronounced as in ground granulated blast furnace slag (GGBFS).

iii. **Solubility**: The solubility of LF slag in water is relatively low, but it can vary depending on the specific mineralogical composition. This affects its leaching behavior and potential environmental impact.

iv. **Trace Elements**: Elements such as chromium (Cr), nickel (Ni), vanadium (V) etc., can be found in LF slag, the amount of which depends on the steel production process. Although their presence is usually minimal it can impact the environmental behavior of the slag.

9.2.6.3 Benefits of Using LF Slag in 3DCP

i. **Environmental Impact Reduction**: Utilizing industrial byproducts like LF slag reduces the need for virgin raw materials and helps to manage waste. Further, replacing traditional cement with LF slag can significantly lower CO_2 emissions associated with concrete production.

ii. **Improved Material Properties**: LF slag can enhance the durability of concrete by providing better resistance to chemical attacks and reducing the permeability of the concrete. In addition, it can improve the compressive and tensile strength of the printed concrete, making it suitable for load-bearing applications.

iii. **Cost-Effectiveness**: As an industrial byproduct, LF slag can be a cost-effective alternative to traditional cementitious materials.

9.2.6.4 Challenges and Considerations

i. **Material Design**: Ensuring compatibility with respect to the printing performance and hardened properties for the mix made of LF slag is still a challenging task.

ii. **Quality Control**: The quality and composition of LF slag can vary depending on its source, which may affect the various properties of the concrete mixture made of LF slag.

iii. **Regulatory and Standardization Issues**: The use of alternative materials like LF slag in construction requires adherence to building codes and standards. The lack of standardized testing methods for 3DCP materials can be a hurdle.

9.2.6.5 Research and Development

Ongoing research focuses on optimizing the mix design, assessing long-term performance, and exploring the potential of LF slag in different applications. Researchers are also investigating the environmental impacts, lifecycle analysis, and scalability of using LF slag in 3D concrete printing.

9.2.7 *LD Slag*

LD slag, also known as Linz-Donawitz slag or steelmaking slag, is a byproduct produced during the conversion of iron into steel in a basic oxygen furnace (BOF). This material has been explored as a potential component in 3D Concrete Printing (3DCP) due to its unique properties. Here's a detailed description of the use of LD slag in 3DCP:

9.2.7.1 Chemical Composition and Physical Properties of LD Slag

LD slag primarily consists of oxides of calcium (calcium oxide:40–60%), iron (iron oxides: 10–15%, silicon (silicon dioxide: 10–20%), magnesium (magnesium oxide: 5–10%), and manganese (manganese oxide: 1–3%). The physical properties of the LD slag, such as its high density and hardness, make it a valuable material in construction.

9.2.7.2 Applications in 3D Concrete Printing: LD slag can be utilized in 3DCP in several ways:

i. As a Cementitious Material

LD slag can partially replace traditional cement in concrete mixes. Its high content of calcium oxide allows it to participate in hydration reactions, forming compounds like calcium silicate hydrates (CSH) that contribute to the strength and durability of the concrete.

ii. **As an Aggregate**

The slag can be crushed and screened to be used as a coarse or fine aggregate. Its angular shape improves interlocking between particles, which can enhance the mechanical properties of the printed structure.

iii. **As a Mineral Admixture**

LD slag can also be used as a mineral admixture to improve the fresh and hardened properties of the concrete mix. It can influence the workability, setting time, and strength development of 3D printed concrete.

9.2.7.3 Challenges and Considerations

i. Variability in Composition

The composition of LD slag can vary significantly depending on the source and steelmaking process. This variability can affect the consistency and quality of the concrete mix.

ii. **Volume Stability**

LD slag contains free lime and magnesium oxide, which can cause expansion due to hydration and carbonation reactions. This can lead to cracking and other durability issues in the printed structure.

iii. **Environmental Concerns**

Although use of LD slag is environmentally friendly in terms of waste utilization, there are concerns about the leaching of heavy metals and other harmful substances. Proper treatment and testing are required to ensure the safety of slag-containing concrete.

Hence, the use of LD slag in 3D Concrete Printing offers promising benefits in terms of sustainability, cost reduction, and enhanced durability. However, challenges related to its variability, volume stability, and environmental impact must be carefully managed. As research continues to advance, the integration of LD slag in 3DCP has the potential to contribute significantly to sustainable construction practices. Research is ongoing to optimize the use of LD slag in 3DCP. This includes developing methods to stabilize its composition, improving its hydration properties, and understanding its long-term performance in printed structures. Studies are also exploring the potential to use LD slag in combination with other supplementary cementitious materials (SCMs) like fly ash and slag cement to enhance concrete properties.

9.2.8 Glass Cullet

Glass cullet, derived from recycled glass, is another promising waste material for 3D printing mortars. It can be used as a fine aggregate replacement or in powdered form as a pozzolanic material.

9.2.8.1 Properties and Benefits:

i. **Durability**: Glass cullet enhances the durability of mortars by providing resistance to chemical attacks and reducing permeability [30].
ii. **Aesthetic Appeal**: The use of glass cullet can improve the aesthetic properties of printed structures, offering unique textures and colors [31].

9.2.8.2 Challenges

i. **Alkali-Silica Reaction (ASR)**: Glass cullet can be susceptible to ASR, which causes expansion and cracking. Proper mix design and the use of supplementary cementitious materials can mitigate ASR risks [32].

9.2.9 *Plastic Waste*

Recycled plastic waste can be used as an aggregate in 3D printing mortars, providing a solution to the growing plastic waste problem.

9.2.9.1 Properties and Benefits

i. **Lightweight**: Plastic aggregates produce lightweight mortars, suitable for applications where weight reduction is critical [33].
ii. **Thermal Insulation**: Mortars with plastic aggregates offer enhanced thermal insulation properties, beneficial for energy-efficient buildings [33].

9.2.9.2 Challenges

i. **Bonding**: The hydrophobic nature of plastic can affect the bonding between the plastic particles and the cement matrix. Surface treatments and the use of bonding agents can improve this interaction.

9.2.10 *Limestone Powder (LP)*

Limestone Powder (LP) is an important supplementary cementitious material used in 3D Concrete Printing (3DCP). It enhances various properties of the concrete mixture, making it suitable for the additive manufacturing process. Here's a detailed look at Limestone Powder in the context of 3DCP:

9.2.10.1 Composition

Limestone Powder primarily consists of Calcium Carbonate (CaCO3) which contributes to its chemical and physical properties. In addition, it also contains small amounts of clay, silt, and other minerals depending on the source of the limestone.

9.2.10.2 Benefits of Using Limestone Powder in 3DCP

i. **Enhanced Workability**: Fine particles of limestone powder help improve the workability of the concrete mix, crucial for smooth extrusion through the 3D printer nozzle.
ii. **Improved rheology**: The thixotropic nature of limestone powder helps the concrete mix to maintain its shape after deposition, reducing deformation and slumping.

iii. **Cost-Effective**: Limestone powder is typically less expensive than other supplementary materials, reducing the overall cost of the concrete mix.
iv. **Reduced Water Demand**: Fine limestone particles reduce the water demand of the mix, leading to a more manageable and consistent extrusion process.

Challenges and Considerations

i. **Consistency of Supply**: Ensuring a consistent quality and supply of limestone powder is essential for maintaining the performance of the 3D printed mix.
ii. **Compatibility with Other Materials**: The interaction between limestone powder and other mix components (e.g., superplasticizers, accelerators) needs to be carefully managed.
iii. **Durability**: Long-term performance and durability of structures printed with limestone powder need thorough evaluation, particularly in aggressive environmental conditions.
iv. **Environmental Impact**: While using limestone powder can reduce the overall carbon footprint, the mining and processing of limestone should also be considered in sustainability assessments.

9.2.11 Calcined Clay (CC)

Calcined Clay (CC) is an increasingly popular supplementary cementitious material in the field of 3D Concrete Printing (3DCP). Calcined clay is produced by heating natural clays to a high temperature, typically between 600 °C and 900 °C, which results in a material with enhanced pozzolanic properties. Here's a detailed look at calcined clay in the context of 3DCP:

9.2.11.1 Chemical and Physical Properties

Calcined clay consists primarily of metakaolin**,** formed by the dehydroxylation of kaolinite clay during calcination. Further, depending on the raw material and calcination process, other phases such as amorphous silica and alumina can be present. On the other hand, the pozzolanic activity of calcined clay, in which it reacts with calcium hydroxide ($Ca(OH)_2$) released during cement hydration to form additional calcium silicate hydrate (C–S–H) and calcium aluminate hydrate (C–A–H), significantly contributes to the refinement of the microstructure and enhancement of mechanical properties.

9.2.11.2 Benefits of Using Calcined Clay in 3DCP

i. **Enhanced Mechanical and Durability Properties**: The pozzolanic reaction between calcined clay and calcium hydroxide (produced during the hydration of Portland Cement) form additional calcium silicate hydrate (C-S–H), which enhances the strength and durability of the concrete.

ii. **Cost-Effective**: Can be sourced locally in many regions, reducing transportation costs and overall material costs.

9.2.11.3 Challenges and Considerations

i. **Consistency of Supply**: Ensuring a consistent quality and supply of calcined clay is crucial for maintaining the performance of the 3D printed concrete.
ii. **Compatibility with Other Materials**: The interaction between calcined clay and other mix components (e.g., superplasticizers, retarders) needs careful management to achieve the desired properties.
iii. **Processing Conditions**: Proper calcination temperature and duration are critical to produce high-quality calcined clay with optimal pozzolanic activity.

9.2.12 Paper Sludge Ash

Paper Sludge Ash (PSA), also known as Paper Mill Sludge Ash, is a byproduct of the paper manufacturing industry. This material is gaining attention in the field of 3D Concrete Printing (3DCP) due to its potential to enhance the sustainability and performance of concrete. Here's a detailed look at Paper Sludge Ash and its application in 3DCP:

9.2.12.1 Composition

The composition of PSA can vary depending on the raw materials used in paper production and the processes employed. Paper Sludge Ash typically contains calcium oxide (CaO), silica (SiO2), alumina (Al2O3), iron oxide (Fe2O3) along with magnesium oxide (MgO) and potassium oxide (K2O).

9.2.12.2 Key Properties for 3DCP

i. **Pozzolanic Activity**: PSA contains reactive silica and alumina, which can react with calcium hydroxide to form additional calcium silicate hydrate (C-S–H), enhancing strength and durability.
ii. **Fineness**: PSA is typically finely ground, which improves the packing density and reduces the porosity of the concrete mix.
iii. **Density**: Lower density compared to traditional cementitious materials, which can be beneficial for reducing the weight of printed structures.
iv. **Sustainability**: Utilizes waste material, reducing the environmental impact and promoting a circular economy.

9.2.12.3 Benefits of Using PSA in 3DCP

i. **Improved Workability**: The fine particles of PSA can enhance the workability of the concrete mix, making it easier to extrude through the printer nozzle.
ii. **Enhanced Strength**: The pozzolanic reaction contributes to higher long-term strength and durability.
iii. **Reduced Carbon Footprint**: Replacing a portion of Portland cement with PSA can significantly lower the overall carbon footprint of the concrete.
iv. **Cost Efficiency**: Using waste materials like PSA can reduce the cost of raw materials in concrete production.
v. **Waste Management**: Provides a valuable use for paper industry waste, reducing the burden on landfills.

9.2.12.4 Challenges and Considerations

i. **Consistency**: Ensuring a consistent quality of PSA can be challenging due to variations in the raw materials and processes used in paper production.
ii. **Processing**: The drying, calcination, and grinding processes must be carefully controlled to produce high-quality PSA.
iii. **Mix Design Optimization**: Finding the right proportion of PSA to achieve the desired properties without compromising the performance of the concrete.
iv. **Performance Evaluation**: Comprehensive testing is required to evaluate the long-term behavior and durability of concrete containing PSA.

9.2.13 Bottom Ash

Bottom Ash is a byproduct of coal combustion in power plants, collected from the bottom of boilers. In 3D Concrete Printing (3DCP), bottom ash can be used as a supplementary material or as a partial replacement for traditional aggregates and cement, promoting sustainability and enhancing certain properties of the printed concrete. Here's a detailed look at the use of bottom ash in 3DCP:

9.2.13.1 Composition

Bottom ash typically contains silica ($SiO2$), alumina ($Al2O3$), iron oxide ($Fe2O3$), calcium oxide (CaO), magnesium oxide (MgO) and unburned carbon.

9.2.13.2 Key Advantages

i. **Enhancement in Properties**: Bottom ash contains reactive silica and alumina that can contribute to the pozzolanic reaction, enhancing strength and durability.

ii. **Workability and Compaction**: Typically, bottom ash particles are angular and porous, which can affect the workability and compaction of the concrete mix.
iii. **Lightweight Structures**: Lower density of bottom ash compared to natural aggregates can reduce the overall weight of the printed structure.
iv. **Thermal Insulation**: The porous nature of bottom ash provides better thermal insulation properties.

9.2.13.3 Challenges and Considerations

i. **Consistency**: Variations in the composition of bottom ash can affect the properties of the concrete mix.
ii. **Water Demand**: Higher water absorption may require increased water content or the use of water-reducing admixtures.
iii. **Quality Control**: Comprehensive testing is required to evaluate the performance and durability of concrete containing bottom ash.

9.2.14 Red Mud (Bauxite Residue)

- **Source**: Byproduct of aluminum production.
- **Properties**: High in iron oxide and alumina, enhances durability, reduces permeability.
- **Usage**: Partial replacement of cement, especially in areas with aluminum production.

9.2.15 Municipal Solid Waste Incineration (MSWI) Ash

- **Source**: Byproduct of municipal waste incineration.
- **Properties**: Contains various oxides, pozzolanic, improves strength.
- **Usage**: Partial replacement of cement or fine aggregates.

9.3 Existing Studies on the Sustainable Utilization of Agro-Industrial Waste in 3DPC

This section presents a summary of the various agro-industrial waste and related other mix details that have been employed in 3D printing of concrete along with their impact on the resultant mechanical and printing characteristics. Firstly, Table 9.1 presents the impact of various agro-industrial waste on the hardened properties of geopolymer binders. The table (Table 9.1) also highlights the particle size gradation and major chemical composition of the employed raw materials used in fabricating the binders.

It is observed that the utilization of various agro-industrial waste in the context of fabricating a sustainable 3D printable binder can produce compressive strength of upto 50 MPa. Table 9.2 further presents in detail about the utilization of various waste in 3DPC along with the major findings with respect to both printing characteristics and mechanical properties. An overall observation of various mentioned studies in Table 9.2 highlights the role of different agro-industrial waste in fabricating a binder with respect to rheological characteristics such as viscosity, yield stress etc., and printability properties such as pumpability, extrudability, and buildability. This signifies the importance of utilizing such waste in 3DPC that promotes circular economy along with reduction in the carbon footprint.

9.4 Experimental Studies

This section presents the usage of agro-industrial waste such as Fly ash (FA) and Bagasse ash (BA) in fabricating sustainable 3D printable binder. In the work, cement (C) is partially replaced with BA and FA and consequently, developed binary (C + BA) and ternary (C + BA + FA) binder is investigated for mechanical properties and printing performance.

In this study, the raw binders comprised 43-grade ordinary Portland cement (OPC) (ASTM C150 [54]), bagasse ash (BA), and Class F fly ash (FA) (ASTM C618 [55]). As the bagasse ash procured from the industry was not fully combusted, it was thermally treated at 650 °C for 2 h [56]. Following thermal treatment and subsequent mechanical processing, the maximum particle size of BA was reduced to approximately 100 μm. X-ray fluorescence (XRF) analysis revealed that BA primarily consisted of SiO_2 (~84%), along with a noticeable amount of unburned carbon, confirming its partially combusted nature. The thermally treated BA exhibited a maximum particle size of 100 μm, with d90 and d10 values of 75 μm and 4 μm, respectively. The fly ash used in this work conformed to ASTM C618 [55] and showed d90 and d10 values of 50 μm and 1.25 μm, respectively. Chemically, FA was mainly composed of SiO_2 (~52%) and Al_2O_3 (~25%). The fine aggregate employed had a fineness modulus of 2.4 and a dry specific gravity of 2.62. Furthermore, to enhance the printability and mechanical performance of the cementitious mixture, a polycarboxylate-based superplasticizer, a viscosity-modifying admixture (VMA), and a retarder were incorporated. The selection of these admixtures was guided by their established functions in improving rheological control, workability, and strength development in cementitious composite systems.

Table 9.1 Influence of different agro-industrial waste on the physical and mechanical properties of geopolymeric binder [34]

Raw constituents	Molarity	Particle size (µm)	Primary chemical phases	Compressive strength (MPa)	Tensile strength (MPa)	Flexural strength (MPa)	Bulk density (kg/m3)	Ref.
Metakao-lin	8	10–50	SiO_2, Al_2O_3	40	4	8	1200	[35, 36]
Fly ash	10	5–20	SiO_2, Al_2O_3, Fe_2O_3	35	3.5	7	1100	[37, 38]
GGBS	12	5–25	SiO_2, Al_2O_3, CaO	45	4.5	9	1250	[38, 39]
Rice husk ash	6	5–40	SiO_2	30	3	6	1000	[40, 41]
Silica fume	8	1–5	SiO_2	50	5	10	1300	[42, 43]
Red mud	10	10–50	Al_2O_3, Fe_2O_3, SiO_2, CaO	35	3.5	7	1200	[36, 44]
Clay	8	5–20	Al_2O_3, SiO_2, MgO, Fe_2O_3	30	3	6	1100	[45, 46]

Table 9.2 Major findings with respect to printing characteristics and mechanical properties of 3D printed binder employing agro-industrial waste [34]

Binders	Inert materials	Activator agents	Testing	Printing	Findings	Ref.
Slag-based geo-polymer powder and plaster-based powder	River sand	Silicate-based activator	• Uniaxial compressive strength, • Apparent porosity, • Inspection of surface quality of 3DPC	Extrusion Printing Method, Fused-based Deposition Modelling	Geopolymer-based cubic shape structures can be developed employing a powdering 3D printed system that can withstand de-powdering and post-processing along with demonstrating higher precision with respect to dimension	[47]
Slag and calcium carbonate	-	Na_2O and SiO_2 based activator	• Fluidity and rheology • Water retention capacity • Compressive strength, flexural strength Drying shrinkage	Extrusion Printing Technique	Calcium carbonate formulations incorporating 4–6% carboxymethyl starch were successfully extruded at a printing speed of 30 mm s^{-1} without exhibiting fracture, discontinuity, or collapse of the deposited layers. Under these conditions, the corresponding plastic viscosity ranged from 10.08 to 10.42 Pa.s, while the yield stress varied between 32.53 and 66.71 Pa, satisfying the requirements for stable extrusion	[48]
Flyash and GGBFS	River sand	Potassium hydroxide and potassium silicate	• Test under compression, tensile and flexural loading • Fe-SEM	Extrusion-Robot assisted concrete	Owing to the anisotropic characteristics of the printing process, mechanical properties of 3D printed geopolymeric material are strongly dependent on loading directions while preserving the material's intrinsic behavior	[49]

(continued)

Table 9.2 (continued)

Binders	Inert materials	Activator agents	Testing	Printing	Findings	Ref.
Flyash, GGBFS and Nano clay	River sand	Hydroxide and silicate solutions of potassium	• Static yield stress • Thixotropy • Setting time • Scanning electron microscope	4-axis gantry concrete printer	Incorporation of nanoclay enhanced the rheological properties of the geopolymer composite, primarily due to flocculation of the clay particles. The developed geopolymer nanocomposite was subsequently employed to fabricate a small-scale bathroom unit, thereby demonstrating the feasibility of the proposed mix design for application in 3D printing technologies	[50]
Flyash, GGBS and Micro silica	River sand	Liquid potassium silicate	• Rheology • Test under compression, tensile and flexural loading • Scanning electron microscope	Extrusion-based Printing Technique, a 4-Axis gantry system	Printer settings can be adjusted to enhance process performance in accordance with the material properties and the size and geometry of the printed sample	[51]
Fly ash, green tow flax fibers and carbon fibers	River sand	Sodium hydroxide and sodium silicate	• Compression and flexural test • Density	Injection method of 3D printing	The results indicate that the inclusion of fibers led to a modest increase in compressive strength while producing a substantial enhancement in flexural strength. In terms of specific performance, specimens reinforced with flax fibers outperformed those containing carbon fibers. Depending on the fiber type, the flexural strength varied from 9.0 to 10.3 MPa, whereas the interlayer bond strength ranged from 2.33 to 2.58 MPa	[52]

(continued)

Table 9.2 (continued)

Binders	Inert materials	Activator agents	Testing	Printing	Findings	Ref.
GGBS and plaster-based powder	Fine sand	Anhydrous sodium metasilicate	• Drop penetration behavior and wettability • Apparent porosity • Dimensional accuracy • Uniaxial compressive strength	3DP inkjet printing system	With respect to both dimensional accuracy and mechanical performance, all printed cubic specimens exhibited anisotropic behavior. The X-direction (direction of binder jetting) demonstrated superior dimensional accuracy and strength compared to the other orientations	[47]
Flyash and GGBS	Silica sand	Saturated anhydrous sodium metasilicate, N Grade sodium silicate and sodium hydroxide (NaOH)	• True/bulk densities, in-process bed density and powder bed Porosity • Linear dimensional accuracy analysis • Compressive strength	Powder-based 3D printer	Accuracy of the post-processed specimens with respect to the linear dimension was also anisotropic and depended on the curing solution. Specimens immersed in Curing Solution III exhibited the highest dimensional accuracy in all directions. This behavior may be attributed to the action of oligomeric silica polymers in the aqueous phase, as well as the viscosity characteristics of the curing solutions	[53]

9.4.1 Role of BA and FA Replacement on Compressive Strength

Figure 9.1 illustrates the effect of varying bagasse ash (BA) content on the compressive strength of cement–BA binary mixtures at a fixed sand-to-binder (S:B) ratio of 1.4 and a superplasticizer (SP) dosage of 1%. As shown in Fig. 9.1, under these optimum conditions, BA can be effectively incorporated up to a 40% replacement level while still achieving the target compressive strength of greater than 38.5 MPa. The increase in compressive strength observed up to the 40% BA replacement level may be attributed to the filler effect of BA, which enhances particle packing and mix uniformity in the presence of SP. In contrast, in the absence of SP, BA particles are more likely to be non-uniformly entrapped within flocculated cement particles, leading to inferior dispersion. However, beyond the optimum replacement level of 40%, the reduction in the proportion of reactive cement becomes significant, resulting in a decline in compressive strength.

Based on the optimum binary mix identified in the previous section (C:BA = 60:40, S:B = 1.4, and SP = 1%), trial mixtures were formulated to develop ternary binders by partially replacing BA with an equivalent amount of fly ash (FA). The cement content (60% of the total binder), S:B ratio (1.4), and SP dosage (1% of the total binder) were maintained constant. Three ternary mixtures were investigated with BA:FA ratios of 30:10, 20:20, and 10:30, collectively constituting 40% of the total binder. Figure 9.2 shows the influence of the BA:FA ratio on the compressive strength of the ternary mixes under fixed cement content, S:B ratio, and SP dosage. The results indicate that the C:BA:FA proportion of 60:20:20 yielded the highest compressive strength, reaching approximately 48 MPa. This enhanced performance of the 60:20:20 ternary mix can be attributed to the combined effects of improved pore structure, synergistic particle packing, and the differing chemical reactivity of

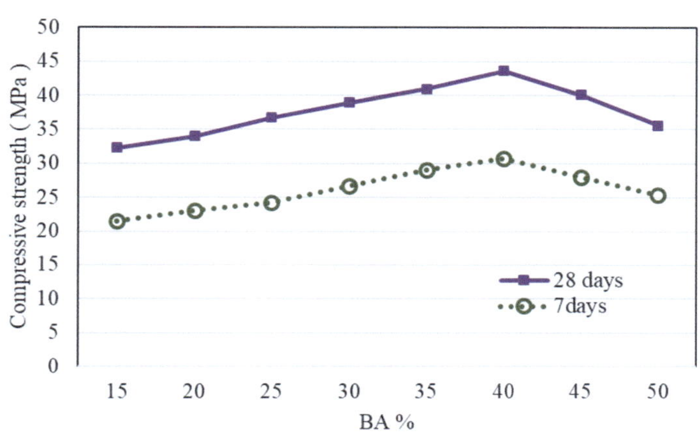

Fig. 9.1 Effect of BA variation on the compressive strength of cement and BA-based binary mix for S:B ratio of 1.4 and SP of 1%

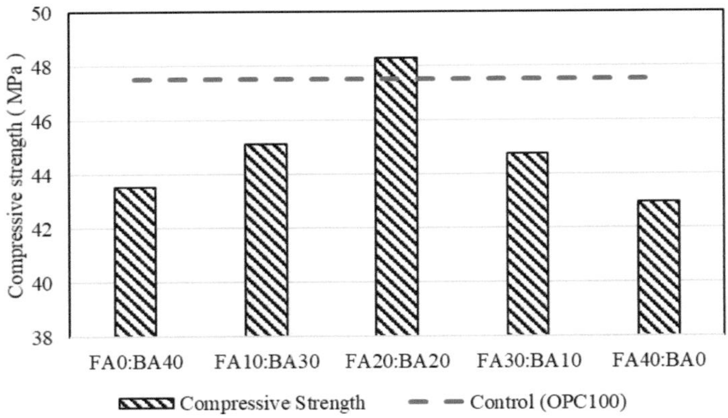

Fig. 9.2 Effect of BA: FA ratio on the compressive strength of ternary mixes for fixed cement content (60% of the total binder), S:B ratio (1.4) and SP content (1% of total binder)

the constituent materials. Accordingly, the ternary binder with a C:BA:FA ratio of 60:20:20 (by mass) was selected as the optimum mix based on compressive strength considerations and was further evaluated in the subsequent section for its suitability in terms of printing characteristics.

9.4.2 Effect of Admixtures on the Compressive Strength and Open Time of the Developed Sustainable Binder

This section investigates the influence of the superplasticizer (SP), viscosity-modifying admixture (VMA), and retarder dosages on the compressive strength and open time of both the binary and ternary mixes. Initially, the SP content was varied for the binary mix (C:BA = 60:40, w/b = 0.4, s/b = 1.4) and the ternary mix (C:BA:FA = 60:20:20, w/b = 0.4, s/b = 1.4) in the absence of VMA and retarder to determine the optimum SP dosage. Subsequently, with the optimum SP content fixed, the effect of VMA dosage on open time and compressive strength was evaluated to identify the optimum VMA content. In the final stage, the retarder dosage was varied, and its influence on open time and compressive strength was examined for mixes containing the previously optimized SP and VMA contents. Based on the targeted hardened properties (compressive strength) and printing requirements (open time), the optimal proportions of the admixtures were established, leading to the final mix designs for both the binary and ternary binders.

Figures 9.3a, b illustrate the effect of SP dosage on the compressive strength and open time of the mixes. The calculated open time corresponds to the duration over which the mix maintains a desired flow value, ensuring pumpability, extrudability, and buildability. From a practical standpoint, a higher open time indicates prolonged printability. As shown in Fig. 9.3, SP exhibits a pronounced influence on open time; increasing the SP dosage led to a corresponding increase in open time for both mixes. This behavior can be attributed to the deflocculation and enhanced dispersion of cement particles induced by SP. However, beyond an SP dosage of 1%, the mixes began to lose shape stability, resulting in a reduction in open time. At an SP content of 1%, the binary mix achieved its maximum open time of approximately 25 min (Fig. 9.3a), along with the highest compressive strength, exceeding the target strength of 38 MPa. Similarly, the ternary mix attained a maximum open time of approximately 30 min at SP dosages between 0.9% and 1% (Fig. 9.3b), while also achieving its peak compressive strength of about 48 MPa at 1% SP. Consequently, an SP dosage of 1% was selected as optimal for both mixes and used in subsequent studies.

Figures 9.4a, b present the effect of VMA dosage on compressive strength and open time at a fixed SP content of 1%. Both binary and ternary mixes exhibited maximum open time at a VMA dosage of 0.5%. However, at this dosage, the compressive strength of the binary mix decreased to approximately 34.84 MPa, representing a reduction of about 8% compared to the design strength of 38.25 MPa. In contrast, for the ternary mix, increasing the VMA dosage from 0.4% to 0.5% reduced the compressive strength from 48 MPa to approximately 39 MPa, which remained above the design strength. Accordingly, a VMA dosage of 0.5% was selected as optimal for the ternary mix and carried forward for retarder optimization.

Figures 9.5a, b illustrate the combined effects of retarder dosage on open time and compressive strength for both mixes with SP = 1% and VMA = 0.5%. For the binary mix, the addition of 0.4% retarder increased the open time to approximately 60 min, while maintaining a compressive strength of about 40.7 MPa, which exceeds the design requirement. Therefore, 0.4% retarder (by mass of binder) was identified as the optimum dosage for the binary mix. In the case of the ternary mix, the maximum open time and a compressive strength of approximately 39 MPa were achieved at a retarder dosage of 0.5%, which was consequently selected as optimal for the ternary system.

Table 9.3 summarizes the final mix design of the developed M30-grade 3D printable binary binder incorporating agro-industrial waste. It is noted that insufficient compaction can significantly influence compressive strength; therefore, identical specimen preparation and compaction procedures were adopted for all mixes. Nevertheless, due to variations in mix composition and constituent proportions, differences in microstructural characteristics—such as hydrated products, unreacted particles, and air voids—are expected. These microstructural variations are responsible for the observed differences in strength properties.

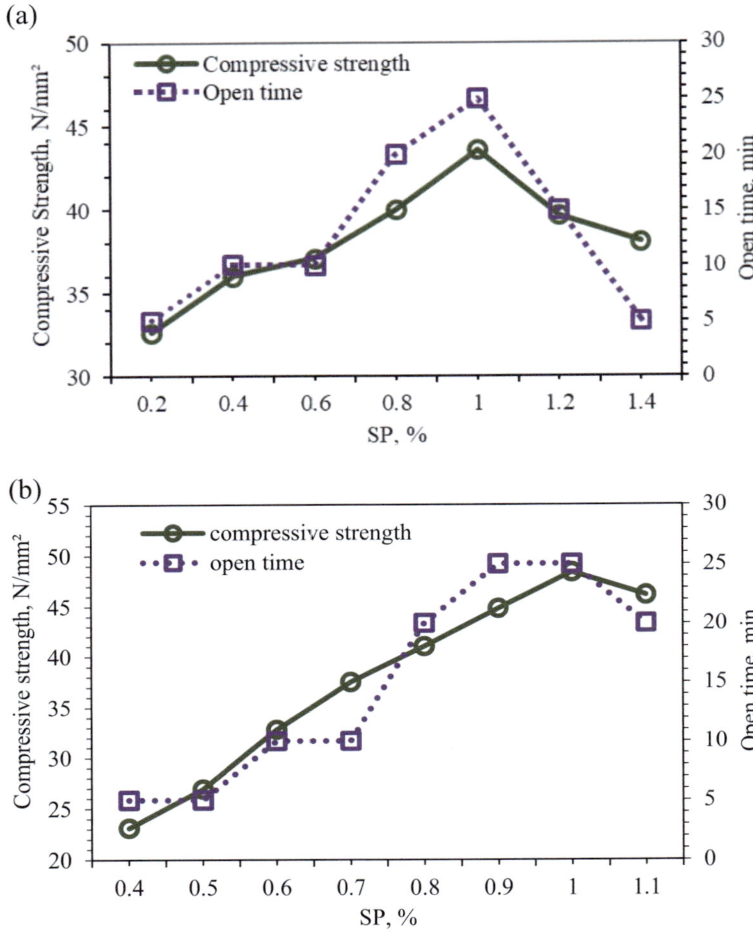

Fig. 9.3 Compressive strength and open time of **a** Binary and **b** Ternary cementitious composite for different dosages of SP

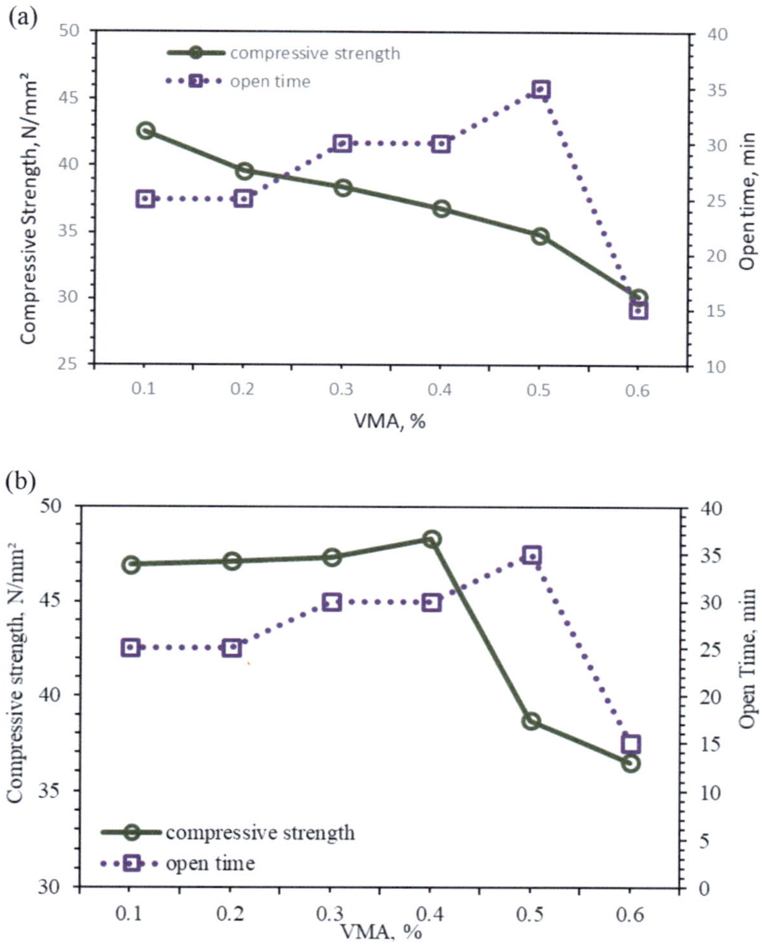

Fig. 9.4 Compressive strength and open time of **a** Binary and **b** Ternary cementitious composite for different dosages of VMA

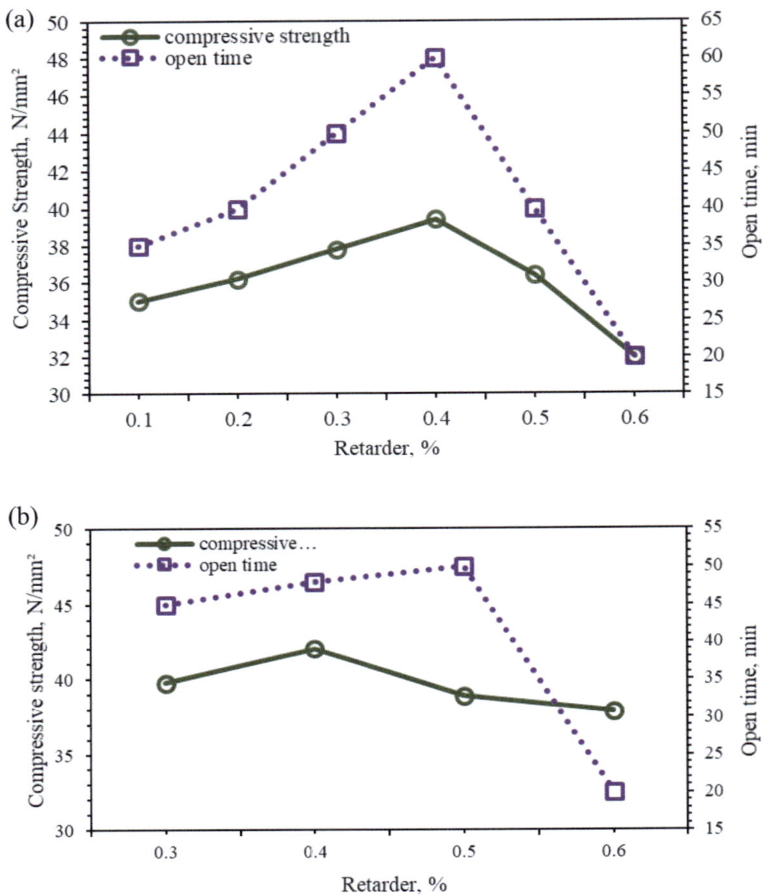

Fig. 9.5 Compressive strength and open time of **a** Binary and **b** Ternary cementitious composite for different dosage of retarder

Table 9.3 Mix-proportion of the developed binary and ternary M30 grade 3D printable cementitious composite

Mix	Cement (%)	BA (%)	FA (%)	Sand/ Binder	Water/ Binder	SP (%)	VMA (%)	Retarder (%)
Binary	60	40	0	1.4	0.40	1	0.5	0.4
Ternary	60	20	20	1.4	0.40	1	0.5	0.5

References

1. Samudrala M, Mujeeb S, Lanjewar BA, Chippagiri R, Kamath M, Ralegaonkar RV (2023) 3D-Printable Concrete for Energy-Efficient Buildings. Energies (Basel) 16:4234
2. Dinh HL, Doh J-H, Liu J, Lu L, Song H, Park D (2023) Comprehensive assessment of geopolymer concrete mechanical and environmental performance with glass cullet fine aggregates. J Build Engineer 107094
3. Shah SFA, Chen B, Oderji SY, Haque MA, Ahmad MR (2020) Improvement of early strength of fly ash-slag based one-part alkali activated mortar. Constr Build Mater 246:118533
4. Reddy MS, Dinakar P, Rao BH (2018) Mix design development of fly ash and ground granulated blast furnace slag based geopolymer concrete. J Build Eng 20:712–722
5. Singh SK, Vashistha P (2021) Development of newer composite cement through mechano-chemical activation of steel slag. Constr Build Mater 268:121147
6. Deng P, Zheng Z (2023). Mechanical properties of one-part geopolymer masonry mortar using alkali-fused lead-zinc tailings. https://doi.org/10.1016/j.conbuildmat.2023.130522
7. de Sande VT, Sadique M, Bras A, Pineda P (2022) Activated sugarcane bagasse ash as efficient admixture in cement-based mortars: Mechanical and durability improvements. J Build Eng 59:105082
8. Jittin V, Bahurudeen A (2022) Evaluation of rheological and durability characteristics of sugar-cane bagasse ash and rice husk ash based binary and ternary cementitious system. Constr Build Mater 317:125965
9. Yurt Ü, Bekar F (2022) Comparative study of hazelnut-shell biomass ash and metakaolin to improve the performance of alkali-activated concrete: a sustainable greener alternative. Constr Build Mater 320:126230
10. Thomas MDA (2007) Optimizing the use of fly ash in concrete, Portland cement association Skokie, IL, USA
11. Malhotra VM, Mehta PK (1996) Pozzolanic and cementitious materials, gordon and breach publishers, advances in concrete. Technology. ISSN1024–50381
12. Reddy PVRK, Ravi Prasad D (2022) A study on workability, strength and microstructure characteristics of graphene oxide and fly ash based concrete, Mater Today Proc 62 2919–2925. https://doi.org/10.1016/J.MATPR.2022.02.495.
13. Li Y, Chai J, Wang R, Zhang X, Si Z (2022) Utilization of sugarcane bagasse ash (SCBA) in construction technology: A state-of-the-art review. J Build Eng 56:104774
14. Chusilp N, Jaturapitakkul C, Kiattikomol K (2009) Utilization of bagasse ash as a pozzolanic material in concrete. Constr Build Mater 23:3352–3358
15. Cordeiro GC, Toledo Filho RD, de Moraes Rego Fairbairn E (2009) Use of ultrafine rice husk ash with high-carbon content as pozzolan in high performance concrete. Mater Struct 42:983–992
16. Nath P, Sarker PK (2014) Effect of GGBFS on setting, workability and early strength properties of fly ash geopolymer concrete cured in ambient condition. Constr Build Mater 66:163–171
17. Shi C, Qu B, Provis JL (2019) Recent progress in low-carbon binders. Cem Concr Res 122:227–250
18. Singh SK, Badkul A, Pal B (2023) Sustainable next-generation single-component geopolymer binders: a review of mechano-chemical behaviour and life-cycle cost analysis. J Mater Cycles Waste Manag 1–27
19. Zhang M-H, Malhotra VM (1996) High-performance concrete incorporating rice husk ash as a supplementary cementing material. ACI Mater J 93:629–636
20. Ganesan K, Rajagopal K, Thangavel K (2008) Rice husk ash blended cement: assessment of optimal level of replacement for strength and permeability properties of concrete. Constr Build Mater 22:1675–1683
21. Givi AN, Rashid SA, Aziz FNA, Salleh MAM (2010) Contribution of rice husk ash to the properties of mortar and concrete: a review. J Am Sci 6:157–165
22. James J, Rao MS (1986) Reactivity of rice husk ash. Cem Concr Res 16:296–302

23. Muthukrishnan S, Kua HW, Yu LN, Chung JKH (2020) Fresh properties of cementitious materials containing rice husk ash for construction 3D printing. J Mater Civ Eng 32:4020195
24. Poon CS, Shui ZH, Lam L, Fok H, Kou SC (2004) Influence of moisture states of natural and recycled aggregates on the slump and compressive strength of concrete. Cem Concr Res 34:31–36
25. Deshpande YS, Hiller JE (2012) Pore characterization of manufactured aggregates: Recycled concrete aggregates and lightweight aggregates. Mater Struct/Mater Constr 45:67–79. https://doi.org/10.1617/s11527-011-9749-2
26. Khan AR, Fareed S, Khan MS (2019) Use of recycled concrete aggregates in structural concrete. In: Proceedings of the fifth international conference on sustainable construction materials and technologies, London, UK, pp 14–17
27. Kou S-C, Poon C-S (2009) Properties of concrete prepared with crushed fine stone, furnace bottom ash and fine recycled aggregate as fine aggregates. Constr Build Mater 23:2877–2886
28. Liu H, Liu C, Wu Y, Bai G, He C, Zhang R, Wang Y (2022) Hardened properties of 3D printed concrete with recycled coarse aggregate. Cem Concr Res 159. https://doi.org/10.1016/j.cemconres.2022.106868.
29. Su C, Shi J, Tambara LUD Jr, Yang Y, Liu B, Revilla-Cuesta V (2024) Improving the mechanical properties and durability of steam-cured concrete by incorporating recycled clay bricks aggregates from C&D waste. Powder Technol 438:119571
30. Ye T, Xiao J, Zhao W, Duan Z, Xu Y (2022) Combined use of recycled concrete aggregate and glass cullet in mortar: Strength, alkali expansion and chemical compositions. J Build Eng 55:104721
31. Qin D, Hu Y, Li X (2021) Waste glass utilization in cement-based materials for sustainable construction: a review. Crystals (Basel) 11:710
32. Khan MNN, Sarker PK (2019) Alkali silica reaction of waste glass aggregate in alkali activated fly ash and GGBFS mortars. Mater Struct 52:93
33. Dai P, Lyu Q, Zong M, Zhu P (2024) Effect of waste plastic fibers on the printability and mechanical properties of 3D-printed cement mortar. J Build Eng 83:108439. https://doi.org/10.1016/J.JOBE.2024.108439
34. Shilar FA, Ganachari SV, Patil VB, Bhojaraja BE, Yunus Khan TM, Almakayeel N (2023) A review of 3D printing of geopolymer composites for structural and functional applications. Constr Build Mater 400:132869. https://doi.org/10.1016/J.CONBUILDMAT.2023.132869.
35. Si R, Dai Q, Guo S, Wang J (2020) Mechanical property, nanopore structure and drying shrinkage of metakaolin-based geopolymer with waste glass powder. J Clean Prod 242:118502. https://doi.org/10.1016/J.JCLEPRO.2019.118502
36. Gonçalves NPF, Olhero SM, Labrincha JA, Novais RM (2023) 3D-printed red mud/metakaolin-based geopolymers as water pollutant sorbents of methylene blue. J Clean Prod 383:135315. https://doi.org/10.1016/J.JCLEPRO.2022.135315
37. Ojha A, Aggarwal P (2023) Durability performance of low calcium Flyash-Based geopolymer concrete. Structures 54:956–963. https://doi.org/10.1016/J.ISTRUC.2023.05.115
38. Ma B, Zhu Z, Huo W, Yang L, Zhang Y, Sun H, Zhang X (2023) Assessing the viability of a high performance one-part geopolymer made from fly ash and GGBS at ambient temperature. J Build Eng 75:106978. https://doi.org/10.1016/J.JOBE.2023.106978
39. Shilar FA, Ganachari SV, Patil VB, Javed S, Khan TMY, Baig RU, Shilar FA, Ganachari SV, Patil VB, Javed S, Khan TMY, Baig RU (2022) Assessment of destructive and nondestructive analysis for GGBS based geopolymer concrete and its statistical analysis. Polymers 14:14. https://doi.org/10.3390/POLYM14153132.
40. Tchakouté HK, Rüscher CH, Kong S, Kamseu E, Leonelli C (2016) Geopolymer binders from metakaolin using sodium waterglass from waste glass and rice husk ash as alternative activators: a comparative study. Constr Build Mater 114:276–289. https://doi.org/10.1016/J.CONBUILDMAT.2016.03.184
41. Tchakouté HK, Rüscher CH, Kong S, Ranjbar N (2016) Synthesis of sodium waterglass from white rice husk ash as an activator to produce metakaolin-based geopolymer cements. J Build Eng 6:252–261. https://doi.org/10.1016/J.JOBE.2016.04.007

42. Yong-Sing N, Yun-Ming L, Cheng-Yong H, Abdullah MMAB, Rojviriya C, Khalid MS, Shee-Ween O, Wan-En O, Yong-Jie H (2023) Interaction of silica fume on flexural properties of 10 mm-thickness geopolymers based on fly ash and ladle furnace slag under the thermal conditions. J Build Eng 69:106331. https://doi.org/10.1016/J.JOBE.2023.106331
43. Khater HM, Gharieb M (2022) Synergetic effect of nano-silica fume for enhancing physico-mechanical properties and thermal behavior of MK-geopolymer composites. Constr Build Mater 350:128879. https://doi.org/10.1016/J.CONBUILDMAT.2022.128879
44. Du Z, Sheng S, Guo J (2023) Effect of composite activators on mechanical properties, hydration activity and microstructure of red mud-based geopolymer. J Mater Res Technol 24:8077–8085. https://doi.org/10.1016/J.JMRT.2023.05.049
45. Liew YM, Heah CY, Mohd Mustafa AB, Kamarudin H (2016) Structure and properties of clay-based geopolymer cements: a review. Prog Mater Sci 83:595–629. https://doi.org/10.1016/J.PMATSCI.2016.08.002
46. Perná I, Hanzlíček T (2016) The setting time of a clay-slag geopolymer matrix: the influence of blast-furnace-slag addition and the mixing method. J Clean Prod 112:1150–1155. https://doi.org/10.1016/J.JCLEPRO.2015.05.069
47. Xia M, Sanjayan J (2016) Method of formulating geopolymer for 3D printing for construction applications. Mater Des 110:382–390. https://doi.org/10.1016/J.MATDES.2016.07.136
48. Sun C, Xiang J, Xu M, He Y, Tong Z, Cui X (2020) 3D extrusion free forming of geopolymer composites: Materials modification and processing optimization. J Clean Prod 258:120986. https://doi.org/10.1016/J.JCLEPRO.2020.120986
49. Panda B, Paul SC, Hui LJ, Tay YWD, Tan MJ (2017) Additive manufacturing of geopolymer for sustainable built environment. J Clean Prod 167:281–288. https://doi.org/10.1016/J.JCLEPRO.2017.08.165
50. Panda B, Unluer C, Tan MJ (2019) Extrusion and rheology characterization of geopolymer nanocomposites used in 3D printing. Compos B Eng 176:107290
51. Panda B, Paul SC, Mohamed NAN, Tay YWD, Tan MJ (2018) Measurement of tensile bond strength of 3D printed geopolymer mortar. Measurement 113:108–116. https://doi.org/10.1016/J.MEASUREMENT.2017.08.051
52. Korniejenko K, Łach M, Chou S-Y, Lin WT, Cheng A, Hebdowska-Krupa M, Gądek S, Mikuła J, Korniejenko K, Łach M, Chou S-Y, Lin W-T, Cheng A, Hebdowska-Krupa M, Gądek S, Mikuła J (2020) Mechanical properties of short fiber-reinforced geopolymers made by casted and 3d printing methods: a comparative study. Materials 13. https://doi.org/10.3390/MA13030579.
53. Xia M, Nematollahi B, Sanjayan J (2019) Printability, accuracy and strength of geopolymer made using powder-based 3D printing for construction applications. Autom Constr 101:179–189
54. ASTM Standard C150/C150M-15 (2015) Standard specification for Portland cement ASTM International, West Conshohocken, PA. https://doi.org/10.1520/C0150-15. www.astm.org
55. ASTM C618–15 (2015) Standard specification for coal fly ash and raw or calcined natural Pozzolan for Use in Concrete; ASTM International: West Conshohocken, PA, USA
56. Hernández JFM, Middendorf B, Gehrke M, Budelmann H (1998) Use of wastes of the sugar industry as pozzolana in lime-pozzolana binders: study of the reaction. Cem Concr Res 28:1525–1536

Chapter 10
Life Cycle Assessment of Building Materials in 3DPC

10.1 Introduction

The construction sector serves as one of the most resource-intensive field that impacts the environment significantly. As per United Nations Environment Programme [1], it covers around 40% of the total energy consumption, produces approximately 38% of the greenhouse gas (GHG) emissions, generates 40% of solid waste, and consumes 12% of freshwater. Owing to the rapid urbanization, especially in the developing countries, it is of utmost importance to shift towards comparatively more sustainable, low-carbon, and building practices that efficiently use the resources.

In this situation, Life Cycle Assessment (LCA) has comes out as a prevailing and scientifically robust method for assessing the environmental impacts of building materials, and systems across all stages of its entire life, from the extraction of the raw material to its processing, transportation, and construction, application, maintenance, and finally to the disposal or recycling at the end of life.

10.1.1 Significance of Life Cycle Assessment (LCA)

Life Cycle Assessment is a systematic, cradle-to-grave approach to assessing the environmental burdens associated with a product, process, or activity. Governed by the international standards **ISO 14040 and ISO 14044**, LCA quantifies multiple environmental indicators, including energy consumption, GHG emissions, water use, acidification potential, eutrophication, and resource depletion. It allows comparisons among alternate materials or components and helps in making decision for procurement to policy development and sustainable design.

A. Chourasia et al., *Advances and Challenges in 3D Printed Concrete (3DPC)*, Materials Horizons: From Nature to Nanomaterials, https://doi.org/10.1007/978-981-95-8557-1_10

10.1.2 Historical Evolution of LCA in the Building Sector

The backgrounds of LCA can be started long back to the 1960s, when it was first employed for energy analysis of industrial systems. However, in the case of building sector, the utilization of LCA acquired momentum in the 1990s with growing awareness about change in climate, resource scarcity, and degradation in ecological aspects. Over the years, LCA has grown from a purely academic means to a critical tool for architects, engineers, manufacturers, and policymakers. The incorporation of LCA into the rating system of **green building** e.g., LEED, BREEAM, and IGBC has further pushed its acceptance in the community and reinforcing its position in taking forward construction practices that are environmentally friendly.

10.1.3 Importance of LCA in Sustainable Construction Practices

In current times, sustainability has become an essential component in construction and in this regard, LCA serves as a crucial factor in performing holistic environmental factors. In comparison to conventional environmental assessments tools that generally consider operational energy, LCA helps to evaluate the embodied effects of building materials and the techniques employed in construction. It is particularly relevant for high-performance based buildings, where operational energy is minimized, and the embodied effects of materials control the life cycle. With the identification of hot spots in the supply chain and endorsing efficiency in the materials, circularity, and low-carbon based alternatives, LCA supports the design of net-zero and climate-resilient buildings.

10.1.4 Emerging Relevance of 3D Printed Concrete (3DPC)

In the context of sustainability in construction, 3D printed concrete (3DPC) has appeared as an advanced and potentially groundbreaking technology. By providing precise material deposition, reduction in formwork, less labor requirement, and flexibility in design, 3DPC helps in reducing the embodied environmental impact in construction. Moreover, the possibility to include agro-industrial waste in fabricating printable concrete mixes unlocks new paths for conservation of natural resources and in achieving circular economy. However, such potential advantages must be validated through comprehensive life cycle assessments, particularly considering the amount of energy used in printing equipment and curing processes.

10.1.5 Objectives and Scope of the Chapter

This chapter aims to:

- Introduce the **principles, methodology, and evolution** of life cycle assessment as applied to the building materials sector.
- Discuss the role of LCA in promoting **sustainable construction, policy formulation, and green rating compliance**.
- Conduct a comparative analysis of LCA among **conventional and alternative building materials**, along with **cementitious materials and composites**.
- Provide an in-depth investigation of the **life cycle performance of 3DPC**, supported by various case studies and evidences from literature.
- Highlight the environmental impact of **agro-industrial waste-based binders** and **supplementary cementitious materials (SCMs)** in 3DPC systems.

By the end of this chapter, readers will gain a solid understanding of how life cycle assessment can be effectively applied to evaluate and optimize building materials for **environmental performance, economic viability**, and **long-term sustainability**, especially in the context of **emerging construction technologies** like 3DPC.

10.2 Fundamentals of Life Cycle Assessment (LCA)

10.2.1 Definition and Purpose

Life Cycle Assessment (LCA) is a comprehensive analytical methodology used to evaluate the environmental impacts associated with all stages of a product's life cycle—from raw material extraction (cradle) through materials processing, manufacture, distribution, use, repair and maintenance, and disposal or recycling (grave). Standardized by the **ISO 14040** and **ISO 14044** frameworks, LCA enables stakeholders to identify environmental hotspots and optimize processes to minimize resource use and emissions.

In the context of building materials, LCA supports environmentally conscious decision-making by quantifying metrics such as **Global Warming Potential (GWP)**, **Cumulative Energy Demand (CED)**, **acidification**, **eutrophication**, and **human toxicity** [2, 3]. The method provides a scientific basis for comparing traditional construction materials (e.g., OPC-based concrete, steel, brick) with innovative materials such as **3D printed concrete (3DPC)** and those incorporating **agro-industrial waste**, facilitating the transition toward low-carbon construction practices.

10.2.2 Key Frameworks and Guidelines

To ensure methodological consistency and comparability, LCA is governed by internationally recognized standards and region-specific guidelines:

- **ISO 14040 and ISO 14044**: These standards outline the principles, framework, requirements, and procedures for conducting a robust LCA. They ensure transparency and reproducibility across industries and geographies.
- **EN 15,804 (European Standard)**: This standard specifies core Product Category Rules (PCRs) for the Environmental Product Declarations (EPDs) of construction products. It is widely used in Europe to benchmark the environmental performance of building materials and systems, particularly for compliance with green building certifications like BREEAM and LEED.
- **ILCD Handbook (International Reference Life Cycle Data System)**: Developed by the European Commission, this guidance harmonizes LCA methodologies by providing detailed technical instructions on goal definition, inventory modeling, impact assessment, and result interpretation.

These standards form the backbone of any credible LCA study and are crucial when applying the methodology to both conventional and emerging materials such as geopolymer cement, biocement, and 3D printed composites.

10.2.3 Main Phases of LCA

The LCA process is composed of four interrelated and iterative phases:

- **Goal and Scope Definition**

This phase defines the purpose of the study, the intended application, target audience, system boundaries (e.g., cradle-to-gate or cradle-to-grave), and the **functional unit**, which is the reference to which all inputs and outputs are normalized.

In building material assessments, common functional units include- 1 cubic meter (m^3) of concrete, 1 square meter (m^2) of wall system, 1 kg (kg) of building product.

- **Life Cycle Inventory (LCI)**

LCI involves compiling and quantifying energy and material inputs (e.g., cement, aggregates, admixtures, electricity, water) and environmental outputs (e.g., CO_2 emissions, wastewater, solid waste) for each life cycle stage. For emerging materials like 3DPC, the LCI phase must also include specialized components such as printing nozzle energy use, robotic arm actuation, and curing techniques (e.g., accelerated carbonation).

- **Life Cycle Impact Assessment (LCIA)**

In this phase, the environmental impacts are assessed using characterization models. For example:

- GWP (in kg CO_2-equivalents)
- Acidification potential (in SO_2-equivalents)
- Human toxicity (in comparative toxic units)
- This stage is crucial for determining whether 3DPC with agro-waste fillers (e.g., rice husk ash, bagasse ash, fly ash) provides net environmental benefits over conventional mixes.

  - **Interpretation**

  This final phase analyzes results, identifies significant issues, evaluates sensitivity and uncertainty, and formulates conclusions and recommendations. It guides material selection, eco-design strategies, and green procurement policies.

10.2.4 Applications of LCA to Conventional and Innovative Building Materials

Conventional Materials

LCA of conventional materials such as Portland cement concrete, fired clay bricks, and structural steel consistently reveals their high environmental burdens, particularly in terms of embodied carbon and energy intensity. For example, cement production alone contributes to nearly 8% of global CO_2 emissions. Traditional LCAs for these materials are often used to develop Environmental Product Declarations (EPDs) and support eco-labeling schemes.

Innovative Materials and Technologies

In response to the environmental impact of traditional materials, attention has turned to low-carbon alternatives, including:

- Geopolymer concrete
- Engineered timber and bamboo
- 3D printed concrete (3DPC)

LCA of **3DPC** is still evolving but has shown promise in reducing material waste, energy use in formwork production, and construction time. Cradle-to-gate assessments have demonstrated that 3DPC structures can achieve up to 30–40% lower material waste compared to cast-in-place concrete, depending on print geometry and mix design.

Agro-Industrial Waste Integration

A key innovation in sustainable construction is the valorization of agro-industrial by-products, which can replace conventional cementitious materials and aggregates.Examples include: **fly ash** (from thermal power plants), **rice husk ash**, **bagasse ash**, and **groundnut shell ash, recycled aggregates** and **plastic fillers.**

The LCA focuses on quantifying and evaluating the environmental performance along with developing a basis for accessing the environmental performance of the process and its byproducts. With a focus on local resources and effective transportation, the process includes materials, trash (10–30%), rework/damage (5–15%), interim buildings, and sustainable alternatives. From embodied energy to the End-of-Life phase, the technique integrates a life cycle study of crucial components in 3D printing for concrete buildings. Fossil depletion, climate change, toxicity, oxidant and particle production, land and water damage, and ozone depletion are among the parameters evaluated. These elements assess how different 3D printing technologies affect the environment in the building industry. The construction experiments of conventional vs 3D printing compare the design mix, thermal and physical properties, and hardening properties that were directly impacted by the aggregate mix of recycled or natural aggregates and also on the transportation distance and cost. As substitutes for natural aggregates, recyclable aggregates such as fly ash, marble sludge, and blast furnace slag are being investigated. Similarly, to lessen dependency on Portland cement, which usually makes up 60% or more of the binder in regular and 3D-printed concrete, alkali-activated binders, recycled fibres, and unusual local materials are being added to concrete formulations [4].

The findings of several studies reported nearly 20% better environmental performance of 3D printing technologies compared to conventional concrete construction methods in the case of singular structures [3]. For large-scale construction, however, the global warming potential and environmental impact were larger of 3D printed concrete due to larger quantities of cement and CO2-producing agents during the manufacturing cycle. This result is consistent in previous studies as the environmental impact of the building materials-acquisition phase plays an important role, regardless of the use of 3D printing or traditional construction techniques [5].

The LCA of such materials demonstrates substantial reductions in **GWP** and **energy demand**, especially when these wastes are locally sourced and integrated into 3D printed concrete. The synergy between **additive manufacturing** and **circular economy principles** positions 3DPC with agro-waste inputs as a frontier in low-carbon construction.

10.3 LCA Methodology for Building Materials

Life Cycle Assessment (LCA) of building materials involves structured methodologies to quantify environmental impacts across the material's life span. Accurate assessment enables designers, policymakers, and industry stakeholders to make informed choices toward sustainable construction. This section elaborates on critical methodological considerations in LCA tailored for building materials.

10.3.1 System Boundaries

System boundaries define the extent of life cycle stages included in the analysis. Common boundary definitions include:

10.3.1.1 Cradle-To-Gate

This boundary encompasses all processes from the extraction of raw materials to the point where the product leaves the manufacturing facility. It **excludes the use phase, maintenance, and end-of-life stages**. This boundary often used for benchmarking and comparing production efficiencies. Particularly suitable for early-stage design decisions or comparing manufacturing impacts of innovative materials.

For example on comparison:

- For **3D printed concrete**, cradle-to-gate captures the production of printable mix, transport of constituents, and on-site robotic printing processes.
- For **conventional concrete**, it includes cement production, batching, and transport to site for casting.
- **Agro-industrial waste concrete** assessed under this boundary includes the sourcing and preprocessing of supplementary cementitious materials (e.g., rice husk ash, fly ash) and their incorporation into the concrete mix.

This boundary is ideal for identifying how alternative material formulations (such as AWC or 3DPC) reduce emissions during manufacturing when compared to traditional concrete.

10.3.1.2 Cradle-To-Grave

This is the most comprehensive boundary, covering **every stage from raw material extraction through product use, maintenance, and final disposal or recycling**. This boundary provides a complete environmental profile over the material's full life cycle, aligning with sustainability assessments. This boundary is generally adapted while comparing long-term impacts of materials used in buildings, especially when durability, maintenance, and end-of-life options vary.

For Example Comparison:

- **3DPC** offers potential savings in maintenance due to precise fabrication and integration of structural and aesthetic features, though its recyclability depends on the specific mix and curing method.

- **CVC** involves recurring maintenance, especially in adverse environments, and generates substantial demolition waste at end-of-life.
- **AWC**, when properly designed, not only improves durability but also imparts more sustainable disposal or reuse owing to the lower clinker content and other additives.

Cradle-to-grave analysis tells how initial manufacturing impacts in 3DPC or AWC and are either compounded or offset at the operational and disposal steps, making it indispensable for developing long-term policy and design decisions.

10.3.1.3 Cradle-To-Cradle

This boundary extends cradle-to-grave by **closing the loop through recycling, reuse, or upcycling**, thereby aligning with circular economy principles. The boundary system promotes regenerative design where materials are viewed as nutrients for future products. This boundary system is critical for evaluating the **true sustainability potential** of materials intended for circular construction systems.

For example comparison:

- **3DPC** can be integrated into cradle-to-cradle models if printed components are designed for disassembly and reuse or if waste print material is reprocessed.
- **CVC** typically ends up in landfills or low-value recycled aggregate, making it less compatible with this model.
- **AWC** is highly relevant here, as many agro-waste binders are inherently recyclable, biodegradable, or amenable to reintegration into cementitious processes.
- **Key Insight**: Cradle-to-cradle boundaries unlock the highest sustainability potential but require materials to be purposefully designed for recovery and reuse—a domain where 3DPC and AWC offer exciting possibilities far beyond traditional systems.

By carefully selecting the system boundary, practitioners can tailor LCA studies to specific decision-making contexts—whether optimizing production (cradle-to-gate), evaluating full life-cycle impacts (cradle-to-grave), or striving toward a regenerative, circular economy (cradle-to-cradle). Each boundary reveals different sustainability dimensions of **3DPC**, **CVC**, and **AWC**, making boundary selection a strategic decision in sustainable material development.

In comparative LCA, especially when evaluating 3DPC vs. conventional concrete, consistent functional units are crucial. For instance, when comparing 1 m^3 of 3DPC using recycled aggregates and fly ash with standard OPC-based concrete, the functional unit ensures that results are normalized for environmental impact per equal service output.

10.3.2 Data Sources and LCA Tools

The quality and granularity of data significantly affect LCA outcomes.

- **Primary Data**: Collected from manufacturers, site-specific operations, and pilot studies. In the context of 3DPC, this includes printer energy use, mix designs, curing methods, and layer deposition rates. For agro-industrial waste concrete, data from agricultural by-product collection, pre-processing, and blending with binders are crucial [3].
- **Secondary Data**: Obtained from reputable databases such as:
  - *Ecoinvent*: Offers granular datasets for global material flows and emissions.
  - *GaBi*: Widely used in industry, supports detailed modeling of construction products.
  - *Athena*: North America–specific database tailored for the built environment.
- **LCA Software Platforms**:
  - *SimaPro*: Powerful and flexible tool suitable for research-grade LCA studies.
  - *OpenLCA*: Open-source, adaptable for academic and industry applications.
  - *OneClick LCA*: Focused on the construction industry, integrates with BIM and supports Environmental Product Declarations (EPDs).

The use of these tools allows for accurate simulation and comparative analysis of innovative materials like 3DPC and hybrid blends with agro-wastes versus traditional Portland cement concrete.

10.3.3 Attributional vs. Consequential LCA

Attributional LCA quantifies environmental impacts associated with a product system as it currently operates. It provides a snapshot of existing practices and is particularly suitable for benchmarking the current environmental footprint of concrete types.

Consequential LCA assesses broader system changes resulting from decisions—such as market shifts to 3DPC or adopting agro-waste binders. It includes indirect effects like altered raw material demand, energy system response, and shifts in waste management.

For example, an attributional LCA may show 3DPC has lower material use per m^3 compared to conventional concrete. However, a consequential LCA might reveal additional emissions if large-scale 3D printing requires energy-intensive robotics or rare additives. Similarly, agro-waste concrete may reduce emissions attributionally but consequentially depend on biomass transport and availability.

10.4 Key Environmental Impact Categories

Understanding the environmental performance of building materials requires quantifying their effects across multiple impact categories. The following indicators, widely recognized in LCA methodology, are instrumental for comparing conventional concrete, 3DPC, and agro-industrial waste–based concrete.

10.4.1 Global Warming Potential (GWP)

Measured in kilograms of CO_2-equivalent emissions, GWP assesses the contribution of greenhouse gases to climate change. Conventional concrete, with high cement content, is a major emitter due to the calcination process and fossil-fuel combustion. In contrast, 3DPC often reduces GWP by optimizing material usage, eliminating formwork, and enabling the use of supplementary cementitious materials (SCMs). Agro-industrial concrete variants show the lowest GWP when they incorporate low-carbon binders like fly ash, rice husk ash, or sugarcane bagasse ash.

10.4.2 Ozone Depletion Potential (ODP)

Expressed in kilograms of CFC-11 equivalent, ODP reflects the contribution to stratospheric ozone layer degradation. While less dominant in concrete manufacturing, ODP may arise from refrigerants used in machinery and curing processes. Innovations in 3DPC must consider equipment refrigerants and curing compounds to minimize this impact.

10.4.3 Acidification and Eutrophication Potential

Acidification Potential: Linked to emissions of SO_2 and NO_x, leading to acid rain that damages ecosystems. Traditional concrete production contributes through fossil fuel combustion. 3DPC may have lower acidification due to efficient energy use, while agro-waste alternatives benefit from using low-sulfur agricultural residues.

Eutrophication Potential: Caused by nitrogen/phosphorus leaching, resulting in water body nutrient overload. This is relevant when agro-waste leachates are poorly managed. Proper pre-processing and mix control in agro-concretes reduce nutrient runoff risks.

10.4.4 Resource Depletion

Fossil Fuel Depletion: Traditional cement production is energy-intensive. 3DPC offers energy savings by reducing onsite labor and optimizing material use. Agro-waste-based concrete significantly offsets fossil resource use when locally sourced residues are used.

Mineral and Water Resource Depletion: Conventional concrete relies heavily on sand, gravel, and water. 3DPC can be engineered for lower water-to-cement ratios and tailored aggregates. Agro-industrial blends further alleviate raw mineral demand, provided the waste material supply chain is sustainable.

10.4.5 Human and Ecotoxicity

Human Toxicity: Includes health impacts from dust, heavy metals, or chemical exposure during production. Cement kiln emissions are notable contributors. 3DPC may reduce on-site dust generation. Agro-waste concretes, if not adequately processed, may introduce biological or chemical contaminants—necessitating proper treatment and testing.

Ecotoxicity: Refers to impacts on aquatic and terrestrial organisms. Runoff from poorly disposed construction waste can be harmful. LCA helps highlight such risks in early-stage material screening, favoring eco-safe additives and waste utilization practices.

10.5 LCA of Common Building Materials

Life Cycle Assessment (LCA) provides a robust framework to evaluate the environmental implications of building materials across all life stages. This section provides a detailed assessment of conventional materials, with a specific focus on the comparative performance of 3D Printed Concrete (3DPC), Agro-Industrial Waste–Based Concrete (AWC), and Conventional Concrete (CC).

10.5.1 Concrete and Cement

Raw Material Extraction and Processing: Cement production emits approximately **0.8–0.9 kg CO_2 per kg of cement**. Quarrying aggregates causes landscape degradation and dust pollution.

Manufacturing Impacts: Ordinary Portland Cement (OPC) has an **embodied energy of ~4.6 MJ/kg** and contributes up to **8% of global CO_2 emissions**. Water and admixtures further contribute to environmental toxicity.

Transportation and Construction: Concrete's bulk density (~2400 kg/m^3) requires significant fuel for transport. Onsite processes generate **up to 10% material waste**, especially through formwork usage.

Use Phase: High thermal mass reduces operational energy by **5–15%** in conditioned buildings. However, shrinkage cracking may lead to **repair frequencies every 15–20 years**.

End-of-Life Scenarios: Crushing for RCA recovery consumes **~0.1 MJ/kg** and emits **0.005–0.01 kg CO_2/kg**.

Comparative Analysis:

- **Embodied carbon: ~ 0.13–0.15 kg CO_2-eq/kg** for conventional concrete.
- **3DPC** eliminates formwork, reducing material waste by **up to 40%**.
- **AWC** with 30–50% cement replacement lowers GWP by **20–60%**.

10.5.2 Steel and Reinforcement

Raw Material Extraction and Processing: Steel production emits **1.85–2.0 kg CO_2/kg** via blast furnace and **0.4–0.6 kg CO_2/kg** using electric arc furnace.

Manufacturing Impacts: Embodied energy of virgin steel: ~ **35 MJ/kg**.

Transportation and Construction: Prefab elements reduce on-site emissions by **10–15%**.

Use Phase: Corrosion risk increases maintenance needs every **20–30 years** in humid environments.

End-of-Life Scenarios: ~90% recycling rate; minimal downcycling.

Comparative Analysis:

- Reinforcement in 3DPC is still under research; robotic placement under development.

10.5.3 Brick and Masonry

Raw Material Extraction and Processing: Kiln-fired clay bricks emit ~ **0.25 kg CO_2 per brick** (~1.7 MJ/brick embodied energy).

Manufacturing Impacts: High particulate emissions and GHG from fossil-fueled kilns.

Transportation and Construction: Laying rate: **400–500 bricks/day/mason**, high labor and water use.

Use Phase: Moisture ingress can reduce insulation and require repair every **10–15 years**.

End-of-Life Scenarios: Low recycling; high fragmentation rates.

Comparative Analysis:

- **3DPC** increases wall construction speed by **200–300%**.
- **AWC bricks** reduce GWP by **30–50%** compared to clay bricks.

10.5.4 Timber and Engineered Wood

Raw Material Extraction and Processing: Sustainably harvested timber stores ~ **1.8 kg CO_2/kg**.

Manufacturing Impacts: Embodied energy: **2–10 MJ/kg** depending on processing.

Transportation and Construction: Lower transport emissions due to density (**~600 kg/m^3**).

Use Phase: Thermal conductivity: **0.12–0.16 W/m·K**, favorable for insulation.

End-of-Life Scenarios: Reusable or incinerated with energy recovery.

Comparative Analysis:

- **3DPC** has limited compatibility with wood but higher fire resistance.

10.5.5 Glass and Aluminum

Raw Material Extraction and Processing: Embodied energy: **glass: ~ 15 MJ/kg, aluminum: 150–200 MJ/kg**.

Manufacturing Impacts: Aluminum smelting emits ~**10–12 kg CO_2/kg** (primary).

Transportation and Construction: Fragility increases transport protection needs.

Use Phase: Glass enhances daylighting, reducing lighting energy by **20–30%**.

End-of-Life Scenarios: Glass recycling saves **25–30% energy**; aluminum saves up to **95% energy** when recycled.

Comparative Analysis:

- Often complements 3DPC facades for energy-efficient designs.

10.5.6 Plastics and Polymers

Raw Material Extraction and Processing: Petroleum-based polymers emit ~ **2–6 kg CO_2/kg**.

Manufacturing Impacts: PVC: ~ **80–100 MJ/kg** embodied energy.

Transportation and Construction: Lightweight materials reduce fuel use by **30–40%** per unit volume.

Use Phase: Longevity ~ **30–50 years**; UV exposure may degrade performance.

End-of-Life Scenarios: Low recycling efficiency (**<20%**) due to contamination.

Comparative Analysis:

- Used in 3DPC conduits and piping; future potential for bioplastics.

10.5.7 Composites and Alternatives (Fly Ash, GGBS, Hempcrete, AAC Blocks)

Table 10.1 presents the LCA of commonly employed construction materials. From the table, the following points can be concluded:

Raw Material Extraction and Processing: By-products like fly ash (0.02 kg CO_2/kg) and GGBFS (0.05 kg CO_2/kg) are low impact.

Table 10.1 LCA comparative summary of materials

Material	GWP (kg CO_2-eq/kg)	Embodied energy (MJ/kg)	Recyclability (%)	Thermal conductivity (W/m·K)	Compatibility with 3DPC	Circular economy potential
Conventional concrete	0.13–0.15	1.1–1.3	20–30	1.4–1.8	High	Medium
Steel	1.85–2.0 (virgin), 0.4–0.6 (recycled)	20–35	85–90	N/A	Medium	High
Brick	0.25–0.3	1.5–1.7	< 10	0.6–1.0	Low	Low
Timber	−1.8 (sequestered)	2–10	60–80	0.12–0.16	Low	High
Glass	1.0–1.2	12–15	30–40	~1.0	Medium	Medium
Polymers	2.0–6.0	80–100	< 20	0.04–0.3	Medium	Low
AWC	0.05–0.08	0.8–1.0	30–60	0.3–0.6	High	High
3DPC	0.10–0.12	1.0–1.2	30–40	0.6–1.2	—	High

Manufacturing Impacts: Embodied energy of AAC: ~ **1.5 MJ/kg**. Hempcrete sequesters **110–165 kg CO_2/m^3**.

Transportation and Construction: Lightweight (~500–700 kg/m^3) reduces transport emissions by **20–40%**.

Use Phase: Good insulation: **thermal conductivity 0.06–0.12 W/m·K**.

End-of-Life Scenarios: Hempcrete can be composted; AAC can be reused.

Comparative Analysis: AWC enhances sustainability of 3DPC with **up to 70% cement replacement**, lowering GWP and enhancing durability.

This section underscores the significant potential of 3DPC and agro-industrial waste–based concrete to transform environmental performance in the construction sector. By reducing resource consumption and leveraging circular economy principles, these materials offer viable pathways to a more sustainable built environment.

10.5.8 Comparative LCA: 3D Printed Concrete Versus Conventional and Agro-Waste–Based Concrete

To contextualize the principles of life cycle assessment (LCA) within real-world applications, this section presents a comparative case study evaluating the environmental impacts of three types of concrete systems: conventional vibrated concrete (CVC); 3D printed concrete (3DPC); agro-waste–based concrete (AWC), incorporating supplementary cementitious materials (SCMs) like rice husk ash (RHA), sugarcane bagasse ash (SCBA), and fly ash. The assessment employs a cradle-to-grave system boundary with a functional unit defined as **1 m^3 of concrete** used in a structural wall application.

The goal is to compare the environmental impacts of three concrete variants and assess their suitability for sustainable construction. While the scope includes raw material extraction, material processing, transport, construction, operational energy requirements (where relevant), and end-of-life treatment. The assumptions made in Life Cycle Inventory (LCI) are summarized in Table 10.2.

Data was modeled using OpenLCA with the Ecoinvent 3.7 database and midpoint indicators from the CML and ReCiPe impact assessment methods. The Life Cycle Impact Assessment (LCIA) results are summarized in Table 10.3.

3D Printed Concrete (3DPC) demonstrates a measurable reduction in most environmental impact categories compared to conventional concrete. The primary drivers include: elimination of formwork; reduced material usage due to geometric efficiency; energy savings from automation, despite increased electricity demand for robotic equipment.

Table 10.2 LCI of three concrete variants

Parameter	CVC	3DPC	AWC
Cement content (kg/ m^3)	300	270	220 (replaced partially by SCMs)
Sand and aggregates	Natural aggregates	Optimized gradation	Recycled or low-carbon local sources
Reinforcement	Mild steel mesh	Minimal or no reinforcement	Similar to CVC or fiber alternatives
Transportation distances	Medium	Short (on-site printing)	Short to medium
Energy use during placement	Vibrators, formwork setup	Pumping, robotic arms	Vibrators and casting equipment
SCM usage	None	Optional (minor)	RHA/SCBA/Fly ash (~30% replacement)

Table 10.3 LCIA of three concrete variants

Impact category	CVC	3DPC	AWC
Global warming potential (kg CO_2-eq)	320	260	180
Ozone depletion potential (kg CFC-11-eq)	1.6×10^{-6}	1.3×10^{-6}	1.2×10^{-6}
Acidification potential (kg SO_2-eq)	1.4	1.2	0.9
Eutrophication potential (kg PO_4^{3-}-eq)	0.12	0.11	0.08
Fossil energy demand (MJ)	4,800	3,900	2,600
Freshwater use (L)	180	160	120

Agro-Waste Concrete (AWC) exhibits the **lowest environmental footprint**, largely due to the partial substitution of cement—a major CO_2 emitter—with industrial by-products. Its performance in GWP and fossil energy use outperforms both CVC and 3DPC.

Notably, **AWC reduces GWP by 44%** compared to CVC and by **30% compared to 3DPC**. Further, **3DPC saves about 19% GWP** over CVC, but its electricity-intensive processes may diminish its relative benefits in regions with high-carbon grids. In terms of **water consumption** which is lowest in AWC due to low water-to-cement ratio and recycled aggregate usage.

Sensitivity analysis showed that 10% increase in cement content in 3DPC increases its GWP by approximately 9%. In addition, substituting 50% of cement with RHA in AWC can further reduce GWP by up to 20%. Uncertainties primarily stem from regional variations in electricity grids, transportation logistics, and SCM availability.

This comparative study reveals that both **3DPC and AWC are viable sustainable alternatives** to traditional concrete, with AWC exhibiting the lowest environmental impact due to its material composition. However, **3DPC holds significant potential** for future gains, particularly when coupled with green power sources and circular design principles. When assessing these technologies, **regional context, material sourcing, and energy infrastructure** must be considered for accurate sustainability evaluation.

10.6 Comparative Life Cycle Assessment: Full Building Case Study

To holistically evaluate the environmental performance of different concrete technologies, a cradle-to-grave life cycle assessment was conducted on a prototype residential building constructed using (a) conventional reinforced concrete, (b) agro-industrial waste–based concrete, and (c) 3D printed concrete. The case study assumes a two-story, 150 m^2 building located in a temperate climate zone and designed for typical residential occupancy over a 50-year service life.

10.6.1 Functional Unit and System Boundaries

- **Functional Unit:** 1 complete residential building (150 m^2) over 50 years of service.
- **System Boundaries:**
- Cradle-to-Grave: Includes raw material extraction, material production, transportation, construction, use phase (maintenance and repair), and end-of-life treatment (demolition, recycling, or landfill). Different key factors associated with different phases are presented in Table 10.4.

Table 10.4 Key inputs and outputs of different phases

Phase	Key inputs and outputs
Raw material extraction	Quarrying of limestone, mining of industrial by-products (ash, slag), fuel use
Material production	Cement kilns, grinding units, admixture blending, digital mix design for 3DPC
Transportation	From plant to site (assumed 50 km average for all materials)
Construction	Concrete mixing, casting or printing, steel reinforcement, scaffolding, curing
Use phase	Minor repairs, energy usage for cooling (assumed equal), negligible emissions
End-of-life	Demolition, recycling of concrete (20% assumed), disposal in landfill or reuse

Table 10.5 Impact assessment

Impact category	Conventional concrete	Agro-waste concrete	3DPC
Global warming potential (kg CO_2-eq)	320,000	260,000	210,000
Cumulative energy demand (MJ)	1,800,000	1,500,000	1,200,000
Water consumption (m^3)	380	320	200
Material waste (kg)	8,000	6,500	2,200
Ozone depletion potential (kg CFC-11 eq)	0.021	0.018	0.016
Human toxicity (kg 1,4-DB eq)	95,000	78,000	62,000
Resource depletion (kg Sb eq)	9.5	8.2	6.1

10.6.2 Scenario Descriptions

- **Scenario A: Conventional Concrete**
  - Portland cement-based concrete with standard river sand, gravel aggregates, and steel reinforcement.
  - Cast-in-place with conventional shuttering, curing, and manual labor.
- **Scenario B: Agro-Industrial Waste Concrete**
  - Concrete mix incorporates 30% fly ash and 20% GGBFS as cement replacements; partial fine aggregate replacement with rice husk ash.
  - Cast-in-place using conventional labor and formwork.
- **Scenario C: 3D Printed Concrete (3DPC)**
  - 3D printable mix with reduced OPC (30% replacement by fly ash and marble dust), tailored for robotic deposition.
  - Eliminates formwork, reduces curing water, and is fabricated on-site using automated gantry printing.

Accordingly, Tables 10.5 and 10.6 present the environmental and economic aspects of three concrete variants. Comparative analysis revealed that AWC and 3DPC has lesser impact than conventional concrete in terms of different environmental factors. In line with environmental factors, utilization of 3DPC resulted in approximately 10% lesser cost than conventional concrete (CC).

10.6.3 Interpretation and Discussion

- **Environmental Benefits**: The 3DPC scenario shows superior environmental performance across almost all impact categories. Material efficiency from design

Table 10.6 Cost comparison (INR, Approx.)

Component	CC (lakh)	AWC (lakh)	3DPC (lakh)
Material cost	₹15.5	₹14.8	₹13.2
Labor cost	₹6.2	₹6.0	₹2.8
Equipment/Setup	₹0.5	₹0.7	₹4.5
Total Cost	**₹22.2 lakh**	**₹21.5**	**₹20.5**

optimization, reduced cement content, and the elimination of formwork contribute to lower embodied carbon and energy.

- **Agro-Waste Concrete as an Intermediate Solution**: Though not as optimized as 3DPC, the agro-waste mix significantly improves GWP and energy performance over traditional concrete, showcasing the potential of industrial by-product valorization.
- **Water and Waste Reduction**: 3DPC uses far less water due to the absence of curing and site mixing. Its automated fabrication also drastically reduces construction waste, especially formwork related.
- **Challenges and Trade-offs**: While 3DPC offers major environmental advantages, barriers such as high capital cost for printing equipment, limited standardization, and skilled workforce requirements may restrict its current scalability. Agro-waste concrete, on the other hand, is relatively easier to adopt within existing infrastructure.

The life cycle assessment (LCA) of building materials reveals critical insights into the environmental trade-offs and sustainability potential of conventional and emerging construction materials. Among all materials analyzed, conventional concrete and steel remain dominant in structural applications but are associated with high embodied energy and global warming potential (GWP). For instance, ordinary concrete emits approximately 0.13–0.15 kg CO_2-eq/kg, and virgin steel production reaches as high as 2.0 kg CO_2/kg, primarily due to energy-intensive processes and raw material extraction.

In contrast, 3D printed concrete (3DPC) and agro-industrial waste–based concrete (AWC) emerge as promising alternatives with significant environmental advantages:

- **3DPC** eliminates the need for traditional formwork, reducing material waste by up to 40%, and enables precise material placement, which improves overall resource efficiency. Its embodied carbon is lower than conventional concrete (approximately 0.10–0.12 kg CO_2-eq/kg), and its compatibility with digital fabrication techniques opens new avenues for sustainable design. **3DPC** exhibits the lowest life cycle emissions, water consumption, and operational waste. Labor savings significantly reduce overall costs despite high initial equipment investment.
- **AWC**, which incorporates industrial by-products like fly ash and GGBFS, can reduce GWP by 20–60% and embodied energy by similar margins, especially with cement replacement levels of 30–70%. It also enhances durability and contributes to a circular material economy. **AWC** provides an environmentally improved and

cost-effective transition, with minor labor savings and the advantage of existing construction practices.

- **CC** remains the least favorable in both emissions and cost due to higher material waste, cement content, and manual labor intensity.

When compared across functional parameters—including recyclability, thermal performance, and end-of-life circularity—both 3DPC and AWC outperform traditional clay bricks, polymers, and even glass in several dimensions. AWC further improves the environmental profile of 3DPC when used in synergy, such as in extrusion mixtures with reduced clinker content.

While certain materials like timber and hempcrete offer exceptional benefits in carbon sequestration and insulation, their structural and fire resistance limitations restrict their universal application. Conversely, the hybrid use of AWC and 3DPC maintains structural integrity while delivering measurable sustainability gains, making it particularly suitable for large-scale deployment in modern construction.

Adopting AWC or 3DPC over conventional concrete reduces carbon emissions by up to 35%, energy use by ~30%, and total lifecycle costs by 3–8%. While 3DPC leads in long-term performance, AWC provides a practical near-term solution, especially in low-tech or resource-constrained settings.

10.7 Conclusions

LCA serves as a vital tool in evaluating the sustainability of both conventional and emerging construction materials. 3D printed concrete, particularly when integrated with agro-industrial waste products, shows considerable promise in reducing environmental burdens while offering economic and technical advantages. However, standardized datasets, methodological consistency, and technology-specific impact factors are essential to fully realize its potential. Future research must focus on scaling up real-world applications, region-specific modelling, and integrating LCA with advanced digital tools.

In summary, the transition to 3DPC and AWC represents a critical step forward in aligning construction practices with climate resilience and circular economy principles. These materials not only reduce environmental burdens but also foster innovation in design and manufacturing, thereby positioning themselves as keystones in the sustainable architecture of the future.

References

1. United Nations Environment Programme (2023) Building materials and the climate: constructing a new future
2. Gursel AP, Ostertag C (2019) Life-cycle assessment of high-strength concrete mixtures with copper slag as sand replacement. Arch Civ Mech Eng 2019:6815348

3. Alhumayani H, Gomaa M, Soebarto V, Jabi W (2020) Environmental assessment of large-scale 3D printing in construction: a comparative study between cob and concrete. J Clean Prod 270:122463
4. Khan SA, Koç M, Al-Ghamdi SG (2021) Sustainability assessment, potentials and challenges of 3D printed concrete structures: a systematic review for built environmental applications. J Clean Prod 303:127027
5. Han Y, Yang Z, Ding T, Xiao J (2021) Environmental and economic assessment on 3D printed buildings with recycled concrete. J Clean Prod 278:123884

Chapter 11
Numerical Modelling of 3D Printed Concrete (3DPC)

11.1 Background

Modern construction techniques, such as additive manufacturing, have given rise to new opportunities in the sector due to their significant advantages over conventional concrete procedures [1]. Although additive manufacturing has the potential to completely transform the industry, before being widely used, it must overcome a number of difficulties and barriers. The absence of thorough research linking from printing properties to mechanical performance is a significant barrier impeding additive manufacturing's wider use in the construction industry [2–4].

In the past few years, significant number of research articles have been published in the field of 3D concrete printing which mostly used experimental approaches to investigate the early-stage printing properties such as extrudability, buildability [5–7]. Along with printing characteristics, mechanical characteristics [8–10] like compressive and flexural strength as well as long-time deformation like shrinkage [11] has also received a lot of attention. Understanding the physics behind the process or to evaluate how small-scale samples behave under mechanical loads require experimental research. However, in order to analyse the mechanical behavior of a large-scale specimen or looking at the wider scale applicability of additive manufacturing in the construction sector, it is imperative to develop a physics-based mathematical model through which the structural response of 3D printed concrete samples can be predicted under mechanical loading. Making use of numerical models could greatly reduce the time and expense involved in conducting experiments. Nevertheless, to create a numerical model that faithfully captures reality, it's essential to represent the governing physics of the problem in a more realistic manner.

In an extrusion-based layering technique, each concrete layer is printed sequentially and the characteristics of each individual layer as well as the interaction between these layers affect the overall mechanical performance of the additively produced composite system. Weaker interface zones between successive printed layers are one of the primary characteristics that makes the difference between a 3D printed

A. Chourasia et al., *Advances and Challenges in 3D Printed Concrete (3DPC)*, Materials Horizons: From Nature to Nanomaterials, https://doi.org/10.1007/978-981-95-8557-1_11

specimen and a traditional mold-cast concrete sample. Hence, investigating the characteristics of the interfilament bond in the case of the 3DPC specimen has attracted the attention of most of the researchers. However, till date, there hasn't been much focus in the literature on identifying interfilament bond characteristics and studying the effect of the same on the mechanical performance of the 3DPC specimen.

With regards to studying the mechanical properties of 3DPC samples, most researchers have concentrated on characterizing the interfilament bond properties among two consecutive printed layers through various experiments [12, 13]. This interlayer bond properties have a strong dependency on the interfilamnet printing time interval [14]. For instance, the experimental results of [15] reported that the interfilamnet bond strength decreases with increase in the interlayer time gap and for a time gap of higher than fifteen minutes, the interface region failed. Analogous outcomes from experimental works have also been reported in [16, 17]. Further, based on the experimental research, researchers [18, 19] also reported that the moisture present in the interfilament region also plays a pivotal role in the bond characteristics.

Apart from the bond characteristics, various other mechanical properties of the 3DPC samples, such as compressive [21], tensile [22], and flexural strength [23], have also been the major focus of research over the years. In this context, the existing work of Kruger and van Zijl [24] highlights that, depending on the interfacial characteristics, the compressive strength of the additively constructed concrete sample can drop by as much as 45% when compared to traditional mold cast samples. Further, through experiments, researchers investigated the effect of different direction of loading on the mechanical properties of concrete i.e., the anisotropic characteristics of printed samples [20–22]. Zareiyan et al. [23] reported that when printed specimens are loaded perpendicular to the printing direction, their compressive strength is higher in comparison to the loading in the other direction. Nonetheless, a contrasting nature in the published data on the compressive strength of the printed specimen has been found. One set of researchers [29, 30] asserted greater compressive strength in a certain printing direction, whilst the opposite pattern in the reported data has been found in [15, 18]. Similar opposite trends in the flexural strength of printed concrete samples have also been reported in [8, 18, 24]. These disparities in the outcomes could be caused by a number of coupled nonlinear characteristics, ranges from the strength properties of each layer of concrete to the bond properties of various interfaces, all of which would interact to affect the system's composite behavior. Moreover, through experiments, researchers also studied the mechanical behavior of 3DPC samples made of geopolymer or other waste-based binders [25, 26].

Contrary to extensive experimental investigations, very few articles have been found that utilized a mathematical framework to examine the mechanical characteristics of 3DPC specimens. Feng et al. [35] adopted a finite element (FE)-based tool with an orthotropic assumption for the material to numerically explore how the printing and loading direction affects the load carrying capability of 3D printed arch-type structures. The orthotropic assumption of the material implicitly incorporates the bond characteristics. Similarly, Shakor et al. [36] studied the mechanical behavior of printed concrete specimen numerically by integrating the experimentally obtained material properties of printed concrete samples that were tested by applying the loading from various direction of printing. However, the primary shortcomings of these models are their applicability in general due to non-considering explicitly the interfilamnet bond properties. Later, Bos et al. [37] examined the flexural performance of fiber-reinforced 3DPC sample using a damage and plasticity-based material behavior for the concrete. However, the work does not account for different bond properties in the numerical model. Following this, a discrete-element method (DEM) was utilized in [27] to study the behavior of printed concrete specimens subjected to a different loading scenario. However, in DEM, the primary challenge lies in accurately defining the contact properties and it entails significant computational costs. Further, Heever et al. [28] applied classical theory of laminate composite to study the mechanical behavior of printed samples subjected to tensile load. In contrast, Xiao et al. [29] employed the CDP model for the concrete layer and traction–separation type constitutive relation to define the interfilamnet bond to investigate the influence of interfacial bond on the variability of the mechanical behavior of the concrete element in different printing directions. Whereas numerical works [29] lack with respect to considering a fixed value of interfilament bond properties in the numerical model i.e., the influence of various bond parameters (presence in the constitutive relation of interfilamnet bond) on the performance of additively constructed concrete specimen under mechanical load has not been investigated. Further, the effect of interlayer pores/gaps (that could arise owing to the usage of circular shape nozzle or could be due to the variability in material type or printing process) on the behavior of the printed samples under mechanical load has not been studied.

Hence, a critical review of literature pertaining to 3D printed concrete highlights a dearth of mathematical model that considers the governing physical mechanism of the extruded layer as well as the interfilamnet bond under mechanical loading and investigates the influence of various printing parameters, ranging from bond characteristics, interfilmanent pores to the loading direction on the behvaior of printed concrete samples under various mechanical loadings. Consequently, there is a need to develop a continuum based mathematical modelling framework considering the variable compressive and tensile strength properties of material like concrete along with the bond strength under tensile and shear load to study the behavior of printed concrete samples under 4-point load. Therefore, this chapter employs a finite-element (FE) framework to numerically capture the mechanical behavior of 3DPC specimen under compression and 4 point loading conditions.

In the developed model, each of the extruded filament of concrete is represented through a constitutive relation that includes both damage and plasticity aspects of concrete. While the interfilamnet bond is represented through a cohesive zone-type constitutive relation that replicates the zero-thickness interactive zone between two consecutive filaments. The required input properties of concrete (e.g., tensile and compressive strength) are obtained through the specific tests [30]. Subsequently, comprehensive sensitivity analyses are performed for different printing related parameters. Within the various types of printing parameters, influence of bond strength, interfilment pores, and the loading direction on the compressive and flexural behavior of 3D printed concrete specimen are investigated. The simulated results under the wide variety of the above-mentioned parameters are compared with that of the traditional mold cast samples. Such exhaustive parametric investigation can aid the researcher in calculating the compressive and flexural strength of 3D printed concrete for different printing conditions and loading directions which ultimately would help in the design of the 3D printed concrete based on load deflection characteristics. Analysis is also extended in the context of a steel reinforced 3DPC building under gravity load. In this regard, suitable strategies with respect to geometrics, reinforcement modelling etc., are highlighted. In addition to the mathematical modelling of hardened properties of 3D printed concrete, different approaches that can be employed to characterize the early age or fresh state behavior of 3DPC are also presented in this chapter.

11.2 Modeling of Hardened Mechanical Properties

11.2.1 Constitutive Behavior of Materials

Any object produced through the technology of 3D printing has highly anisotropic mechanical performance. The anisotropic nature of the 3D printed specimen is created from the interaction of the printed layers through the adhesive bond. With respect to the material behavior of 3D printed concrete, till date, there is no well-established material model that has been developed by the researchers [39], particularly examining the current viewpoint in which different agro-industrial waste generally replace Ordinary Portland Cement (OPC). The employed material in this work consists of OPC along with two waste, bagasse ash (BA) and fly ash (FA) in a mass proportion of 60:20:20. The chosen mix proportion of the three primary constituents are decided among the trial designs of various mixes that satisfy the printing performance and compressive strength. Since, the main goal of this work is to investigate how different aspects of concrete printing, such as bond strength, stiffness of the bond, post-peak or plastic bond displacement, interfilment pores, and the number of extruded layers affects the uniaxial compressive behavior of 3D printed concrete specimen, thus, the details of comprehensive trial mixes are not provided in this chapter.

In this work, the mechanical behavior of the extruded concrete filament is represented through the plasticity and damage-based model present in Abaqus®, known as concrete damage plasticity (CDP) model [29, 31]. In the CDP model, the failure of concrete is considered to occur either through crushing due to compression or due to cracking in the case of tensile loading. Within each type of failure mode, initiation of damage or cracks and its propagation is described through a separate criterion, described in detail later in this section. Numerous studies, e.g., [31], have confirmed the effectiveness of the CDP model among the variety of plasticity and damage-based material models existing in Abaqus® for capturing the softening behavior of concrete under diverse loading scenarios. By utilizing the stress–strain behavior under uniaxial compressive and tensile loading, the model computes the needed inelastic or plastic deformation and cracking strains that are required for numerical analysis. As a result, the model automatically accounts for any rapid decrease in the stress values when peak stress is reached in a uniaxial stress–strain relationship, a characteristic frequently seen in brittle materials like concrete. Consequently, to describe the mechanical behavior of either plain concrete or concrete reinforced with steel, the Concrete Damage-Plasticity (CDP) model has been extensively employed in literature [32, 33], along with the current work.

In the CDP model, a scalar damage variable is used to represent the continuum-based damage in concrete, and a non-associated Drucker–Prager type yield function with predetermined cutoffs under the tensile loading characterizes concrete's plastic behavior. Additionally, the Mohr–Coulomb model defines the flow behavior in the plasticity. Whereas, the cohesive zone technique is employed to describe the adhesive contact between two subsequent extruded layers of the concrete.

The behavior of a system subjected to quasi-static mechanical loading can be characterized by the linear momentum balance, expressed as [34]

$$div(\boldsymbol{\sigma}) + \rho\overline{b} = 0 \tag{11.1}$$

where, $\boldsymbol{\sigma}$ *is* the stress tensor**,** *div* represents the operator "divergence" that acts on $\boldsymbol{\sigma}$, ρ stands for density, $\underline{b}$ represents the body force. Appropriate boundary conditions must be given to solve the governing equation stated above. A mechanical system's boundary conditions can be defined as follows: traction type boundary conditions represented as $\underline{t} = \boldsymbol{\sigma}.\underline{n} = \underline{t}^{Specified}\left(\underline{X}\right)$ on surface S_t and displacement based conditions as $\underline{u} = \underline{u}^{specified}\left(\underline{X}\right)$ on surface S_u. In the CDP model, the description of plasticity and damage through the constitutive relation is represented as

$$\boldsymbol{\sigma} = (1-d)\boldsymbol{D}_0^{el} : \left(\boldsymbol{\varepsilon} - \boldsymbol{\varepsilon}^{pl}\right) = (1-d)\overline{\boldsymbol{\sigma}} \tag{11.2}$$

Here, $\boldsymbol{\sigma}$ and $\overline{\boldsymbol{\sigma}}$ present the Cauchy stress and effective stress tensors, d is the isotropic scalar damage component and depends on $\overline{\boldsymbol{\sigma}}$, $\overline{\epsilon}_{p_l}$ i.e., $d = d(\overline{\sigma}, \overline{\epsilon}_{p_l})$, where $\overline{\epsilon}_{p_l}$ represents the equivalent plastic strain tensor. In the CDP model, $\overline{\epsilon}_{p_l}$ is decomposed into equivalent tensile and compressive plastic strain $(\varepsilon_{t,pl}, \varepsilon_{c,pl})$ that separately controls the tensile and compressive damages $(D_c andD_t)$ evolution. The failure surface $F(\overline{\sigma}, \overline{\epsilon}_{p_l})$ in the CDP model is represented as [35, 36]

$$F\left(\bar{\sigma}, \bar{\epsilon}_{p_l}\right) = \frac{1}{1-\alpha}\left(\bar{q} - 3\alpha\bar{p} + \beta\left(\bar{\epsilon}_{p_l}\right)\sigma'_{max} - \gamma\sigma'_{max}\right) - \bar{\sigma}_c\left(\bar{\epsilon}_{p_l}\right) \tag{11.3}$$

where, $\overline{p} = -\frac{1}{3}trace(\overline{\sigma})$, $\overline{q} = \sqrt{\frac{3}{2}(\overline{\boldsymbol{\sigma}} + \overline{p}I) : (\overline{\boldsymbol{\sigma}} + \overline{p}I)}$.

Here, $\sigma_{max}\prime$ is the maximum value of principal stress and α, β,γ are represented as:

$$\alpha = \frac{(f_{by}/f_{cy}) - 1}{2(f_{by}/f_{cy}) - 1}; \beta = \frac{\overline{f}_{cc}\left(\boldsymbol{\varepsilon}_{c,pl}\right)}{\overline{f}_{tc}\left(\boldsymbol{\varepsilon}_{t,pl}\right)}(1-\alpha) - (1+\alpha); \gamma = \frac{3(1-K_r)}{2K_r - 1} \tag{11.4}$$

Table 11.1 Values of the different parameters of the plasticity damage model [37]

Dilation angle	Eccentricity	f_{by}/f_{cy}	K	Viscous parameters
31	0.1	1.16	0.67	0.01

In Eq. (11.4), f_{by} and f_{cy} denote the biaxial and uniaxial compressive yield stresses, respectively. K_r represents the ratio of the second stress invariant on the tensile meridian to that on the compressive meridian at the initial yield point. $\overline{f}_{cc}$ and $\overline{f}_{tt}$ correspond to the effective compressive and tensile cohesion stresses. The coupled plasticity–damage model is formulated based on a non-associated flow rule, expressed as

$$G = \sqrt{(\epsilon f_{tg} \tan\phi)^2 + \overline{q}^2} - \overline{p} \tan \phi \tag{11.5}$$

where ϕ signifies the dilatation angle in the p-q plane, $f_{tg} = f_t(\varepsilon_{t,pl} = 0, \dot{\varepsilon}_{t,pl} = 0)$, ϵ stands for eccentricity parameter utilized in plastic flow. Compressive damage (D_c) is related to the plastic component of strain through Eq. (11.6)

$$\varepsilon_{c,pl} = \varepsilon_{c,in} - \frac{D_c}{(1 - D_c)} \frac{f_{c0}}{E_m} \tag{11.6}$$

Here, ε_c^{in} denotes the inelastic component of strain due to compressive load, f_{c0} stands for compressive yield stress, whereas, the intact elastic modulus is represented by E_m. To improve the mesh size facor, the damage D_t under tensile load is estimated as

$$u_{t,pl} = u_{t,ck} - \frac{D_t}{(1 - D_t)} \frac{f_t l_0}{E_m} \tag{11.7}$$

Here, $u_{t,pl} and u_{t,ck}$ symbolize the plastic and cracking displacement due to tension, f_t signifies the tensile stress. Input values of different parameters of the damage model [37] are specified in Table 11.1.

11.2.2 Modeling of Bond/Interaction Between Successive Layers

The zero-thickness cohesive element is characterized by a traction and separation rule with softening, governing its constitutive response in both the normal as well as shear directions, as illustrated in Fig. 11.1.

Damage (d_b) in the cohesive interlayer is considered to initiate in the cohesive zone model when the quadratic stress condition is met

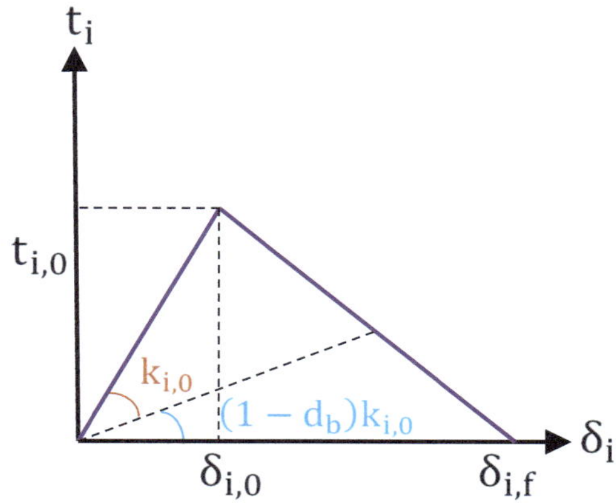

Fig. 11.1 Constitutive relation (traction–separation) of cohesive elements in normal ($t_n - \delta_n$) and shear loading ($t_s - \delta_s$)

$$\left\{\frac{\langle t_n \rangle}{t_{n,0}}\right\}^2 + \left\{\frac{t_s}{t_{s,0}}\right\}^2 = 1 \tag{11.8}$$

$\langle \cdot \rangle$ denotes the Macaulay bracket. The stiffness degradation due to damage (d_b) in the cohesive element is expressed as $k_n = (1 - d_b)k_{n,0}$ and $k_s = (1 - d_b)k_{s,0}$ [38]. The damage evolution (d_b) is governed by the effective relative displacement (δ_m) as:

$$d_b = \frac{\delta_{mf}(\delta_{m,max} - \delta_{m0})}{\delta_{m,max}(\delta_{m,f} - \delta_{m0})} \tag{11.9}$$

where, $\delta_m = \sqrt{\langle \delta_n \rangle^2 + \delta_s^2}$, $\delta_{m,max}$ signifies the maximum value of relative slip attained during the loading history, δ_{m0} and δ_{mf} are the effective displacements which belong to $\delta_{n,0}$ and $\delta_{s,0}$; and $\delta_{n,f}$ and $\delta_{s,f}$ (see Fig. 11.1).

11.2.3 Modeling Details in Abaqus

11.2.3.1 Loading and Boundary Conditions

Figure 11.2 presents a schematic illustration of the boundary conditions and loading adopted in the simulation. The loading conditions are four points with simply supported ends. At the left support, there are constraints on the x, y, and z axes of displacement. Conversely, the right support has zero displacement along the y axis. The loading method used in the analysis is gradual displacement-controlled, and the corresponding reactions are computed. The two displacement applied points in the beam are connected to a "reference point (RP)" in order to assess the reaction using the "Equation" interaction features in Abaqus. Value of the reaction measured at RP is then computed and stated as the load for a given displacement.

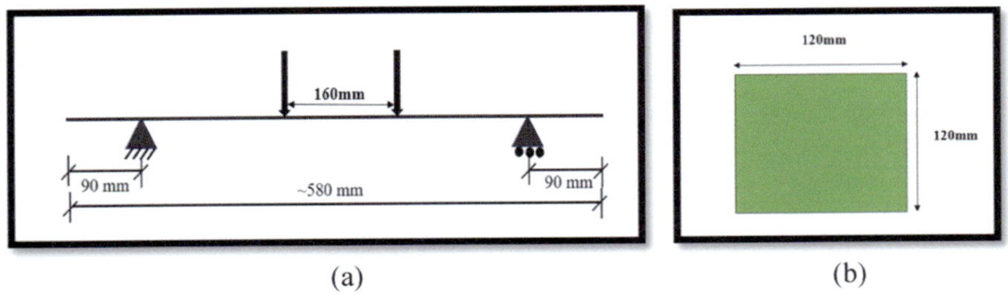

Fig. 11.2 **a** Schematic representation of boundary and loading in 4-point loading, **b** cross section of the beam

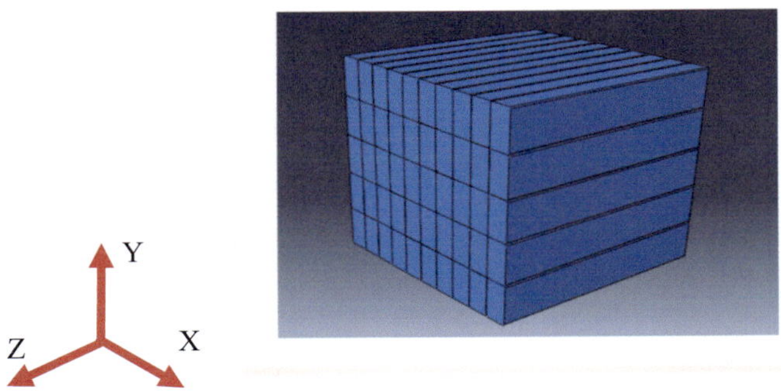

Fig. 11.3 Printed cube specimen of side 150 mm with extruded layer of size 30 × 15 mm

To simulate the compressive behavior, the bottom face of the cube sample of size 150 mm (see Fig. 11.3) is restrained for all the three-translation movements. However displacement-controlled scheme of loading is applied incrementally in the vertical direction. Here, the "Equation" interaction features in Abaqus is used to connect the top surface loading with a "reference point (RP)." As a result, calculation of the reaction only at the reference point reveals the sum of the total reaction at different nodes in the top surface of the cube.

11.2.3.2 Geometry Construction and Interlayer Interaction Definition

In the context of different types of samples simulated for compression and flexural loading condition, two categories of specimen are analyzed: conventional (Fig. 11.4a) or mold cast, printed specimen (Figs. 11.4b, c). Within the category of the printed specimen, two cases are considered: printed sample made of straight extruded layers without any cross-sectional curvatures (Fig. 11.4b) and with pore of radius r mm (Fig. 11.4c). The printed layer has a width of 30 mm and thickness of 15 mm extruded from a circular nozzle. Based on the geometric information of the layers, a 3D printed concrete specimen is constructed in Abaqus®. Here, the "solid extrusion" feature available in Abaqus is used to build the geometry. Along with that, the geometrical coordinates are utilized to represent the cross-sectional curvatures. In our previous study [30], where similar specimen height and layer cross-section was used, the interlayer pore radius was found to vary from 0.3 mm (at bottom) to 3 mm (top). Hence, in this work three different average interlayer pores of radius 1, 2, and 3 mm are considered. To do it, the geometric coordinates of the modelled layers are suitably adjusted. Whereas, the interaction between two layers are described through a cohesive zone interaction feature available in Abaqus®. Python scripting is used to define the interaction between two layers, which either interacts horizontally or vertically through one or more faces, see Fig. 11.5. The constitutive behavior of the interacting surfaces is defined through the cohesive zone-based interaction feature available in Abaqus®

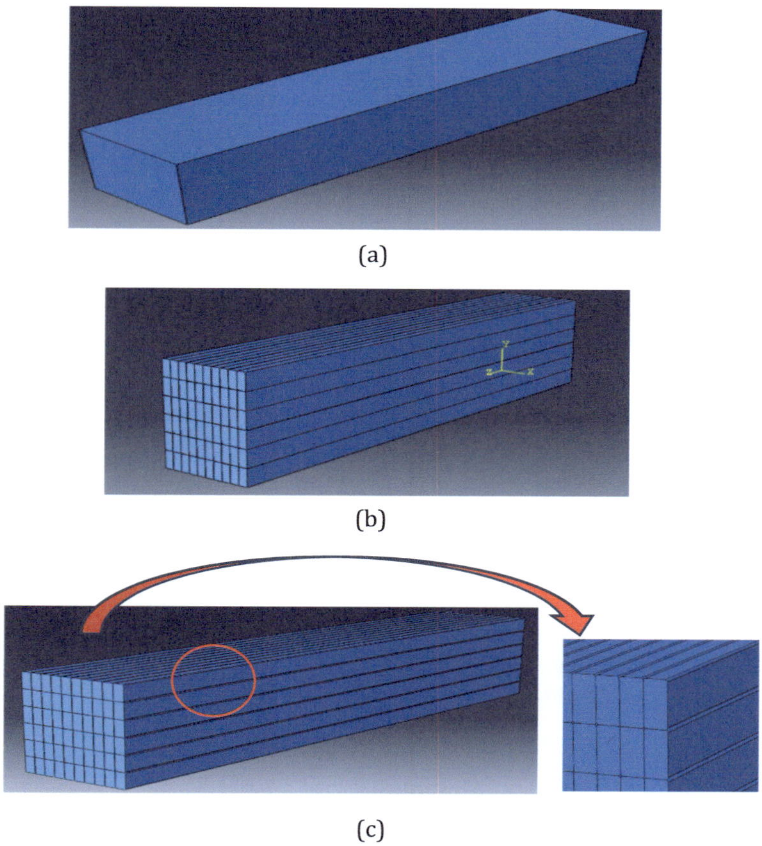

Fig. 11.4 Three different types of simulated beam specimens: **a** mold cast, **b** printed with no pore, **c** printed with pore of radius 1 mm along with enlarged view

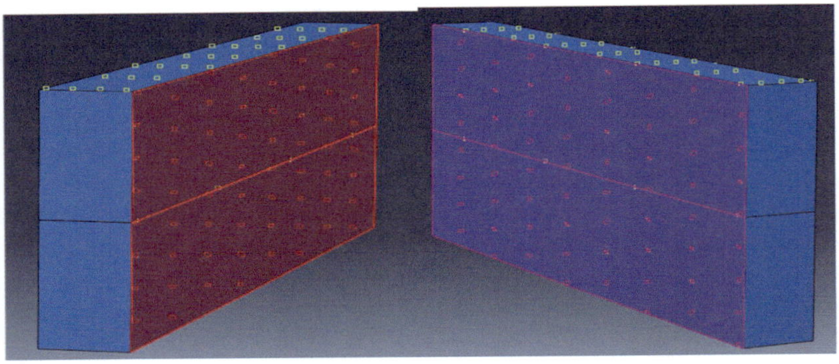

(a)

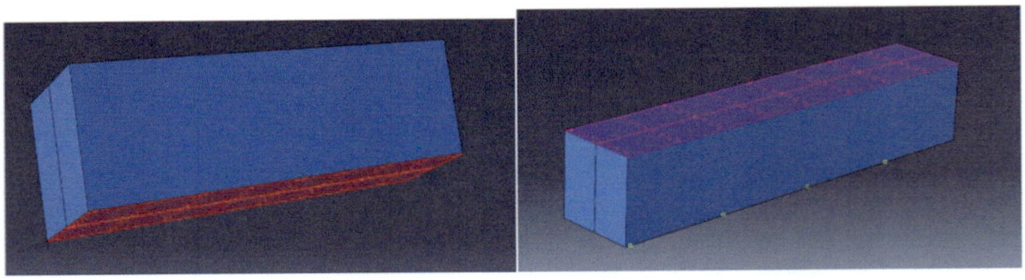

(b)

Fig. 11.5 Defining interaction between two adjoining printed layers in **a** vertical and **b** horizontal direction

11.2.3.3 Material and Interlayer Bond Values

Uniaxial properties of employed concrete are derived from testing under uniaxial compression and tension on the mold cast specimen. A cube 150 mm side is used for the compressive strength test [39], and a cylinder of 150 mm diameter and 300 mm height is used for the split tensile strength test [40]. The measured tensile and compressive strengths of concrete are 1.78 MPa and 32 MPa, respectively. However, parametric variation of the interlayer bond properties (such as bond strength, stiffness, post peak bond parameters) are conducted to investigate the effect of the same on the resultant homogenised compressive strength of the 3D printed specimen. Such parametric variation of the bond properties is inspired by the typical ranges reported in previous studies [28, 30].

11.2.3.4 Element Types and Meshing

Tetrahedral elements are used to discretize the domain because of the complex geometrical features, which include cross-sectional curvatures. In the selected element type, a quadratic displacement-based function is used. For different mesh sizes, a mesh convergence investigation is carried out in the finite element analysis. The pre-peak region's homogenized reaction versus displacement response is compared to establish the optimum mesh size, which is afterwards kept constant for further simulations of the specific sample.

11.3 Results and Discussions

On the basis of the methodology outlined in Sect. 11.2, numerical analysis is carried out to investigate the different printing parameters that affect the mechanical behavior of the 3DPC sample. Figure 11.6 illustrates the flowchart that describes the different steps in the FE analysis. The developed FE model is validated in terms of simulated compressive strength of the 3DPC specimen in different directions with the experimental work carried out. Afterwards, for a wide range of key printing related parameters, the relative variation in the strength properties is simulated from the developed model. Key printing parameters considered in this work include tensile bond strength, the presence of gaps between layers, number of layers, and the direction of loading. These parameters are individually varied while others are held constant, and their impact on the mechanical behavior is assessed. The relative drop in the load carrying capacity due to each parameter, compared to a mold cast specimen, is also analyzed.

The feasible range of the different bond characteristics in the cohesive model is inspired by various existing works. Nonetheless, a value outside of the specified range for any of the aforementioned constitutive or geometric characteristics may be selected. However, a range of the parameters that spans the potential scientific fluctuation of those parameters is chosen based on the practical application point.

11.3.1 Simulation Under Compressive Loading

At first, the developed model is validated in terms of the simulated results with the experimental work conducted. The specimen considered is a cube sample of side 150 mm with an average interlayer pore of size 1 mm, see Fig. 11.3. Whereas for the purpose of experimental work, cube specimens of same dimensions, as of simulation, are extracted from the printed bed. Afterwards, the extracted cubes are tested in the compression testing machine from three different directions, such as shown in

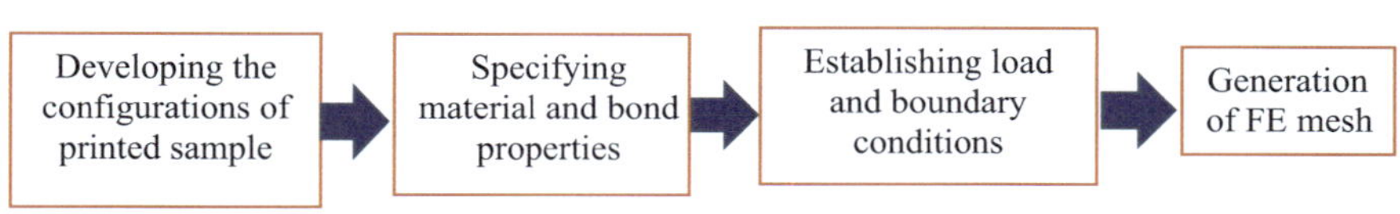

Fig. 11.6 Different steps employed in the FE simulations

Fig. 11.7 Uniaxial compressive strength assessment of 3D printed specimen

Fig. 11.7 in one direction. The experimentally found compressive strength values are then compared with the simulated strength values in Fig. 11.8. The interlayer bond strength in this simulation is taken as 0.43 MPa from our previous work [30], where the interlayer bond strength was evaluated from the experiment. It is observed (Fig. 11.8) that the simulated results in all three directions are approximately closer to the experimental values with roughly 10% higher prediction obtained from the model. Such higher prediction with the model could be due to the approximate considerations of the interfilamentous pore, in which an average pore throughout the thickness of the specimen is considered in comparison to the real 3D printed sample. Whereas, in reality, the interlayer pore has a varying size across the height of the sample with higher pore in the top and lower value is at the bottom due to the weight of the above printed layers. Hence, the over-estimation of compressive strength from the model could be due to the smaller consideration of the interlayer pore. Further, Fig. 11.9 presents the simulated damage pattern of the cube sample under compression load. Typically a diagonal type failure in the specimen is observed which is directly linked as perpendicular to the principal tensile stress direction that arises in the specimen under compressive load with end friction.

11.3.1.1 Influence of Tensile Bond Strength

In the present section, tensile bond strength is varied keeping the bond stiffness ($k_{i,0}$) and failure displacement ($\delta_{i,f}$) constant. The shear to tensile bond-strength ($t_{s,0}/t_{n,0}$) is maintained at 2 throughout the current section. Based on the tensile and shear bond strength test, authors in [28] reported that for the additively built concrete samples, shear bond strength would be higher than the corresponding tensile bond strength. Furthermore, the bond-stiffness ($k_{i,0}$) and failure displacement of bond ($\delta_{i,f}$) are assumed to be 1000 N/mm^2/mm and five times of the peak bond displacement respectively, in the current section.

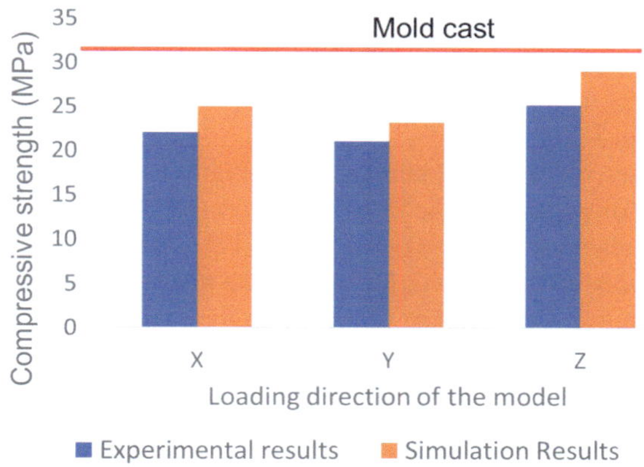

Fig. 11.8 Comparison of simulated and experimental results of compressive strength for a 3D printed specimen in different directions

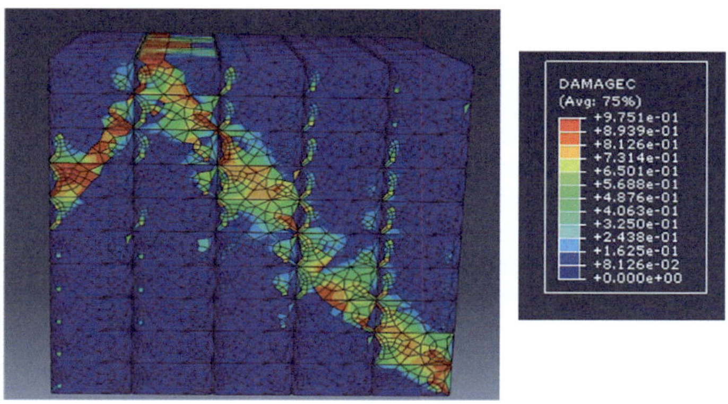

Fig. 11.9 Simulated damage pattern of printed specimen under compressive loading along with the compressive damage (d_c) value

Figure 11.10(a–c) presents the variation in compressive strength for different tensile bond strengths and interlayer pore sizes when loaded from three different directions. In all the three plots, where loads are applied in three different directions, an increase in the interlayer pore resulted in the drop of compressive strength of the printed specimen. Such an observation of decreasing trend in compressive strength with increase in pore is in line with the physical understanding. As more weak zones are introduced in the model, it naturally results in the reduction in an effective load bearing area. A comparative assessment reveals that it is the Y direction of loading in which the drop in strength is highest and the respective drop amount are 30%, 65%, and 75% respectively for the introduction of 1, 2, and 3 mm pore as compared to the mold cast sample. Whereas, it is the z direction of loading i.e., parallel to the direction of the printed layer, in which the influence of the interlayer pore is minimum with 8%, 13%, and 18% respectively for the 1, 2, and 3 mm pore than the mold cast sample. Consequently, the relative drops in compressive strength in the Y direction (the lowest strength) with respect to Z direction is 12% and 70% respectively for 1 and 3 mm pore size.

Similar outcomes of highest compressive strength (i.e., less effect of printing parameters) along or parallel to the layer direction (i.e., z direction) from experimental work have also been observed in [30, 41, 42], where different material type and printing technique were utilized. Moreover, the phenomena of lowest compressive strength (i.e., maximum effect of printing parameters) in the case of Y direction loading have been reported in [29, 30]. Such similarities of highest compressive strength in a particular printing direction (Z direction) in the case of the 3DPC specimen highlights the potential impact of printing direction on the strength and resultant anisotropy. Based on such comparative results, irrespective of the material type and printing process, one could choose that optimum printing path which results in the highest compressive strength. The elevated strength "Perpendicular" to the direction of printing could be due to the considerable compaction resulting from the applied pressure while extrusion [41] occurs. A similar explanation for the maximum strength along the "Perpendicular" direction in comparison to other two directions has also been reported in [41]. However, opposite trends in the reported studies, with respect to the directional dependency of strength properties, were observed. Further, unlike the behavior of the printed specimen under flexural load [30], the effect of interlayer bond on the compressive strength is negligible with the highest drop in compressive strength bieng 12% for a bond strength value of 0.25 MPa with respect to the value at bond strength of 1.78 MPa. Such dependency of strength of the printed specimen in a particular direction depends on multiple reasons ranging from various printing process-related parameters to material type amd compaction of layers. Hence, the reported outcome in the present work could also be problem specific and hence, necessitate future study.

Moreover, analysis of compressive strength with respect to interfilament pores reveals that the compressive strength of the 3D printed concrete is significantly affected when loaded transverse to the printing layer direction (*X* or *Y* direction). However, inter-layer pores have comparatively negligible effect on strength when loaded along the printing layer direction (Z-direction). For the interlayer pore of radius 1-, 2-, and 3 mm, the drop of compressive strength is upto 25%, 65%, and 75% respectively with respect to mold cast sample considering a different direction and bond strength. Consequently, such analysis would be helpful for the material designer in deciding the optimum printing path of printing direction with respect to the relative drop of strength in different directions and for different interlayer pore radii.

Fig. 11.10 Compressive strength of printed specimen for different bond strengths and interlayer pore radii for load applied in **a** X, **b** Y and **c** Z- direction (refer Fig. 11.3 for details of direction) along with mold cast specimen (red line)

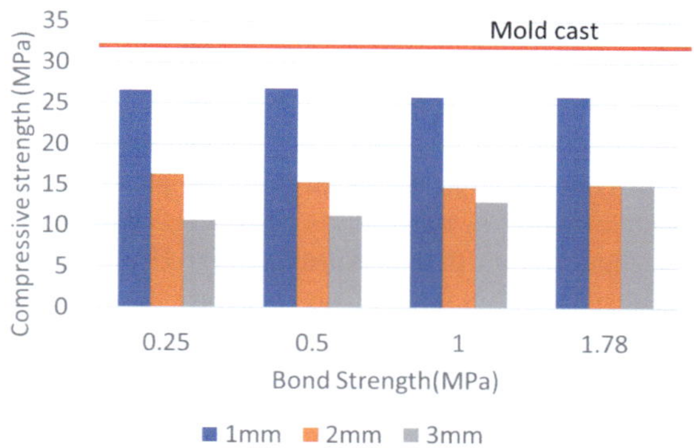

(a)

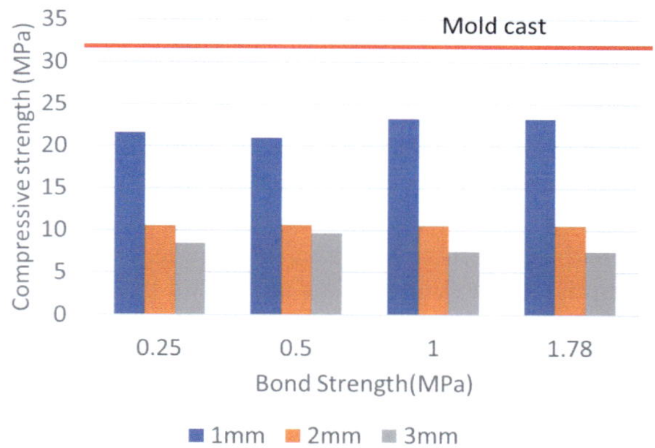

(b)

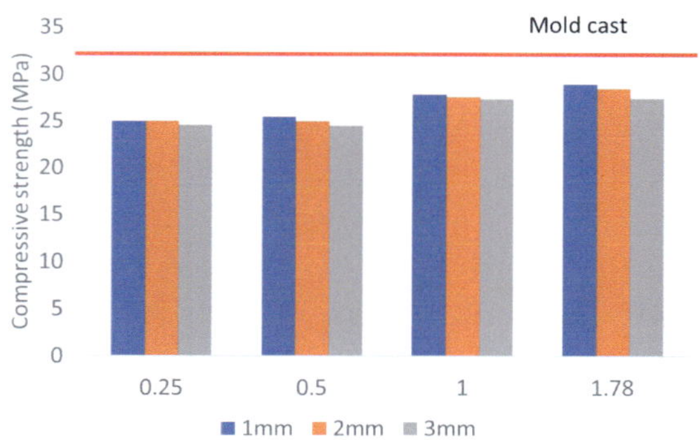

(c)

11.3.1.2 Effect of Number of Extruded Layers on the Compressive Strength

In the last section, the influence of different number of extruded layers on the compressive strength of the printed sample is examined. Consequently, the cube specimen with 150 mm side is considered while taking into account two different cross-sections (width x thickness) of extruded layers: 30 × 15 mm and 15 × 7.5 mm. As a result, the number of layers along the width and thickness direction of the extruded layers become 5 and 10 for the case of extruded layers with 30 × 15 mm cross section, see Fig. 11.11a. On the other hand, for the case of extruded layers of 15 × 7.5 mm, the respective number of layers are 10 and 20, see Fig. 11.11b. Since the effect of varying amount of pore is already investigated in the previous section, hence, the simulation conducted in this section is the one where there is no pore in the sample, which typically can occur in the case of a rectangular type nozzle. In the simulation, the tensile bond strength is kept as 1 MPa and the bond stiffness as 1000N/mm^2/mm. Figure 11.12 presents the compressive strength of the printed sample when the load is applied from different directions for the two kinds of extruded layers. In the same plot, the strength of a mold cast specimen is also shown. It is observed that by increasing the number of extruded layers by 2 times (i.e., from sample with extruded layer of size 30 × 15 mm–15 × 7.5 mm), the compressive strength of the cube decreases in all three directions. The relative reduction amount of compressive strength with the rise in the number of layers by 2 times in the X, Y, and Z directions are 14.8%, 34.4%, and 22.6%, respectively. This drop in the compressive strength could be due to the increase in the number of comparatively weak interlayer bond regions in the case of specimens made with the extruded layer of size 15 × 7.5 mm. Consequently, such analysis would be helpful for the material designer in deciding the appropriate nozzle size or the relative drop of strength that could occur for smaller size of the nozzle.

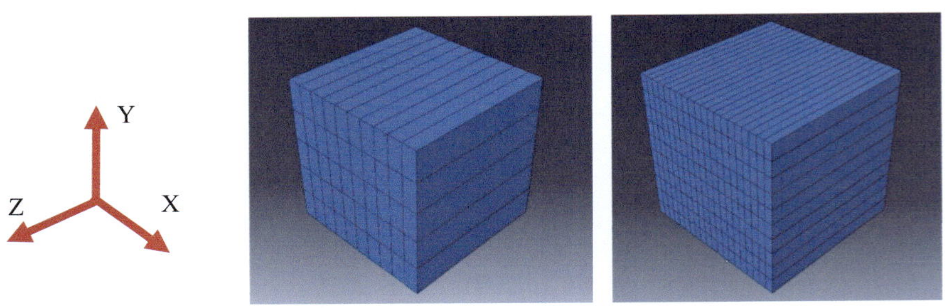

(a) (b)

Fig. 11.11 Printed cube sample of size 150 mm with extruded layers of two different sizes: **a** 15 × 30 mm, **b** 15 × 7.5 mm

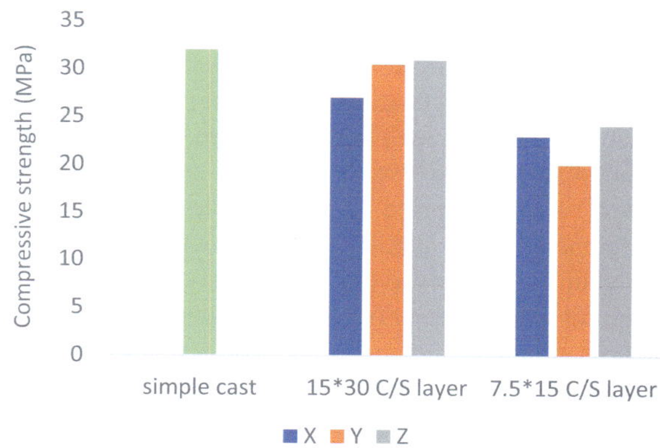

Fig. 11.12 Comparison of compressive strength results for two different layer sizes: 30 × 15 mm and 15 × 7.5 mm with the mold cast sample when loaded in three different directions with no interlayer pore

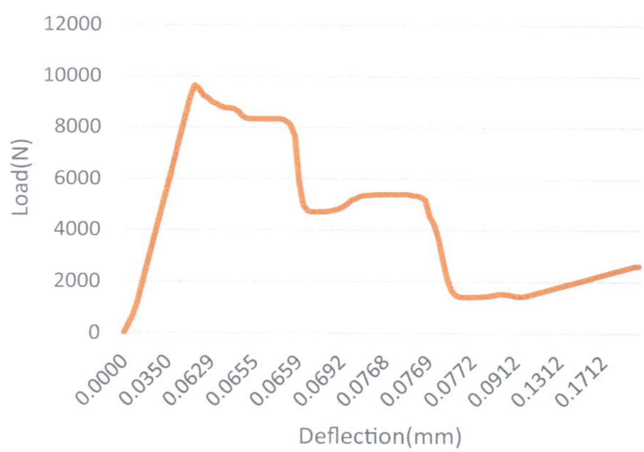

Fig. 11.13 Load deflection characteristics of mold cast beam

11.3.2 Simulation Under Flexural Loading

11.3.2.1 Simulation of Mold Cast Specimen

Figure 11.13 presents the load deflection behavior of the mold cast specimen, from which the peak load capacity of the beam is found to be approximately 9273 N. Beyond the peak load value, the post peak behavior of concrete typically belongs to the material which is brittle to quasi-brittle in nature.

11.3.2.2 Simulation of Printed Specimen

Figure 11.14 presents the load deflection behavior of the printed specimen loaded in the y direction for two pores condition: one with no pore (Fig. 11.14a) and another with an average interfilament pore of radius 1 mm (Fig. 11.14b). Similarly, Fig. 11.15a, b shows the load deflection characteristics of the same specimen when loaded in the X direction. In the case of both conditions, two extreme values of tensile bond strength are considered, 0.25 and 1.78 MPa. Here, the highest value of bond strength is taken as the split tensile of the monolithic concrete layer. The lowest value is inspired by the possible reported value of bond strength by previous researchers such as in [30] that could occur if the printed time gap between the layers is higher. It is observed that the interlayer bond strength has a stronger effect on the resultant load capacity of the beam when the load is applied in the Y direction than in the X direction. In the case of Y direction loading, without any inter filament pore, the possible load capacity can range between 7959 and 5847 N for the two extreme values of tensile bond strength. However, with interfilamnet pore of 1 mm radius in the case of Y direction loading, the same load capacity ranges between 7359 and 3806 N. This implies for the two pores conditions (no pore and 1 mm pore), the load value can drop upto 26.5 and 48% with the drop in tensile bond strength from 1.78 to 0.25 MPa. Lowering the interlayer bond strength introduces additional weak regions in the printed beam that resulted in the drop in the load capacity of the printed beam when the bond strength decreases. Further, introduction of the interfilamnet pore of 1 mm radius resulted in the drop in load capacity by 7.5% and 35% for the interfilament bond strength values of 1.78 and 0.25 MPa. Similar to the interlayer bond strength values, introduction of pores or gaps between the extruded layers effectively weaken the structure which causes the peak load value to drop with the introduction of inter filament pores. Figure 11.16 shows the damage pattern of the 3DPC specimen under 4-point loading, in which damage or crack occur at the centroid of the specimen.

On the other hand, in the case of X direction loading, the relative difference in peak load for the two bond strength values is insignificant in the case of no pore and 1 mm pore conditions. However, the introduction of the pore has an effect on the resultant peak load value in the case of X-direction loading condition also. However, the same effect is minimal with approximately 4.5% drop in the peak load value with the introduction of 1 mm pore in comparison to the load value in the case of no pore condition. Such variable effect of the different loading direction on the peak load value of 3D printed beam implies the resultant anisotropic characteristics that could arise due to the effect of printing and loading direction. Consequently, a designer can choose the optimum printing direction of the beam based on the required performance and other constraints. Further, the comparison of the peak load of the printed specimen (loaded in X and Y direction) with respect to the mold cast sample shows a relative drop in the load value upto 60%. Incorporation of a different degree of weakness in the system either through lowering the bond strength or through pore results in such drop in the load carrying capacity of the printed beam in comparison to the mold cast sample.

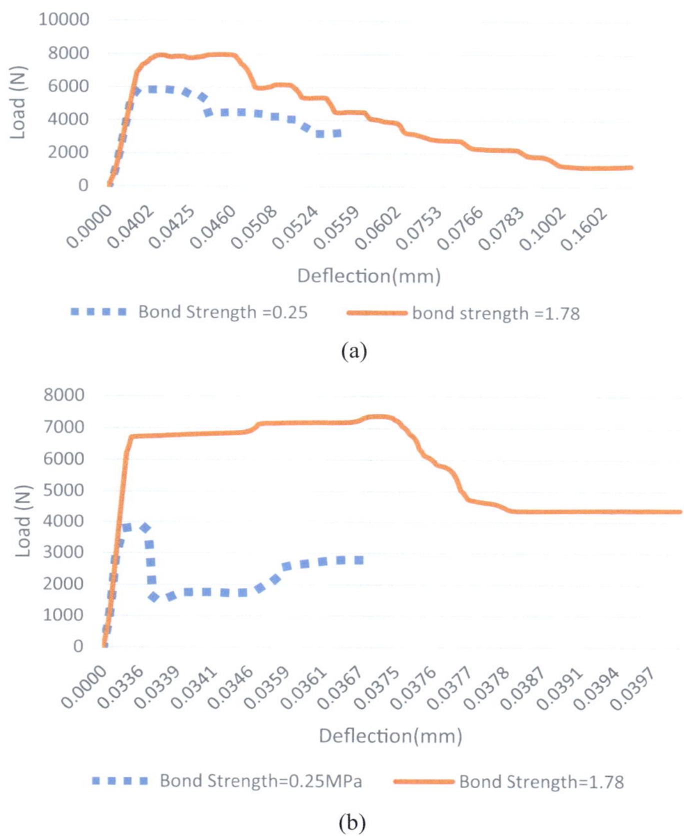

Fig. 11.14 Load–deflection behavior of printed sample with each extruded layer of size 20 × 10 mm for load applied in Y direction in the case of **a** without pore, **b** with average pore of 1 mm

11.3.3 Simulation of 3D Printed rReinforced Concrete Building

This section presents the analysis of a 3D printed steel reinforced concrete building under gravity load. The building is made of wall of thickness of 180 mm and height of 3 m. Details of the geometry are presented in Fig. 11.17. Based on the geometrical configurations, the developed printed building from two different sides is shown in Fig. 11.18. Along with that, an enlarged portion of the 3D printed building from two different angles is also shown in the same figure. Each the extruded layer has a width of 60 mm. Here it is required to mention that the outer and inner walls of 60 mm thickness have been printed. While the centroid portion of the wall of thickness 60 mm is filled with concrete at a later time. The overall arrangement of the reinforcement is shown in Fig. 11.19. Whereas Fig. 11.20 presents the placement of reinforcements in

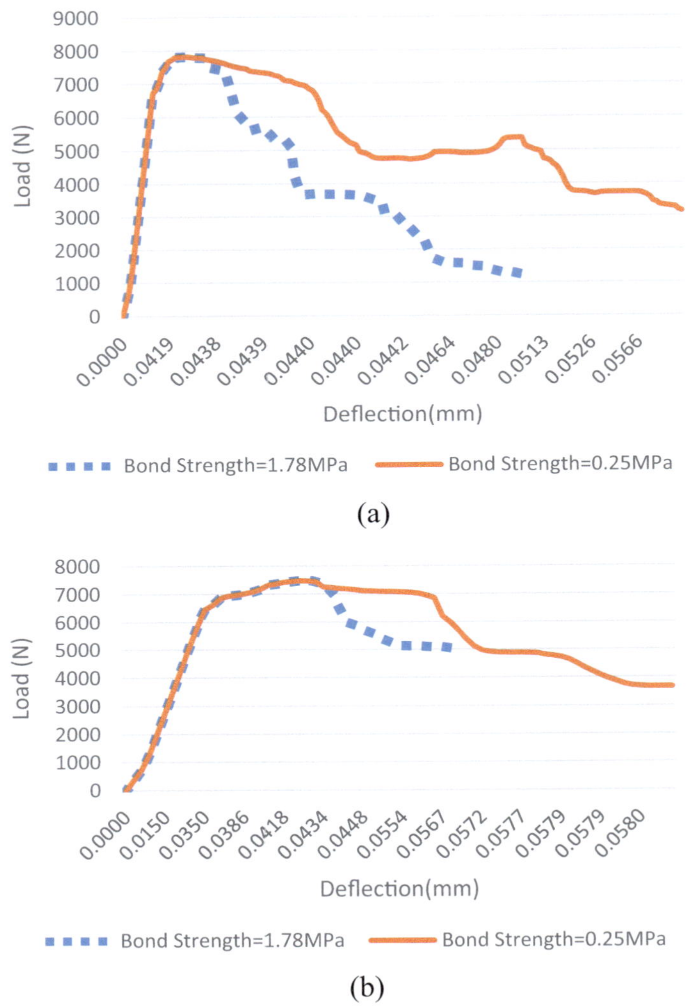

Fig. 11.15 Load–deflection behavior of printed sample with each extruded layer of size 20 × 10 mm for load applied in X direction in the case of **a** without pore, **b** with average pore of 1 mm

a particular printed layer. The vertical reinforcement in the wall is 8 mm @ 150 mm c/c. Horizontal main reinforcement in wall is of 8 mm @ 120 mm c/c. Horizontal stirrups in wall has a diameter of 8 mm and of 120 mm length placed at distance of 250 mm (see Fig. 11.19). On the other hand, owing to the difficulty in printing the slab, a precast slab of 150 mm thickness with 10 mm dia bar placed @150 mm c/c in both horizontal directions is taken. In building the reinforced concrete building, the reinforcements are modelled using the "embedded elements" features available in Abaqus®. Afterwards, the analysis is conducted for gravity loading conditions

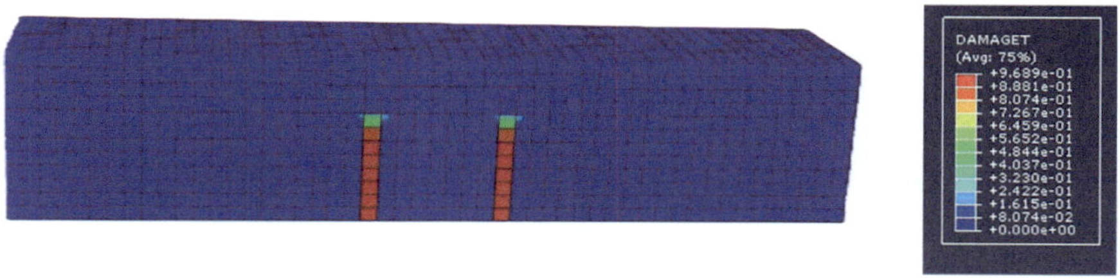

Fig. 11.16 Typical tensile damage pattern of printed beam under 4-point load along with the tensile damage value (d_t)

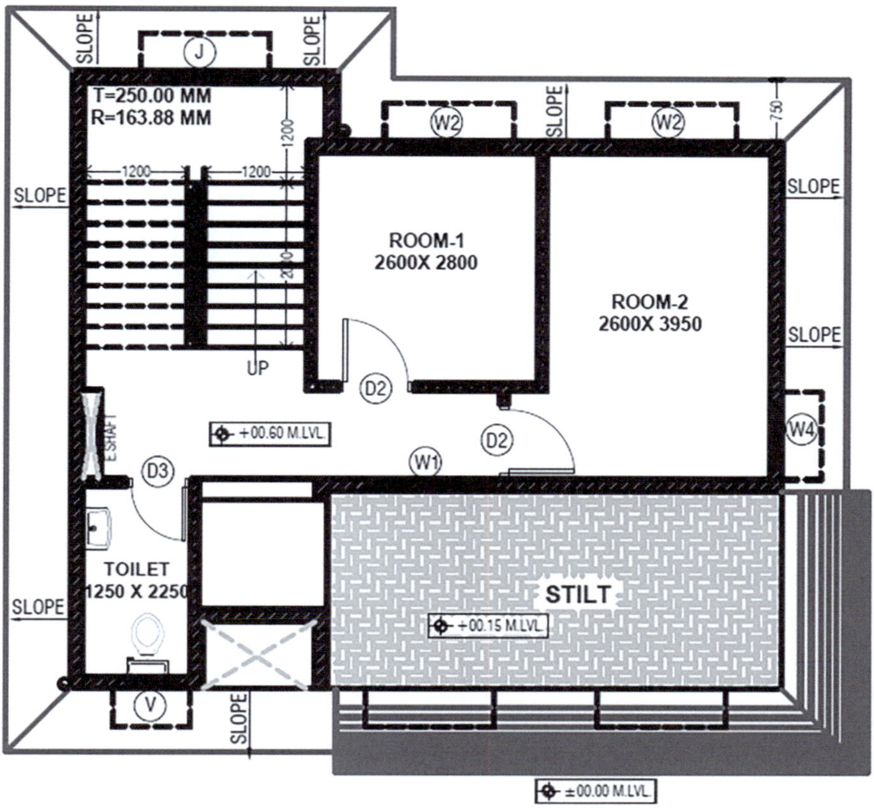

Fig. 11.17 Detailed plan of 3DPC building used for simulation

and the resultant vertical stress distribution is shown in Fig. 11.21. It is observed that the bottommost layer of the building experiences the highest vertical stress due to gravity. Since the primary aim of the chapter is to introduce the readers about the various modelling methodologies that can be employed to simulate the

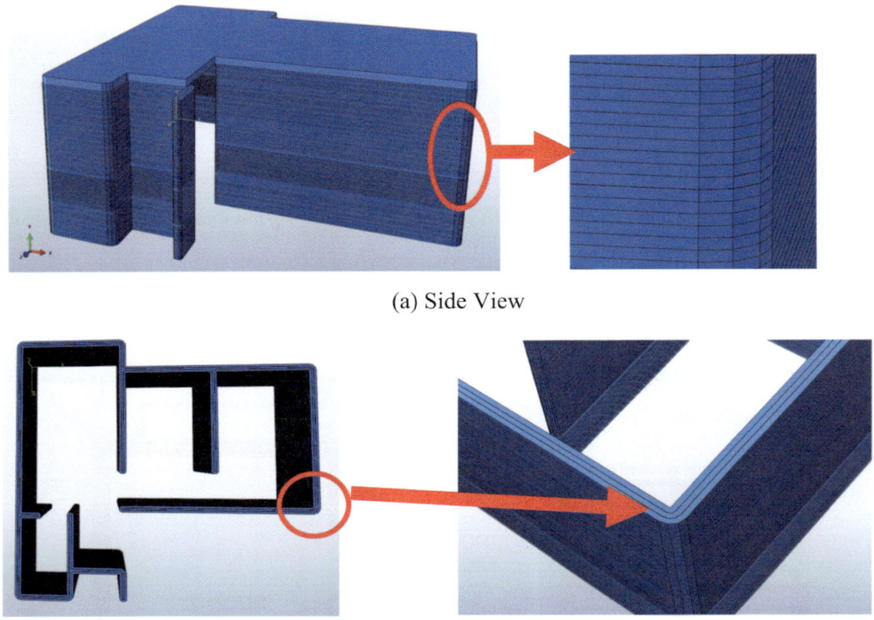

(a) Side View

(b) Top view

Fig. 11.18 Different views of 3DPC building along with enlarged view of extruded layers

mechanical behavior of 3D printed structural elements, hence, further analysis under different loading conditions has not been shown. Consequently, readers can extend such methodologies to characterize the behavior of 3DPC components under other loadings conditions.

11.4 Closure

This chapters delves into the various approaches of numerically predicting the mechanical behavior of 3D printed concrete element ranging from cube to beam and building of 3 m height. The model predicted observations are compared with existing studies or experimental works that highlighted the importance of the described modelling methodologies in the context of 3DPC in terms of reduction of time and cost associated with experimental activities. However, looking at the various complexities and rheological requirements, more advanced and robust modelling methodologies are required to be developed for 3DPC.

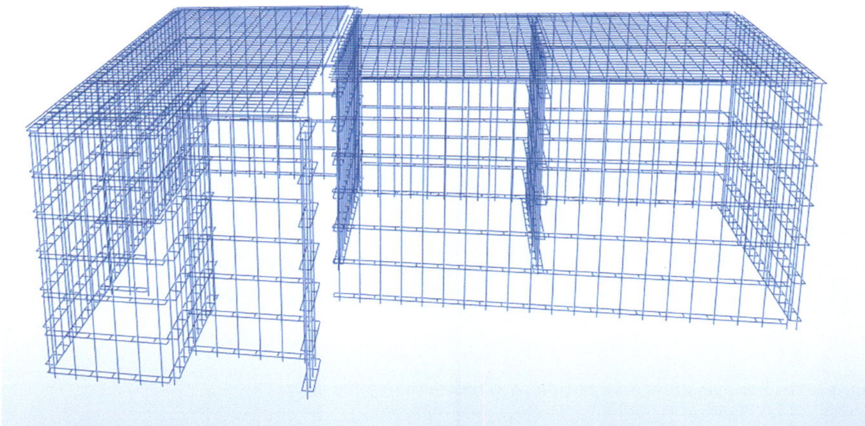

Fig. 11.19 View of overall reinforcement employed in 3DPC building

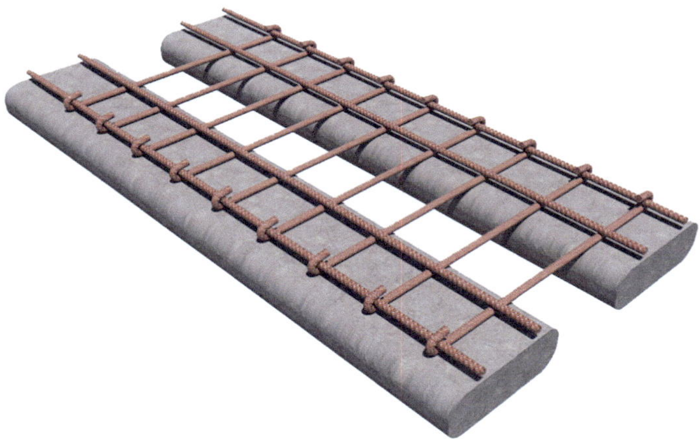

Fig. 11.20 Details of different reinforcements with respect to one extruded layer

Fig. 11.21 Vertical stress distribution of 3DPC building due to gravity

References

1. Kaikai X, Yadong G, Qiang Z (2023) Comparison of traditional processing and additive manufacturing technologies in various performance aspects: a review. Arch Civ Mech Eng 23:1–28
2. Buswell RA, De Silva WRL, Jones SZ, Dirrenberger J (2018) 3D printing using concrete extrusion: a roadmap for research. Cem Concr Res 112:37–49
3. Paul SC, Van Zijl GPAG, Tan MJ, Gibson I (2018) A review of 3D concrete printing systems and materials properties: current status and future research prospects. Rapid Prototyp J 24:784–798
4. El-Sayegh S, Romdhane L, Manjikian S (2020) A critical review of 3D printing in construction: benefits, challenges, and risks. Arch Civ Mech Eng 20:1–25
5. Nguyen-Van V, Nguyen-Xuan H, Panda B, Tran P (2022) 3D concrete printing modelling of thin-walled structures. In: Structures, Elsevier, pp 496–511
6. Duarte G, Duarte JP, Memari A, Brown N, Gevaudan JP (2023) Towards a model for structural performance in concrete printing based on buildability and toolpath design. J Build Eng 69:106325
7. Xu Z, Zhang D, Li H, Sun X (2023) Effects of the distribution of solid particles on the rheological properties and buildability of 3DPM fresh pastes with different FA/GGBFS content. Arch Civ Mech Eng 23:82
8. Wolfs RJM, Bos FP, Salet TAM (2019) Hardened properties of 3D printed concrete: the influence of process parameters on interlayer adhesion. Cem Concr Res 119:132–140
9. Liu H, Liu C, Wu Y, Bai G, He C, Zhang R, Wang Y (2022) Hardened properties of 3D printed concrete with recycled coarse aggregate. Cem Concr Res 159:106868
10. Sun B, Li P, Wang D, Ye J, Liu G, Zhao W (2023) Evaluation of mechanical properties and anisotropy of 3D printed concrete at different temperatures. Structures 51:391–401. https://doi.org/10.1016/j.istruc.2023.03.045
11. Zhu L, Zhang M, Zhang Y, Yao J, Yang G, Guan X, Zhao Y (2023) Research progress on shrinkage properties of extruded 3D printed cement-based materials. J Build Eng 107394
12. Li Q, Gao X, Su A, Lu X (2023) Interlayer adhesion strength of 3D-printed cement-based materials exposed to varying curing conditions. J Build Eng 74:106825
13. Wang Y, Qiu L, Hu Y, Chen S, Liu Y (2023) Influential factors on mechanical properties and microscopic characteristics of underwater 3D printing concrete. J Build Eng 107571
14. Roussel N, Cussigh F (2008) Distinct-layer casting of SCC: the mechanical consequences of thixotropy. Cem Concr Res 38:624–632

15. Le TT, Austin SA, Lim S, Buswell RA, Law R, Gibb AGF, Thorpe T (2012) Hardened properties of high-performance printing concrete. Cem Concr Res 42:558–566
16. Panda B, Paul SC, Mohamed NAN, Tay YWD, Tan MJ (2018) Measurement of tensile bond strength of 3D printed geopolymer mortar. Measurement 113:108–116
17. Tay YWD, Ting GHA, Qian Y, Panda B, He L, Tan MJ (2019) Time gap effect on bond strength of 3D-printed concrete. Virtual Phys Prototyp 14:104–113
18. Sanjayan JG, Nematollahi B, Xia M, Marchment T (2018) Effect of surface moisture on inter-layer strength of 3D printed concrete. Constr Build Mater 172:468–475
19. Keita E, Bessaies-Bey H, Zuo W, Belin P, Roussel N (2019) Weak bond strength between successive layers in extrusion-based additive manufacturing: measurement and physical origin. Cem Concr Res 123:105787
20. van den Heever M, du Plessis A, Kruger J, van Zijl G (2022) Evaluating the effects of porosity on the mechanical properties of extrusion-based 3D printed concrete, Cem Concr Res 153. https://doi.org/10.1016/j.cemconres.2021.106695
21. Singh A, Liu Q, Xiao J, Lyu Q (2022) Mechanical and macrostructural properties of 3D printed concrete dosed with steel fibers under different loading direction, Constr Build Mater 323. https://doi.org/10.1016/j.conbuildmat.2022.126616
22. Paul SC, Tay YWD, Panda B, Tan MJ (2018) Fresh and hardened properties of 3D printable cementitious materials for building and construction. Arch Civ Mech Eng 18:311–319
23. Zareiyan B, Khoshnevis B (2017) Effects of interlocking on interlayer adhesion and strength of structures in 3D printing of concrete. Autom Constr 83:212–221
24. Ma G, Li Z, Wang L, Wang F, Sanjayan J (2019) Mechanical anisotropy of aligned fiber reinforced composite for extrusion-based 3D printing. Constr Build Mater 202:770–783
25. Chen W, Pan J, Zhu B, Ma X, Zhang Y, Chen Y, Li X, Meng L, Cai J (2023) Improving mechanical properties of 3D printable 'one-part'geopolymer concrete with steel fiber reinforcement. J Build Eng 107077
26. Yu Q, Zhu B, Li X, Meng L, Cai J, Zhang Y, Pan J (2023) Investigation of the rheological and mechanical properties of 3D printed eco-friendly concrete with steel slag. J Build Eng 72:106621
27. Wang L, Jiang H, Li Z, Ma G (2020) Mechanical behaviors of 3D printed lightweight concrete structure with hollow section. Arch Civ Mech Eng 20:1–17
28. van den Heever M, Bester F, Pourbehi M, Kruger J, Cho S, van Zijl G (2020) Characterizing the fissility of 3D concrete printed elements via the cohesive zone method. In: Second RILEM international conference on concrete and digital fabrication: digital concrete 2020 2, Springer, pp 489–499
29. Xiao J, Liu H, Ding T (2021) Finite element analysis on the anisotropic behavior of 3D printed concrete under compression and flexure. Addit Manuf 39:101712
30. Pal B, Chourasia A, Kapoor A (2024) Intricacies of various printing parameters on mechanical behaviour of additively constructed concrete. Arch Civ Mech Eng 24:41
31. Xiao J, Li W, Corr DJ, Shah SP (2013) Effects of interfacial transition zones on the stress-strain behavior of modeled recycled aggregate concrete. Cem Concr Res 52:82–99. https://doi.org/10.1016/j.cemconres.2013.05.004
32. Huang Y, Yang Z, Ren W, Liu G, Zhang C (2015) 3D meso-scale fracture modelling and validation of concrete based on in-situ X-ray Computed Tomography images using damage plasticity model. Int J Solids Struct 67:340–352
33. Pal B, Ramaswamy A (2020) Meso-scale study of notched and un-notched plain concrete beam under three-point bending. In: IABSE congress, Christchurch 2020: resilient technologies for sustainable infrastructure-proceedings, international association for bridge and structural engineering (IABSE), pp 890–897
34. Jog CS (2015) Continuum mechanics: volume 1: foundations and applications of mechanics, Cambridge University Press
35. Lubliner J, Oliver J, Oller S, Oñate E (1989) A plastic-damage model for concrete. Int J Solids Struct 25:299–326

36. Lee J, Fenves GL (1998) Plastic-damage model for cyclic loading of concrete structures. J Eng Mech 124:892–900
37. Hafezolghorani M, Hejazi F, Vaghei R, Bin Jaafar MS, Karimzade K (2017) Simplified damage plasticity model for concrete. Struct Eng Int 27: 68–78
38. Wang X, Yang Z, Jivkov AP (2015) Monte Carlo simulations of mesoscale fracture of concrete with random aggregates and pores: a size effect study. Constr Build Mater 80:262–272. https://doi.org/10.1016/j.conbuildmat.2015.02.002
39. IS:516–2013 (2013) Methods of tests for strength of concrete. Bureau of Indian Standards, New Delhi. (n.d.)
40. I.S. BIS (1999) 5816–1999 Specification for splitting tensile strength of concrete—method of test, Bureau of Indian Standards, New Delhi, India
41. Luo S, Lin Q, Xu W, Wang D (2023) Effects of interval time and interfacial agents on the mechanical characteristics of ultra-high toughnessnn cementitious composites under 3D-printed technology. Constr Build Mater 374:130936
42. Pasupathy K, Ramakrishnan S, Sanjayan J (2021) Formulating eco-friendly geopolymer foam concrete by alkali-activation of ground brick waste. J Clean Prod 325:129180

Chapter 12
Applications of Artificial Intelligence and Machine Learning in 3D Concrete Printing

12.1 Introduction

The construction sector is undergoing a key revolution by the adoption of leading-edge digital tools, like three-dimensional concrete printing, Artificial Intelligence (AI), and Machine Learning (ML). Combiningly, these innovations are opening newer potentials in the construction sector that allows simulating the fresh and hardened state behavior through the optimization of the process parameters and thus improves quality control, along with offering superior flexibility in design. AI replicates human cognitive abilities, e.g., reasoning, learning, and self-correction, through the utilization of various computational schemes. Machine Learning (ML), one of the key branches of AI, allows the systems to learn from the input data and gradually the predictive accuracy of the model improves without the need of explicit coding. Thus, when ML integrated with 3DPC, the combined tool can redefine the path through which various construction materials and processes can gainfully be utilized.

The present chapter focuses on the deployment of AI and ML in the field of 3D concrete printing, in order to predict performances such as fresh-state characteristics, mechanical, and durability properties. In general, prediction of such properties through experimental process can be costly and requires extensive time. In fact, by choosing suitable ML algorithms and adjusting its parameters to optimum, such properties can be projected with greater accuracy that could significantly reduce the cost and time associated with the experimental works.

Utilization of an ML algorithm involves sequence of stages, starting from data acquisition, to preprocessing, formulation of the model, evaluating the performance accuracy, and predicting the output. In this regards, various ML algorithms, such as linear regression, fine decision tree, support vector machine, Gaussian process regression, etc., were described and subsequently employed to simulate the desired outcomes. In addition, the chapter also presents the execution process of these ML algorithms in MATLAB, a widely utilized programming language. The collected

A. Chourasia et al., *Advances and Challenges in 3D Printed Concrete (3DPC)*, Materials Horizons: From Nature to Nanomaterials, https://doi.org/10.1007/978-981-95-8557-1_12

input data was divided into two subsets: one part for training and another for validation. Afterward, the accuracy of the models in predicting the output was assessed through RMSE and R^2 values. The predicted output results were validated with the actual values obtained from experiments to authenticate reliability and predictive capacity of the models. Analysis showed that optimized ML models can significantly reduce the requirement for extensive experiments and provide the precise predictions quickly, and thus make them extremely valued for designing material and quality control.

The chapter concludes by examining the practical applications of various predictive models in real-world constructions and suggests future research pathways to enhance and expand their usefulness. In essence, this chapter advances concrete technology through the integration of ML into 3DPC, supporting the creation of smarter, more efficient, and sustainable construction practices.

12.2 Kinds of ML Algorithms

Machine Learning (ML), a core subcategory of Artificial Intelligence (AI), encompasses a wide range of computational algorithms aimed at recognizing patterns within data and making independent decisions or predictions. In contrast to conventional rule-based programming, ML models develop predictive capabilities by learning from real-world data, removing the necessity for explicitly programmed instructions for each task. This data-driven approach is especially advantageous in complex and evolving fields, such as 3D-printed concrete (3DPC), where the properties and behavior of construction materials and systems often challenge traditional deterministic modeling methods.

In the context of 3DPC and advanced construction automation, ML enables process optimization, real-time defect detection, adaptive control of printing parameters (e.g., extrusion speed and nozzle trajectory), and prediction of material properties based on composition and environmental factors. ML is typically organized into four categories: supervised, unsupervised, semi-supervised, and reinforcement learning [1]. This section discusses each of these in detail, highlighting their underlying concepts and how they differ with respect to training data and learning goals.

12.2.1 Supervised Learning (SL)

Among various ML techniques, supervised learning is the most widely applied, as it depends on labeled data for training. In this framework, every input data point is clubbed with a conforming output value. The main objective of supervised learning is to construct a function that maps inputs to outputs, thereby uncovering the relationship between features and the target variable. As a result, the model can use input features to generate predictions for the "desired" outputs, as illustrated in Fig. 12.1.

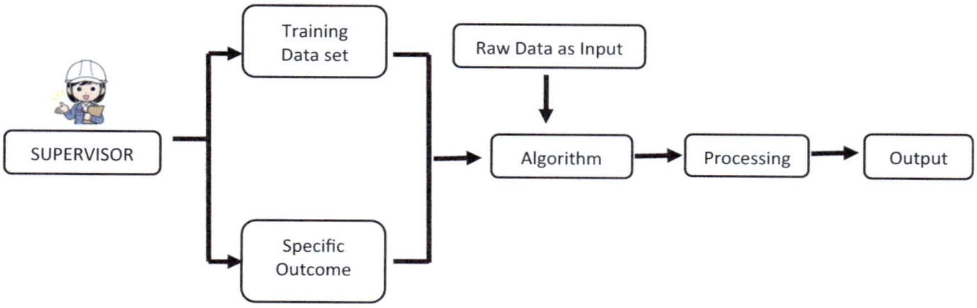

Fig. 12.1 Supervised ML

Various types of supervised learning algorithms are.

- **Logistic Regression**—Widely applied in binary classification scenarios.
- **Support Vector Machine**—Accomplishes well in high-dimensional feature spaces and excels in margin-based classification tasks.
- **k-Nearest Neighbors**—A simple instance-based approach that relies on distance measures to perform classification or regression.
- **Decision Trees**—Models that organize data hierarchically by repeatedly dividing it into subsets according to feature attributes.
- **Random Forests**—A collection of decision trees combined to improve predictive accuracy while decreasing the overfitting risk.
- **Neural Networks**—Can handle complex, nonlinear patterns and serving as the foundation of deep learning methods.

Applications of supervised learning:

- **Email Spam Detection**—Identifying whether incoming emails are spam or genuine by analyzing their content and metadata.
- **Disease Diagnosis**—Predicting the likelihood or presence of a disease using patient data.
- **Sentiment Analysis**—Analyzing textual data, including customer feedback or social media content, to identify sentiment polarity.
- **In 3DPC (3D-Printed Concrete)**—Predicting the compressive strength of printed structures using parameters such as material mix design, nozzle configuration, and curing conditions; real-time classification of print quality based on sensor-derived data.

12.2.2 Unsupervised Learning (USL)

In comparison to SL, USL algorithms operate without requiring data to be labeled by a human expert. These methods automatically extract features from unlabeled input data and classify it based on patterns discovered through self-generated rules. As a

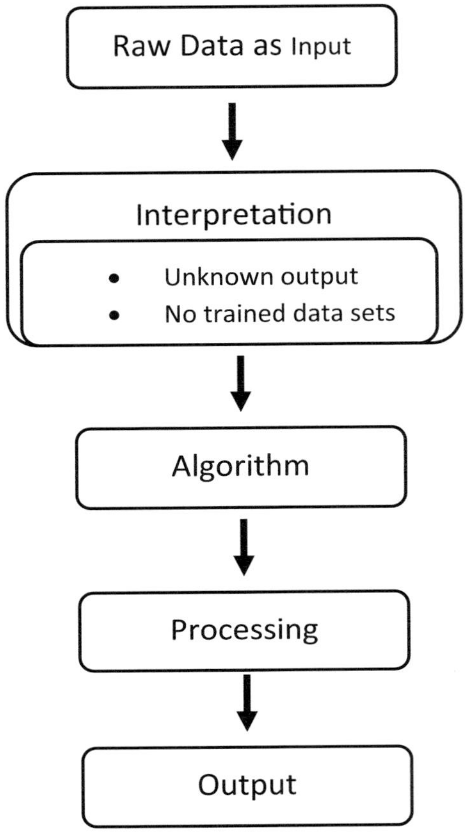

Fig. 12.2 Unsupervised machine learning

result, they are often applied to uncover hidden or previously unknown relationships within datasets, as illustrated in Fig. 12.2 [1].

USL is especially advantageous when labeling data is costly, subjective, or impractical—challenges that are frequently encountered in large-scale construction projects and materials research.

Examples of unsupervised learning algorithms:

- **K-Means Clustering**—Partitions data into *k* clusters by reducing the variance within each group.
- **Hierarchical Clustering**—Forms nested clusters based on similarity measures, often visualized using dendrograms.
- **Principal Component Analysis**—Decreases the dimensionality of data while preserving the greatest amount of variance.

Unsupervised learning in practice:

- **Customer Segmentation**—Categorizing consumers according to buying patterns.
- **Anomaly Detection**—Recognizing abnormal patterns or deviations in datasets such as sensor outputs or financial transactions.
- **Topic Modeling**—Identifying hidden themes in large collections of text.
- **In 3DPC (3D-Printed Concrete)**—Clustering material batches with similar characteristics, detecting anomalies in print quality records, and reducing high-dimensional sensor data to critical indicators for informed decision-making.

12.2.3 Semi-supervised Learning

Semi-supervised learning incorporates characteristics of both supervised and unsupervised paradigms. It is especially beneficial for large datasets where labeling every data point is costly and impractical. In this approach, the input to the algorithms consists of a combination of labeled and unlabeled data. By utilizing a small labeled subset together with the larger unlabeled data, these models usually deliver improved performance compared to purely unsupervised methods, while remaining cost-effective and easier to train than fully supervised approaches.

This paradigm is increasingly relevant in construction and civil engineering, where obtaining complete labels—such as performing mechanical tests on every printed specimen—can be labor-intensive and costly.

12.3 Examples of Semi-supervised Learning Algorithms

- **Semi-Supervised SVM**—An extension of traditional SVMs that utilizes unlabeled data by maximizing the margin not only between labeled samples but also across the unlabeled data space.
- **Graph-Based Methods**—Models data as a graph in which nodes (representing data points) are connected based on similarity, with labels propagated throughout the network structure.

12.4 Applications of Semi-supervised Learning

- **Speech Recognition**—Training models with a limited amount of transcribed audio data alongside a large collection of unlabeled speech recordings.
- **Text Classification**—Categorizing documents or articles using only a limited number of labeled samples.

- **In 3DPC**—Building predictive models for structural failure or deformation using a small number of experimentally tested samples alongside extensive unlabeled process logs or sensor readings; classifying material behavior based on partially labeled experimental datasets.

12.4.1 Reinforcement Learning

In contrast to SL, which depends on labeled data, reinforcement learning operates in environments where the training data only indicates whether an action is correct or not. In this paradigm, algorithms iteratively acquire "good" performance through continuous interaction with its surroundings. While the learning process shares similarities with supervised learning, it does not depend on large labeled datasets. Instead, the model engages with its environment, getting response of positive rewards or negative consequences type. Such feedback loop reinforces desirable actions, which is the basis for the term *reinforcement learning*, as illustrated in Fig. 12.3.

Reinforcement learning strategies often involve two key concepts: exploration and exploitation. *Exploitation* involves selecting actions known to yield the highest possible reward, while *exploration* focuses on trying new actions to gather more information about the environment. By balancing these two approaches, the model progressively learns which inputs lead to positive outcomes, ultimately achieving optimal solutions.

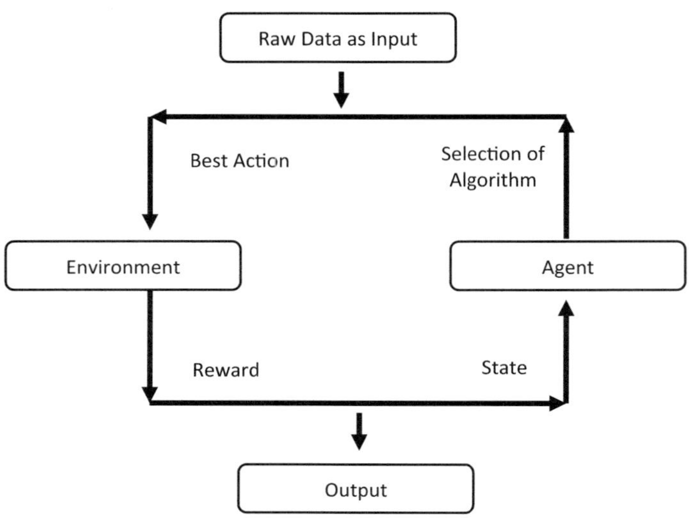

Fig. 12.3 Reinforcement machine learning

Examples of reinforcement learning algorithms:

- **Q-Learning**—A value-created reinforcement learning method that estimates the optimum action plan.
- **Temporal Difference (TD) Learning**—An approach that integrates concepts from Monte Carlo methodology and dynamic training to enable incremental learning.
- **Deep Adversarial Networks**—Uses deep neural architectures to improve decision-making in complex environments.

Grasping the fundamental groups of ML algorithms is essential for deploying effective data-based approaches in 3DPC and other cutting-edge construction technologies. Each type provides distinct benefits that depends on features, e.g., availability of data, project goals, and the desired level of model interpretability. By selecting an ML approach that aligns with the unique characteristics of the construction process, researchers and industry professionals can improve decision-making, streamline quality control, and drive innovation in digital fabrication.

12.5 Methodology

The section presents the procedure for applying ML algorithms to predict the behavior of 3DPC, as illustrated in Fig. 12.4. Ensuring the correctness and completeness of the dataset begins with essential steps such as data collection, selecting features, and cleaning of data. Following this, normalization of the data is performed to eliminate dimensional inconsistencies and enhance the robustness of the models.

Next, an appropriate model—such as Linear Regression (LR), SVR, DT, Random Forest (RF), or Artificial Neural Network (ANN)—is selected depending on the specific problem requirements. Afterward, the models are then trained using the training set, while the test set is reserved for performance evaluation.

During training, the K-fold method of cross-validation is applied to fine-tune hyperparameters and improve efficiency. Model performance is assessed using different evaluation metrics, including Root Mean Squared Error (RMSE), Mean Squared Error (MSE), Mean Absolute Error (MAE), and the Coefficient of Determination (R^2). Through this structured process, the most effective model is identified for accurate prediction of 3D-printed concrete performance.

12.5.1 Data Preprocessing

To satisfy accurateness and robust features of the ML algorithms, the current work meticulously preprocessed on a set of data before conducting a detailed analysis [2]. These preprocessing measures are indispensable, as they improve data reliability and subsequently improve the training efficacy of the models.

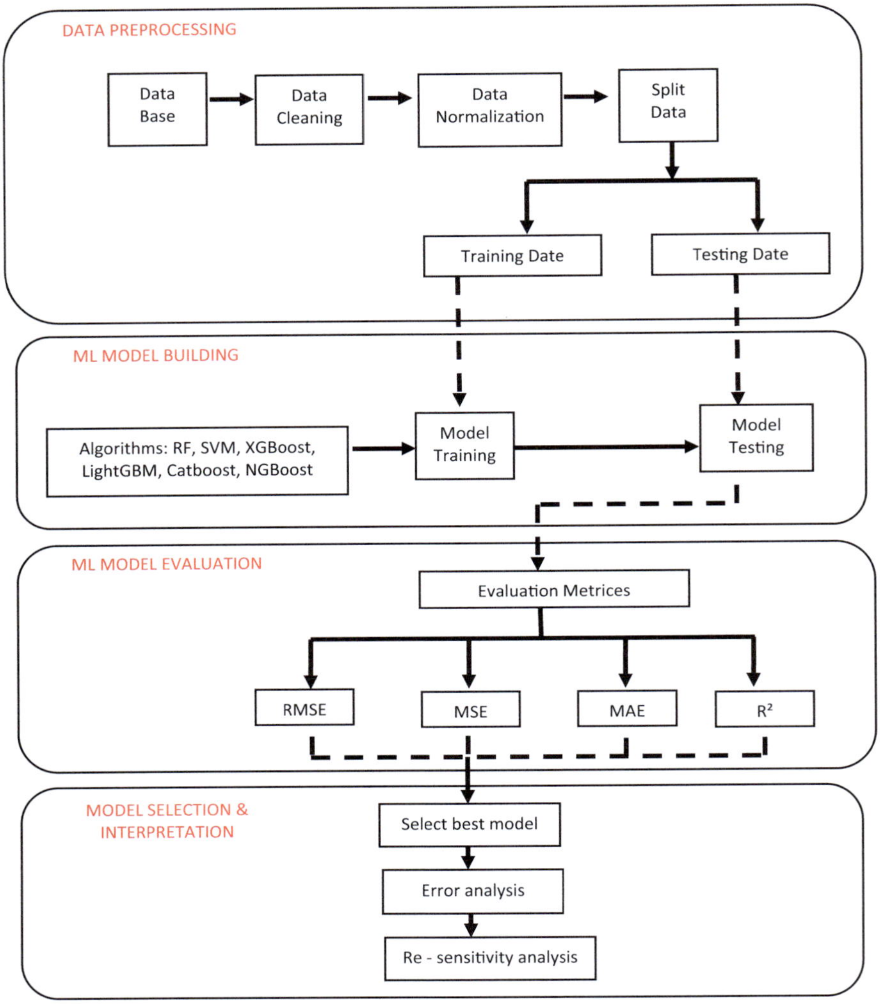

Fig. 12.4 Overview of the methodology of the application of ML algorithms in 3DPC

Further, normalization was performed to remove analytical bias caused by differences in the dimension between different characteristics. All feature values were scaled to the range of 0 to 1, bringing them to the similar value range and confirming that each feature carried identical weight in the models. This process preserves impartiality and objectivity in model training. Normalization formula used was [2]

$$\text{Normalized value} = \frac{\text{x} - \text{min value}}{\text{max value} - \text{min value} + 0.001} \tag{12.1}$$

where x stands for the original data, whereas *min_value* and *max_value* stand for minimum and maximum values in a given data. A constant of 0.001 was added to the denominator to avoid dividing it by 0, for the case when *max_value* matches with *min_value*, thus circumventing computational errors.

Feature selection combined advanced correlation analysis within the domain. The study investigated key material properties—such as water–binder ratio, content of aggregate, and admixture type—and their influence on compressive and flexural strengths. Repeated trials confirmed that cement content influences both fluidity and setting time, prompting an examination of the effects of different cement types on the mechanical characteristics of 3DPC.

The uniformity, size gradation, and impurity content of aggregates were found to significantly affect strength and stability. The role of admixtures and additives in improving workability and durability was also explored, with particular attention to their synergistic effects in enhancing overall performance. Fiber addition was analyzed for its contribution to resistance due to crack.

Water–cement ratio emerged as a critical factor influencing both workability and strength, while printing parameters, e.g., printing speed, interlayer time gap, were recognized as impacting the final mechanical properties. However, due to limited available data on printing process parameters in existing literature, which was insufficient for machine learning needs, the study prioritized raw material properties as primary input features, drawing from both literature and experimental data. To improve scientific rigor, advanced correlation analysis techniques were applied to identify features with the strongest influence on compressive and flexural strength.

12.5.2 Dataset Partitioning

To improve the generality of the models and achieve higher accuracy in the predicted values, dataset was divided into a training and a testing dataset in an 80%–20% ratio, respectively, following the approach described in [2].

Multiple partition ratios were evaluated [2], including 90–10%, 85–15%, 75–25%, and 70–30%. Among these, the 80–20% split consistently produced the highest correlation and predictive performance [2]. The decision to allocate 80% of the data for training was based on its ability to provide sufficient information for the models, yielding higher R^2 values and enabling better learning and generalization of underlying patterns. Although a 70–30% split is commonly adopted, the nature of this study—where high-quality data in the area of 3D-printed concrete is limited—favored maximizing the training data proportion for efficiently using the available resources.

In this framework, training dataset was used for model training and parameter fine-tuning, allowing the models to find the core features of the dataset. Testing dataset was reserved completely for evaluating predictive performance on unseen data, ensuring that the models remain reliable for practical applications. Such segmentation strategy emphasizes not only strong performance on the trained dataset but, more importantly, exceptional predictive performance when exposed to new, previously unseen data.

12.5.3 Model Training

Selecting the right machine learning model is critical when predicting the mechanical properties of 3DPC. In current study, five representative models were examined in detail, **LR, SVR, DT, RF,** and **ANN** [3]. Each model offers unique capabilities for extracting deep insights from data across multiple dimensions, enabling more accurate and comprehensive forecasts of 3D-printed concrete properties.

Multiple Linear Regression (MLR) adopts a straightforward linear association between input and output variables. While computationally efficient and easy to interpret, it falls short in modeling complex nonlinear interactions between different variables. On other side, nonlinear approaches, e.g., DT, RF, and ANN, offer advantages in treating such complexity.

DT captures nonlinear dependencies by recursively splitting the dataset into smaller, more homogeneous groups. RFs, as groups of DTs, improve robustness and predictive accuracy by averaging the outputs of multiple trees. ANNs, equipped with several hidden layers along with nonlinear activation type functions, excel at modeling intricate and highly nonlinear patterns.

Support Vector Regression can operate as either a linear- or a nonlinear-type model, conditional on the kernel applied. Through the transformation of data into a higher-dimensional space, SVR identifies an optimally separated hyperplane, making it particularly effective for complex nonlinear problems.

While linear models are valued for their simplicity, interpretability, and speed, they are less suited to materials like 3D-printed concrete, which exhibit highly nonlinear characteristics. Nonlinear models are better equipped to capture and represent these complex patterns.

Comparative analysis in this research showed that nonlinear methodologies consistently delivered higher predictive precision and greater flexibility than linear models when estimating the mechanical behavior of 3DPC.

12.5.3.1 Linear Regression

Linear Regression (LR) [4] stands as a fundamental and intuitive modeling approaches to predict the response. It operates by estimating the target variable through the determination of a fitted straight line, as illustrated in Eq. (12.2) and Fig. 12.5.

In the work, the amount of constituents and other experiment-related parameters were used as independent entities to predict the strength of 3DPC under compression

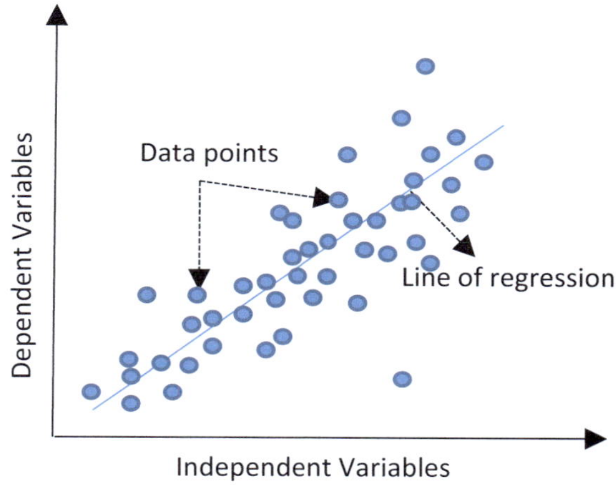

Fig. 12.5 General schematic diagram of machine learning models: linear regression

and flexure loading. Selection of LR as a base model was motivated by its simple nature, interpretability, and by the capability to explicitly express the linear variability among input and output features.

Although the mechanical performance of 3DPC is often affected by a range of nonlinear entities, a linear model provides a valuable start point, for initial knowledge of the primary tendencies in the dataset. The predictive function for LR is mathematically expressed as shown in Eq. (12.2) [5]:

$$\hat{y} = ax + b \tag{12.2}$$

Here, x denotes independent entities, ax represents the weight coefficient, and $\hat{y}$ stands for predicted value of the dependent entity. In the most general scenario—accounting for an intercept term and multiple regression coefficients, model computes the coefficients a and b directly through the **Generalized Least Squares (GLS)** method, which optimizes the fit by the minimization of sum squared errors among the actual and estimated outputs.

$$a = \left(x^T x\right) + x^T y \tag{12.3}$$

$$b = \overline{y} - a\overline{x} \tag{12.4}$$

In this context, x refers to independent entity, $\overline{x}$ represents the mean of x, and $\overline{y}$ denotes the mean of the actual dependent entity, while T indicates the transpose operator. Model parameters were determined employing **Least Squares Method**, which aims to find best-fitted line (hyperplane in the case of multiple dimensions) by minimizing residuals of evaluated and actual values. The objective was to determine the parameter set that produces the smallest possible sum of squared errors. The preprocessed dataset was fed into the algorithm, and the corresponding slope and intercept parameters were iteratively optimized using numerical algorithms to reduce prediction errors to a minimum.

12.5.3.2 Support Vector Machines (SVM)

Support Vector Machines (SVM) are SL algorithms intended to classify data by determining the optimum hyperplane which best splits various classes of algorithms. In the context of 3D-printed concrete, SVMs have been applied to assess the printability of material mixtures and to classify quality control results.

Support Vector Machine Regression (SVR) extends this concept to regression tasks by using a **kernel** to project the input observations into a high-dimensional space. When a **linear kernel** is applied, SVR operates similar to linear regression. However, when a **nonlinear kernel**—such as a polynomial or Gaussian (RBF) kernel—is used, the predictors are transformed into a larger set of nonlinear characteristics. This transformation allows the model to capture complex relations by utilizing similarities among observations with comparable predictor values.

SVM is inherently a binary type method. For linearly separated problems, it determines the linear decision boundary with the maximum margin between classes by resolving a **convex and quadratic problem** [6]. Thus, the resulting separation is illustrated as

$$\omega^{T}x + b > 0, y = +1(\text{positive}) \tag{12.5}$$

$$x + b < 0, y = -1(\text{negative}) \tag{12.6}$$

Here, x represents independent entity, while ω and b denote the model parameters. For the cases where the problem cannot be not linearly separated, the model applies **kernel approach** to project the sample data points to a higher-dimensional space, thereby transforming a nonlinear problem to a linear in that space.

SVR operates using the same underlying principles as SVM but is tailored for regression tasks. The prediction function for SVR can be expressed as follows:

$$\hat{y} = \omega^{T}x + b \tag{12.7}$$

Here, x denotes the independent variable and $\hat{y}$ represents predicted value of the dependent variable. Unlike Linear Regression (LR), **Support Vector Regression (SVR)** incorporates an ε-insensitive tube to tolerate small deviations in the model's predictions. The final prediction model is obtained by simultaneously minimizing the total loss and maximizing the margin, resulting in a more reliable estimation. However, SVM-based methods can be computationally rigorous to train, particularly for large datasets or when employing nonlinear kernels.

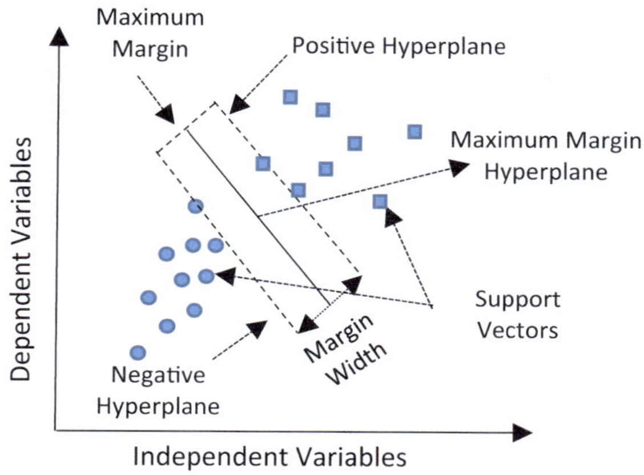

Fig. 12.6 General schematic diagram of machine learning models: SVM

SVR [7] is a part of SVM [8] for regression applications. Its goal is to determine an optimum hyperplane which reduces the distance of data points in the hyperplane, while applying an ε-insensitive function that allows a predefined margin of error. This approach ensures that the difference among estimated and actual values is kept as small as possible, making SVR highly effective for applications such as predicting the strength of concrete (Fig. 12.6).

12.5.3.3 Decision Trees

DT [9] is a predictive modeling approach that categorizes data to a tree-type structure through recursive partitioning. At each of the internal node, a decision rule defined by a specific attribute is employed; the branches denote the outcome of a decision. Whereas, each of the leaf node contains either a labeled class (for classification) or a numerical value (for regression). A schematic representation is shown in Fig. 12.7.

In regression trees, observations are passed through a sequence of binary decision nodes, referred to as nodes. The path from the root to a terminal point is known as a branch, and the leaf nodes at the ends of branches assign fixed prediction values to all observations that arrive at them.

Decision Trees can be constructed with varying levels of granularity. Fine Trees involve numerous splits, allowing them to capture intricate patterns within the dataset, but they also carry a higher risk of overfitting, which may reduce their effectiveness on new data. In contrast, Coarse Trees use fewer splits, making them simpler and more generalizable, though they may overlook subtle yet significant relationships. The complexity of a Decision Tree can be controlled by adjusting parameters such as the minimum leaf size—where a smaller value increases the number of leaves and overall complexity—or the maximum number of splits, which extends branch length

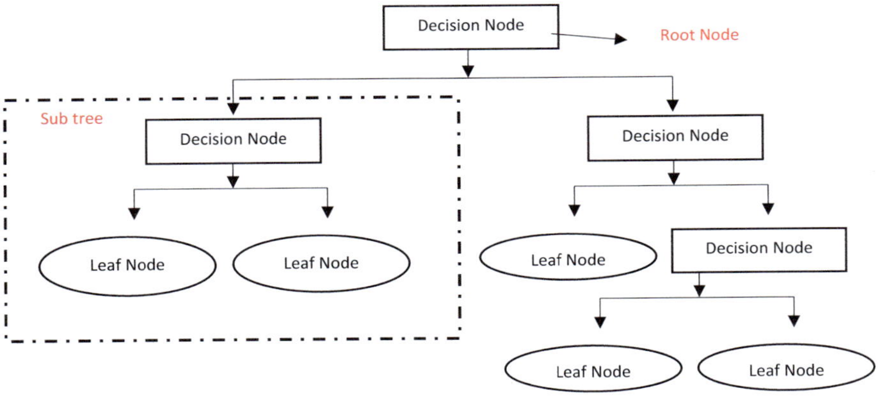

Fig. 12.7 General schematic diagram of machine learning models: Decision Tree

and enables the model to capture more detailed patterns. Unlike algorithms such as Support Vector Regression (SVR), Decision Trees do not require standardization of predictor variables.

Among the various DT algorithms, the Classification and Regression Tree (CART) method is widely used [10]. CART employs a least squares regression-type tree for regression tasks and follows a four-step procedure:

1. For a dataset with x that comprises N entities, each having M features, select the optimum variable j and split point s.
2. Iterate over $j \in M$, and for each fixed j, iterate through possible s values to estimate the pair (j, s) which minimizes the target cost function:

$$\min_{j,s}\left[\min_{C_1}\sum_{x_i \epsilon R_1(j,s)}\left(y_i-\widehat{C}_1\right)^2+\min_{C_2}\sum_{x_i \epsilon R_2(j,s)}\left(y_i-\widehat{C}_2\right)^2\right] \tag{12.8}$$

The method of division is

$$R_1=\{y_i|x_{i,j}\leq s\}, R_2=\{y_i|x_{i,j}>s\} \tag{12.9}$$

$\hat{c}_1$ and $\hat{c}_2$ are

$$\hat{c}_1=\text{aver}(y_i|x_i\epsilon R_1(j,s)), \hat{c}_2=\text{aver}(y_i|x_i\epsilon R_2(j,s)) \tag{12.10}$$

1. Split the dataset using the selected optimal pair (j, s) according to Eq. (12.9) and compute the corresponding predicted values $\hat{c}_1$ and $\hat{c}_2$ for the two resulting subsets.
2. Recursively repeat Steps 1 and 2 on each subset generated in Step 2 till a specified stopping condition (exit condition) is satisfied, e.g., reaching a minimum leaf size or maximum tree depth.

3. Once the process is complete, the dataset will be divided into *K* distinct regions, each associated with a predicted value, thereby forming the final regression tree.

$$f(x) = \sum_{i=1}^{k} C_i I(x \epsilon R_i) \tag{12.11}$$

Key principles of Decision Trees involve gain of information, gain ratio, along with the Gini index, which are applied to determine the most suitable feature for splitting the dataset. In cases where datasets contain numerous features, these techniques have proven highly effective for predicting concrete strength. During implementation, cross-validation is employed to fine-tune model parameters, reducing the likelihood of overfitting issues and confirming that the method retains strong predictive capabilities when applied to new set of data. As Decision Trees can handle both classification and regression tasks, it holds significant possibility in the domain of predicting the performance of material.

Decision Tree regression offers high interpretability, is easy to visualize, and effectively models nonlinear relationships between variables. However, it can be sensitive to minor variations in data and may overfit if the tree is excessively deep or not pruned. To address these challenges, ensemble approaches such as Random Forest and Gradient Boosting are frequently used, as they combine multiple trees to deliver improved predictive accuracy and greater robustness.

12.5.3.4 Gaussian Process Regression (GPR)

GPR predicts response values by evaluating their similarity to the training data. Unlike traditional models that assume a fixed functional form for the distribution of observations, GPR makes no such assumption. Instead, it is a probabilistic model, meaning that predictions for new inputs are expressed not as a single deterministic value but as a probability distribution, specifically, a Gaussian distribution. Thus, it allows the model to compute prediction intervals, providing a measure of uncertainty alongside the prediction.

GPR can be implemented using different kernel functions, with rational quadratic, squared exponential, and exponential kernels being among the most common. In the case of the rational quadratic GPR, the kernel can be taken as a mix of Gaussian kernels with variable length scales, formulated as follows:

$$k_{RQ}\left(x, x^{'}\right) = \sigma^2\left(1 + \frac{\|x - x^{'}\|^2}{2\alpha l^2}\right)^{-\alpha} \tag{12.12}$$

where σ^2 denotes signal variance, ℓ represents the length scale, and α governs the relative weightage of large- and small-scale variations. When $\alpha \rightarrow \infty$, the RQ kernel approaches the squared exponential kernel. Effective for data that varies at multiple scales. It is useful for modeling data with varying degrees of smoothness.

On the other hand, squared exponential GPR, also called Radial Basis Function (RBF) kernel, is commonly utilized kernel:

$$k\left(x, x'\right) = \exp[-2l^2\left(x - x'\right)^2] \quad (12.13)$$

It assumes that the function being modeled is smooth. Suitable for smooth and continuous functions. Similarly, exponential GPR, also known as the Laplacian kernel. It models functions that are less smooth than those modeled by the squared exponential kernel. Suitable for functions with sharp changes or discontinuities.

$$k\left(x, x'\right) = \exp(-1\left|x - x'\right|) \quad (12.14)$$

12.5.3.5 Random Forest

Random Forest (RF) [11] categorized as a learning method that aggregates the predictions of several decision trees to enhance predictive accuracy. Each tree is constructed from a bootstrap sample of the trained data, and at every node, the split is determined using only a randomly selected subset of features [12].

For regression, the final estimate is obtained by averaging the outputs of all trees, while for classification, it is determined by the majority vote (mode) among the trees. A schematic depiction of this process is shown in Fig. 12.7.

Key principles underlying RF comprise of bootstrap sampling, random feature, and ensemble aggregation. This approach efficiently manages datasets with a large number of characteristics, excels in capturing nonlinear relations, and inherently performs feature selection. Furthermore, Random Forest is highly robust to outliers and noisy data, offering stable and accurate predictions. These qualities make it particularly well suited for complex predictive works, e.g., estimating the mechanical performance of 3DPC.

12.5.3.6 Artificial Neural Network (ANN)

ANNs [13] are adaptive computational systems designed to learn from input data, enabling them to detect patterns, classify information, or generate predictions.

An ANN comprises multiple interconnected layers: an input and output layer along with single or more than one hidden layers. Among different layers, each contains neurons connected through weighted links, which are adjusted during the learning process [14].

The operation of ANNs is driven by two fundamental processes:

- **Forward Propagation**—Input data passes sequentially through the network's layers, undergoing transformations at each neuron until the final output is produced.
- **Backpropagation**—The network weights are updated depending on the difference among the predicted and actual outputs. It involves calculation of error gradients and accordingly the application of optimization algorithms, e.g., gradient descent, to iteratively minimize prediction errors and improve accuracy.

Key concepts within ANN include neurons, activation functions (e.g., Sigmoid, ReLU), and gradient descent optimization. Due to their adaptableness, nonlinear modeling capability, and parallel processing power, ANNs are extensively used for recognizing the pattern, classification, regression, and predictive modeling.

In 3DPC, ANNs serve as a powerful tool for analyzing and forecasting material behavior. By integrating ANNs into predictive frameworks, this study harnesses their innovative learning abilities to effectively solve complex and multivariable interactions impacts the mechanical behavior of 3DPC, thereby improving both the accuracy and reliability of predictions.

12.5.3.7 K-Nearest Neighbors (k-NN)

k-NN algorithm is a straightforward yet powerful non-parametric, instance-based learning methodology applicable to both classification and regression works. In this approach, predictions are derived from the k most similar training instances in the feature space, with similarity typically measured using distance metrics such as Euclidean distance.

For classification, the predicted label is determined by the majority class among the k-nearest neighbors. In regression, the algorithm outputs the average value of these neighbors. One of k-NN's strengths is that it makes no prior assumptions regarding the data distribution, making it highly adjustable to diverse problem types.

However, the model's performance is strongly influenced by features, e.g., choice of k, the selected distance metric, and scaling of features. While k-NN is generally effective for small- to medium-sized datasets with low dimensionality, its efficiency and accuracy can degrade in high-dimensional spaces owing to the curse of dimensionality, and the computational cost increases significantly as the dataset grows.

12.5.3.8 Genetic Algorithms (GA)

Genetic Algorithms (GAs) are a kind of optimization and search method encouraged by the natural selection principles and genetics. Introduced by John Holland in the 1970s, GAs simulate the evolutionary process to generate approximate solutions for complex, nonlinear, multimodal, or analytically challenging problems.

At the heart of a GA lies a population of candidate solutions, commonly represented as strings or "chromosomes"—most often using binary encoding, though real-valued or alternative encodings are also possible. Each chromosome represents to a possible solution, and its quality, or fitness, is determined by a fitness function that evaluates how well it meets the problem's objectives.

The evolutionary process is driven by three genetic operators: selection, crossover (recombination), and mutation.

- Selection operates by choosing individuals from the population according to their fitness, with higher-fitness individuals having a greater probability of being selected for reproduction. Common strategies include roulette wheel selection, tournament selection, and rank-based selection.
- Crossover then combines pairs of selected parents by swapping segments of their chromosomes, generating offspring that get traits from both parents. This step is crucial for exploring new areas of the search space.
- Mutation leads random changes to certain genes in the offspring, preserving genetic diversity and reducing the danger of early convergence to local optima.

The process—evaluation, selection, crossover, and mutation—is repetitive over multiple generations. As evolution progresses, the population ideally moves toward optimal or near-optimal solutions. Stopping criteria can comprise of reaching a maximum generation count, attaining a target fitness level, or observing no significant improvement over successive generations.

Genetic algorithms are particularly well suited for large, complex, and poorly understood search spaces. They have been successfully applied in scheduling, engineering design, machine learning hyperparameter optimization, robotics, bioinformatics, and financial modeling. While GAs are valued for their robustness and flexibility, they can be computationally demanding and require careful parameter tuning—e.g., population size, mutation rate, and crossover probability—to achieve the best performance.

12.5.3.9 Deep Learning (DL)

Deep Learning (DL), a subfield of machine learning, utilizes multi-layered artificial neural networks to capture and model complex patterns and hierarchical data representations. Drawing stimulation from the structure and functioning of the human brain, deep learning systems can learn directly from raw data, automatically extracting hierarchical features. This capability has driven significant advancements across domains, e.g., computer vision, natural processing of language, recognition of speech, robotics, and many others.

At the core, DL relies on deep neural networks (DNNs), architectures composed of multiple layers of interrelated nodes. Each of the nodes processes incoming signals by employing a weighted sum afterward an activation function, and then transmits the result to subsequent layers. The depth of these networks, which indicates the number of layers, allows us to learn abstract features progressively. For instance, in the case of an image classification task, early layers generally sense textures, while deeper layers recognize shapes, objects, or even faces.

Among deep learning architectures, CNNs stand out for image and video analysis due to their capability to detect spatial hierarchies in data. Recurrent Neural Networks (RNNs), along with advanced variants like Long Short-Term Memory (LSTM) and Gated Recurrent Units (GRUs), are well suited for sequential or time-dependent data, such as language and time series analysis. In recent time, Transformer-based architectures (e.g., BERT, GPT) have transformed natural language understanding and generation through attention mechanisms and parallel computation, achieving remarkable efficiency and accuracy.

Training DL models typically demands large, high-quality datasets and significant computational resources, often utilizing Graphics Processing Units (GPUs) or specialized accelerators such as Tensor Processing Units (TPUs). The learning process is driven by backpropagation, where model parameters are updated to minimize a loss function via gradient descent. Despite their complexity, DL models have achieved human-level or superhuman performance in tasks, e.g., image recognition and machine translation.

However, deep learning also presents notable challenges: it often requires extensive labeled data, incurs high computational costs, is prone to overfitting, and can lack interpretability in decision-making. To address these issues, ongoing research explores methods such as data augmentation, transfer learning, unsupervised and self-supervised learning, and model distillation to improve efficiency, generalization, and transparency.

12.6 Steps to Develop ML Models Using MATLAB

Machine Learning (ML) has become an important part across various engineering disciplines, including civil engineering, for predicting the desired output employing the algorithm, along with the classification, and data analysis. In this regard, MATLAB provides a flexible and powerful medium for developing, training, and employing ML models that are combined with an extensive variety of functions and toolboxes, which provides the complete application process.

12.6.1 Steps to Develop Machine Learning Models in MATLAB

The process of developing an ML model in MATLAB starts with data preparation in a structured workflow. Data preparation comprises assembling related datasets from various sources that were measured in different sensors or from historical records of experimental results. Afterward, the data were preprocessed for addressing the missing values or to normalize it, and subsequently to extract key entities. As mentioned before, the dataset is next divided into training, validation, and testing subsets, with a split ratio of 70% for training, 15% for validation, and 15% for testing.

Next comes feature engineering, which involves selecting the most influential features to boost the model's predictive capability and, where needed, creating new features to enhance performance. Once the features are ready, the model selection phase begins. MATLAB offers a variety of machine learning models, such as linear regression, decision trees, support vector machines (SVM), neural networks, and ensemble techniques, and the choice depends on the type of the problem, be it regression, classification, or clustering—and the characteristics of the dataset.

The model training phase includes fitting the chosen model to the training data using MATLAB's built-in functions, while applying cross-validation to fine-tune hyperparameters and decrease the overfitting risk. Model evaluation follows, where performance is measured on the validation dataset employing metrics, e.g., MSE for regression tasks, along with visual aids like confusion matrices and ROC curves.

In the testing phase, the finalized model is applied to the test dataset to evaluate its generalization capability, and results are compared with validation performance to ensure overfitting has been avoided. Finally, model deployment enables the trained model to be integrated into real-time prediction systems or embedded into larger applications. MATLAB supports deployment across various platforms, including standalone software, web services, and embedded hardware.

By following these steps, MATLAB provides a robust, end-to-end environment that empowers civil engineers and other professionals to develop and implement ML models tailored to their specific project needs, leveraging its extensive toolboxes and user-friendly interface.

12.6.2 Various Functions Used for Different ML Algorithms in MATLAB

12.6.2.1 Support Vector Machines (SVM)

In MATLAB, SVM can be implemented using the fitrsvm function, which allows customization through several options such as Kernel, BoxConstraint, Epsilon, and KernelScale. Different variations of SVM can be applied based on the nature of the data and the complexity of relationships it exhibits.

A Linear SVM fits a straight-line decision boundary and works best for data that is linearly separable or close to linear, implemented in MATLAB via fitrlinear. For more complex patterns, a Quadratic SVM employs a second-degree polynomial kernel (fitrsvm with "Polynomial" kernel and Polynomial Order set to 2), while a Cubic SVM applies a third-degree polynomial kernel (fitrsvm with "Polynomial" kernel and Polynomial Order set to 3) for capturing higher-order relationships.

Gaussian kernel-based models offer further flexibility. The Fine Gaussian SVM uses a small kernel scale to capture intricate details (fitrsvm with "Gaussian" kernel and a small Kernel Scale), the Medium Gaussian SVM uses a moderate kernel scale to balance fine detail recognition and general pattern detection (fitrsvm with "Gaussian" kernel and a medium Kernel Scale), and the Coarse Gaussian SVM applies a large kernel scale to focus on broad, general patterns in the data (fitrsvm with "Gaussian" kernel and a large Kernel Scale).

This range of SVM types enables MATLAB users to tailor model complexity and flexibility to match the data's structural characteristics, from simple linear trends to highly nonlinear relationships.

12.6.2.2 Decision Trees

In MATLAB, Decision Tree models can be trained using the fitrtree function. Key configurable options for this function include MinLeafSize, which specifies the minimum number of observations per leaf, and MaxNumSplits, which determines the maximum number of decision splits allowed in the tree.

12.6.2.3 Gaussian Process Regression

For Gaussian Process Regression (GPR) in MATLAB, the training is performed using the fitrgp function. One of the main customizable parameters for this model is the Kernel option, which allows the selection of an appropriate kernel function to model data relationships effectively.

12.7 Results and Discussion

The current section presents the outcomes from training different ML models on the concrete dataset and provides a detailed analysis of their performance. The models evaluated comprise of DT, SVM, GPR, and Tree Ensemble methods. To compare their effectiveness, the performance was assessed using several evaluation metrics: R^2, RMSE, MSE, MAE, MAPE.

12.8 Evaluation Metric

After completing the training phase, the predictive accuracy of the models was thoroughly evaluated utilizing an independent dataset. To achieve a well-rounded performance assessment, multiple evaluation metrics were employed [15]. Each of the metrics plays a distinct role in measuring and understanding model performance. The selection of these specific metrics was intentional, aiming to provide a comprehensive view of the models' predictive abilities.

MSE and RMSE both assess the squared differences between predicted and actual values, with greater penalties assigned to larger deviations. MSE offers a cumulative measure of these squared errors, whereas RMSE presents an interpretable value representing the average magnitude of prediction errors, thus providing a clearer picture of error characteristics [16, 17]. MAE determines the absolute difference among the predicted and actual values, making it less sensitive to outliers and providing an unbiased perspective on overall model accuracy. R^2 presents as an indicator of the model's explanatory power, expressing the variance in the target variable which can be accounted for by the predictors [18].

By utilizing these metrics together, the evaluation captures various dimensions of predictive performance. Sensitivity of MSE and RMSE to larger deviations is particularly valuable for finding substantial prediction errors, while MAE's uniform treatment of all deviations allows for a balanced accuracy measure. R^2 complements these by offering a holistic view of model fit and reliability.

12.8.1 Mean Squared Error (MSE)

MSE specifically measures the average squared difference between predicted outputs and actual observed values. A lower MSE signifies stronger predictive accuracy, indicating that the model's valuations are meticulously allied with real-world results.

$$MSE = \frac{1}{n}\sum_{i=1}^{n}\left(y_i - \hat{y}_i\right)^2 \tag{12.15}$$

where n stands for sample numbers, and y_i and $\hat{y}_i$ present the true and predicted values.

12.8.2 Root Mean Squared Error (RMSE)

RMSE is derived by taking the square root of the Mean Squared Error (MSE) and represents the standard deviation of the prediction errors. This metric allows for an intuitive evaluation of model performance, as it expresses the prediction deviations in the same units as the actual observed values. By doing so, RMSE provides a straightforward and interpretable means of understanding the typical magnitude of errors made by the model.

$$\mathrm{RMSE} = \sqrt{\frac{1}{n}\sum_{i=1}^{n}\left(y_i - \hat{y}_i\right)^2} \tag{12.16}$$

12.8.3 Mean Absolute Error (MAE)

The Mean Absolute Error (MAE) measures the average of the absolute differences between predicted and actual values, disregarding the direction of the errors. By removing the influence of error signs, MAE provides a clear and unbiased representation of the typical magnitude of prediction errors, offering a straightforward measure of model accuracy.

$$MAE = \frac{1}{n}\sum_{i=1}^{n}\left|y_i - \hat{y}_i\right| \tag{12.17}$$

12.8.4 Coefficient of Determination (R^2)

The coefficient of determination, denoted as R^2, serves as a key indicator of how well a model accounts for the variation present in the dataset. A value of R^2 closer to 1 signifies that the model provides a stronger fit to the observed data, effectively capturing a greater proportion of its variability. This interpretation aligns with the findings reported by

$$R^2 = 1 - \frac{\sum_{i=1}^{n}\left(y_i - \hat{y}_i\right)^2}{\sum_{i=1}^{n}\left(y_i - \overline{y}_i\right)^2} \tag{12.18}$$

12.9 Prediction of Strength Properties of 3DPC

12.9.1 Comparative Analysis

Figure 12.8 illustrates the performance indicators used to evaluate the regression models. Multiple machine learning techniques demonstrated strong capabilities in predicting both the strengths of 3DPC under compression and flexure. Among them, LR model showed notable efficiency for datasets with linear correlations due to its simplicity and effectiveness. In predicting flexural strength, LR achieved an MSE of 0.006, RMSE of 0.08, MAE of 0.001, and an R^2 of 0.777. In the case of compressive strength estimation, although RMSE (0.228) and MSE (0.052) were slightly higher, the MAE remained low at 0.003, and R^2 reached 0.947. These results confirm LR's capability to effectively find linear relations between selected features and strength values.

In contrast, Support Vector Machines (SVM) proved less effective for regression tasks. When predicting flexural strength, SVM produced an MSE of 0.0129, RMSE of 0.113, MAE of 0.1, and an R^2 of 0.549. For compressive strength, the model recorded an MSE of 0.07, RMSE of 0.263, MAE of 0.102, and R^2 of 0.93. This comparatively weaker performance is likely due to SVM's stronger suitability for classification problems and its sensitivity to data scaling and feature distribution.

Decision Tree (DT) models, on the other hand, delivered excellent predictive accuracy, benefiting from their capability to model nonlinear relations while requiring minimal preprocessing. In the case of strength under flexure, DT attained an MSE of 0.0021, RMSE of 0.0462, MAE of 6e-3, and R^2 of 0.923. To estimate the strength under compression, it obtained an MSE of 0.0129, RMSE of 0.1165, MAE of 1e-3, and an R^2 of 0.984. Furthermore, the straightforward interpretability of DT models enhances their practical applicability.

The Random Forest (RF) model, which integrates multiple DTs, further improved stability. To estimate the strength under flexure, RF recorded an MSE of 1e-3, RMSE of 32e-3, MAE of 5e-4, and an R^2 of 0.963. To estimate strength under compression, it achieved an MSE of 14e-3, RMSE of 118e-3, MAE of 1e-3, and R^2 of 0.985. These results demonstrate RF's robustness and high predictive capability.

Artificial Neural Networks (ANN) also showed strong learning potential, especially for complex nonlinear relationships, but required more rigorous preprocessing and hyperparameter optimization. For flexural strength, ANN achieved an MSE of 2e-3, RMSE of 52e-3, and R^2 of 0.905. However, the estimation capability of strength under compression decreases, with MSE at 31e-3, RMSE at 176e-3, and MAE at 5e-3, values that were comparatively higher than those of the top-performing models. This suggests that additional tuning could be necessary to unlock ANN's full potential.

Among all tested algorithms, the Bagged Tree Model delivered the best overall results, attaining the peak R^2 value of 0.959 inline with the lowermost RMSE, MSE, MAE, and MAPE scores. This combination indicates exceptional predictive accuracy and model fit.

In summary, each algorithm exhibited specific strengths in estimating the mechanical performance of 3DPC. Models such as LR, DT, and RF excelled due to their capability to recognize linear and nonlinear patterns in the data. Conversely, SVR was less effective for these regression problems, reflecting its limitations in this context despite advantages in handling high-dimensional, nonlinear datasets. ANN models, though powerful, may have underperformed here due to data size constraints or suboptimal tuning. RF emerged as the best performer for both compressive and flexural strength predictions, largely because its ensemble approach aggregates predictions from multiple DTs, reducing variance and improving stability. Although LR is

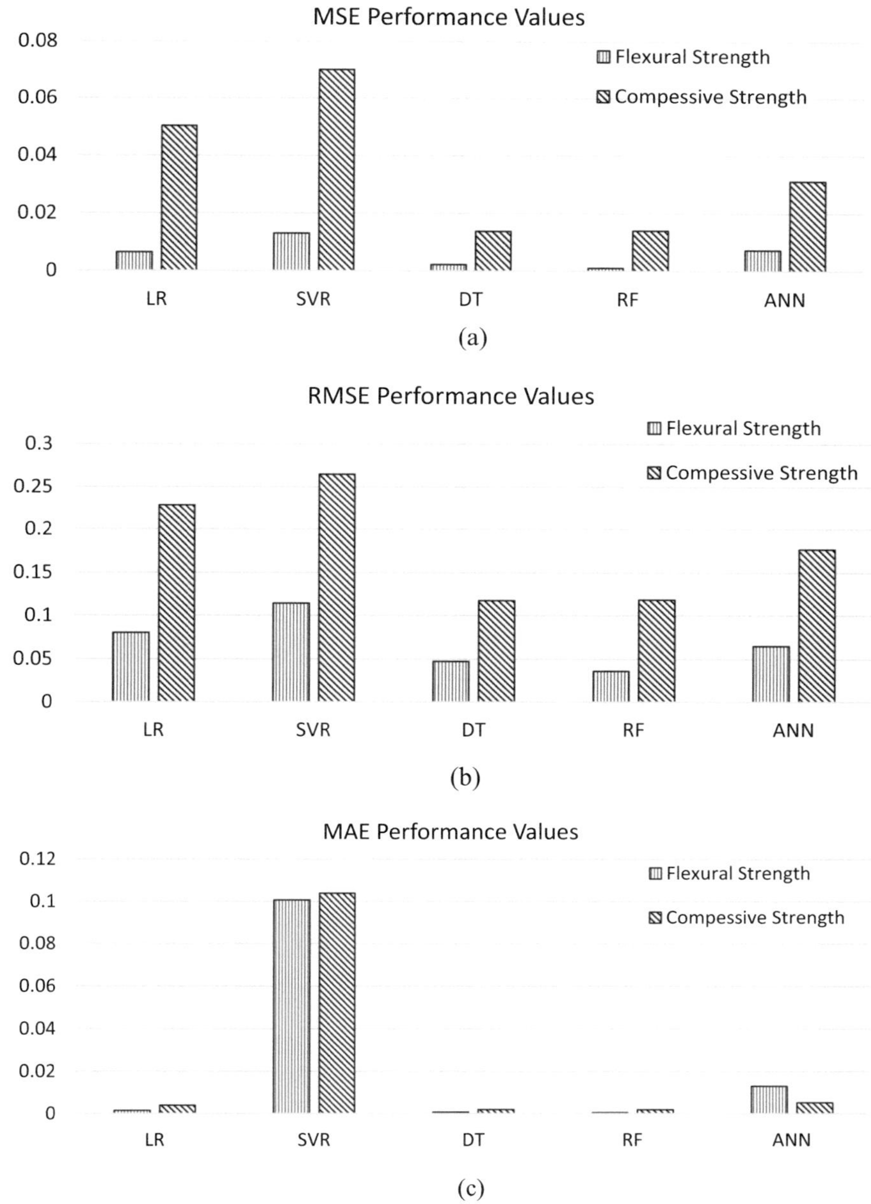

Fig. 12.8 Evaluation of regression model performance values

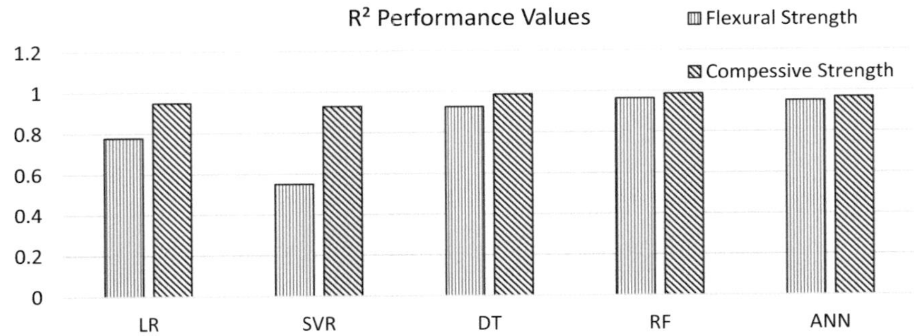

(d)

Fig. 12.8 (continued)

simple in nature but it's effectiveness diminishes with complex nonlinearities, while DTs can capture nonlinear relationships yet remain prone to instability and noise sensitivity.

When selecting a predictive model, considerations should extend beyond accuracy to include interpretability, computational cost, and susceptibility to overfitting. For real-world use, the most effective choice is often one that balances precision with clarity and ease of maintenance.

12.9.2 Prediction of Fresh-State Properties

Figures 12.9a, b presents a summary of the outcomes for different models under their respective optimal hyperparameters for each evaluation metric, showing how each ML approach performed in estimating plastic viscosity (PV) and yield stress (YS). Majority of models achieved prediction correlation coefficients above 0.85, indicating strong predictive accuracy. Among all, the XGBoost proven to be the most comprehensive and consistent performance.

Across the board, nearly all models performed better on PV than on YS, suggesting that PV data contains less overall error. Since R^2 is not influenced by scale differences, PV data can be considered more reliable. In contrast, the Linear Regression (LR) model showed the weakest results, with the highest MSE, RMSE, and MAE values, and the lowest R^2. As depicted in both figures, many LR predictions fell outside the $\pm 20\%$ error range, particularly for YS outputs.

The Support Vector Regression (SVR) model performed well for PV predictions, achieving low MSE, RMSE, and MAE scores. However, its performance for YS was even worse than LR's, showing high error metrics and a low R^2. Figures 12.9a, b reveals that many SVR predictions for YS also exceeded the $\pm 20\%$ error range.

Fig. 12.9 **a** Testing of different models of plastic viscosity. **b** Testing of different models of yield stress

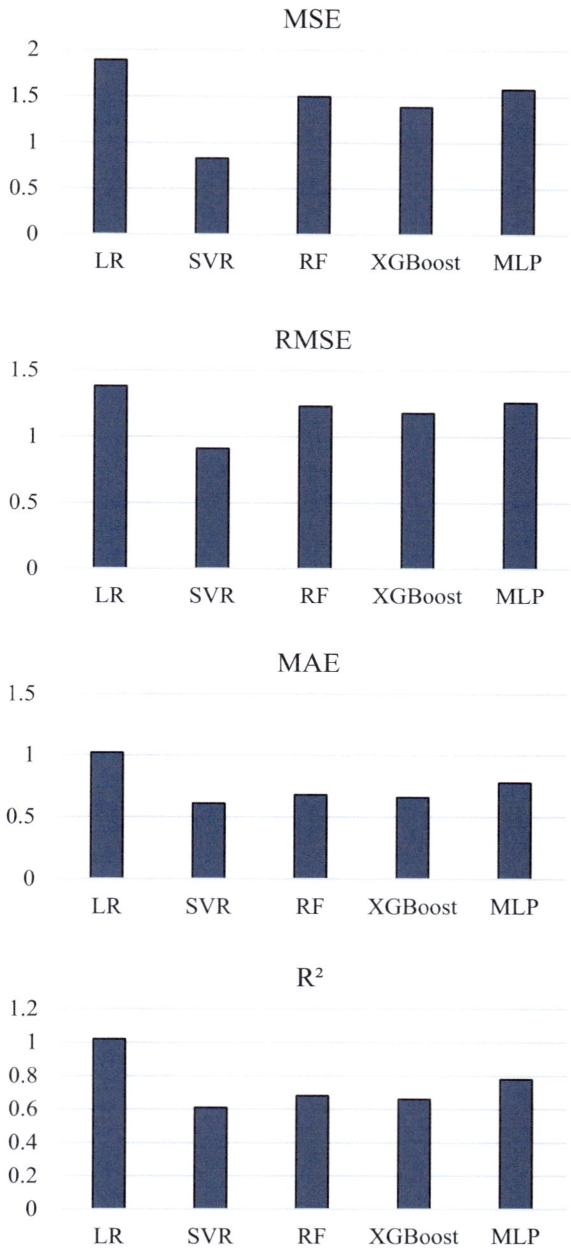

Fig. 12.9 (continued)

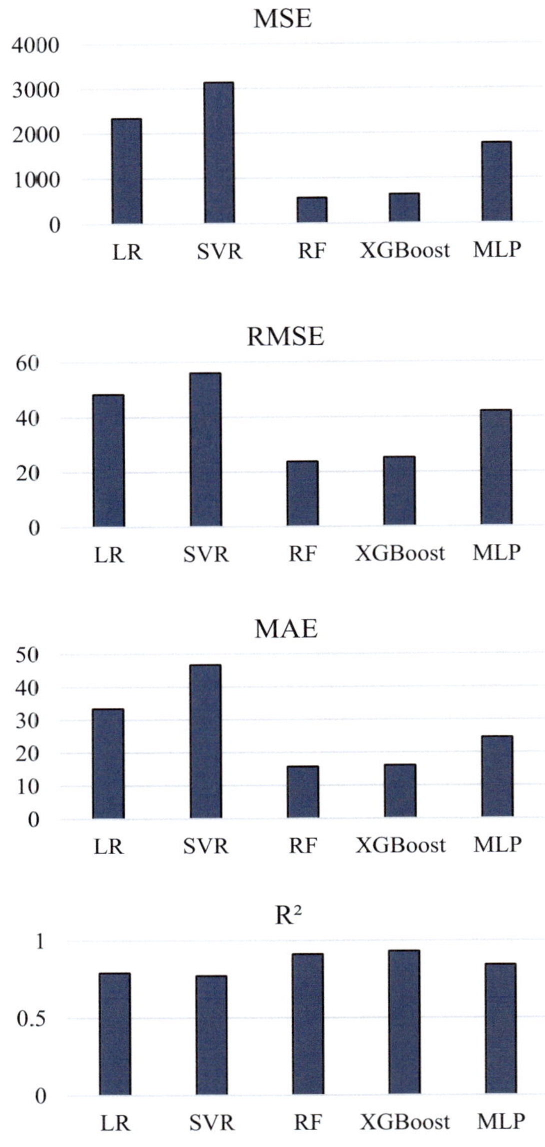

Both Random Forest (RF) and XGBoost models achieved test R^2 values close to 0.9 for PV and around 0.95 for YS, confirming that ensemble learning enhances stability of single-model approaches. Within these, XGBoost produced the lowest error metrics and the highest R^2, marking it as the top-performing model when stability is also considered.

The Multi-Layer Perceptron (MLP) model outperformed LR and SVR but was slightly inferior to RF. As seen in the figures, MLP's interquartile range (IQR) was similar to RF's, but with a higher median value. It also showed lower RMSE, MSE, and R^2 values along with a higher MAE compared to RF, indicating weaker performance. Furthermore, MLP produced the largest number of outliers for YS, highlighting its instability and poorer prediction accuracy for that target. Nonetheless, when compared to SVR, MLP's predictions were more centralized in distribution, which demonstrates better fitting ability despite similar overall accuracy levels.

Additional sets of mortar mix designs, illustrated in Figs. 12.10a, b, were fed into the model for validation dataset to generate the rheological parameters and assess printability. This assessment was carried out using the previously identified best-performing trained XGBoost model. Figure 12.10c shows the corresponding error of the prediction. It is observed that the relative error across all the samples lies within $\pm 8\%$ that implies the predictive capacity of the employed ML model.

The findings reveal that the Bagged Tree Model delivers superior performance compared to the other models in predicting UHPC compressive strength. Its high R^2 value demonstrates that it explains for a substantial proportion of the variance in the dataset, while the low RMSE, MSE, MAE, and MAPE values provide additional evidence of its precision and dependability.

12.10 Challenges and Future Outlook

Although the results are promising, several challenges remain. These include limitations in the availability and quality of data, the necessity for models to generalize effectively across various 3D printers and material types, and the integration of real-time sensor data with machine learning frameworks. Future studies should prioritize the creation of standardized datasets, the adoption of transfer learning techniques, and the development of hybrid physics-informed machine learning models.

12.11 Closure

The application of artificial intelligence (AI) and machine learning (ML) in 3D concrete printing is transforming the design and construction of concrete structures. By using predictive models, it becomes possible to improve process control and evaluate material performance, leading to faster, more efficient, and sustainable construction methods. With the continuous growth of available data and advancements in modeling techniques, the influence of AI and ML in 3DCP is expected to increase substantially, fostering innovation within construction technology.

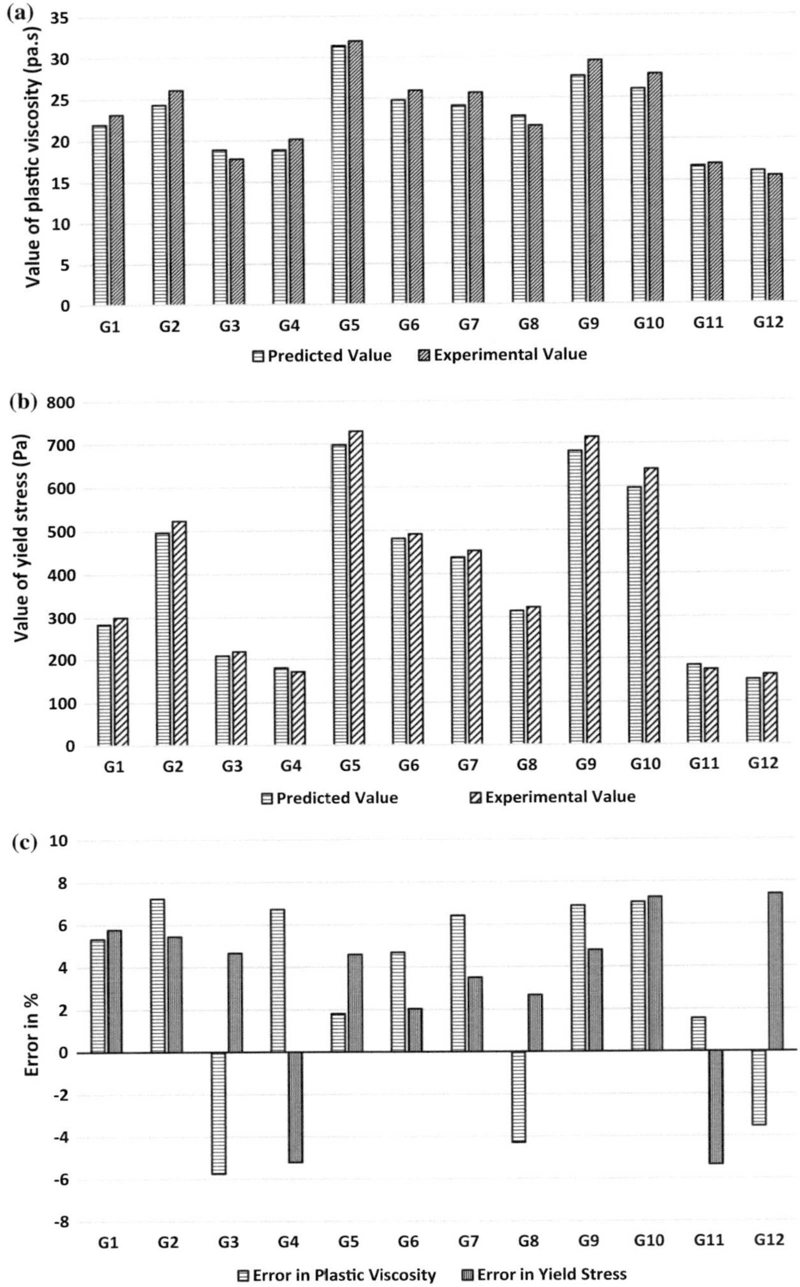

Fig. 12.10 **a** Comparison between predicted and experimental rheological properties and printing outcomes for plastic viscosity (PV). **b** Comparison between predicted and experimental rheological properties and printing outcomes for yield stress (YS). **c** Comparison between experimental and predicted values of plastic viscosity and yield stress

References

1. Goh GD, Sing SL, Yeong WY (2021) A review on machine learning in 3D printing: applications, potential, and challenges. Artif Intell Rev 54:63–94
2. Zhang Y, Cui S, Yang B, Wang X, Liu T (2025) Research on 3D printing concrete mechanical properties prediction model based on machine learning. Case Stud Constr Mater 22:e04254
3. Ge W, Zhang F, Wang Y, Ashour A, Luo L, Qiu L, Fu S, Cao D (2024) Machine learning predictions for bending capacity of ECC-concrete composite beams hybrid reinforced with steel and FRP bars. Case Stud Constr Mater 21:e03670
4. Cardoso RA, Oliveira GAB, Almeida GMJ, Araújo JA (2024) A simple linear regression strategy for fretting fatigue life estimates. Tribol Int 198:109852
5. Gao H, Jin L, Chen Y, Chen Q, Liu X, Yu Q (2024) Rheological behavior of 3D printed concrete: Influential factors and printability prediction scheme. J Build Eng 91:109626
6. Zhang J, Wang P, Gao RX (2019) Deep learning-based tensile strength prediction in fused deposition modeling. Comput Ind 107:11–21
7. Liu C, Qian Q (2024) Twin proximal support vector regression with heteroscedastic Gaussian noise. Expert Syst Appl 250:123840
8. Zhang X, Chu D, Zhao X, Gao C, Lu L, He Y, Bai W (2024) Machine learning-driven 3D printing: a review. Appl Mater Today 39:102306
9. Kioumarsi M, Dabiri H, Kandiri A, Farhangi V (2023) Compressive strength of concrete containing furnace blast slag; optimized machine learning-based models. Clean Eng Technol 13:100604
10. Geng S, Luo Q, Liu K, Li Y, Hou Y, Long W (2023) Research status and prospect of machine learning in construction 3D printing. Case Stud Constr Mater 18:e01952
11. Dabbagh SR, Ozcan O, Tasoglu S (2022) Machine learning-enabled optimization of extrusion-based 3D printing. Methods 206:27–40
12. Sirisena G, Jayasinghe T, Gunawardena T, Zhang L, Mendis P, Mangalathu S (2024) Machine learning-based framework for predicting the fire-induced spalling in concrete tunnel linings. Tunn Undergr Space Technol 153:106000
13. Shahmansouri AA, Yazdani M, Hosseini M, Bengar HA, Ghatte HF (2022) The prediction analysis of compressive strength and electrical resistivity of environmentally friendly concrete incorporating natural zeolite using artificial neural network. Constr Build Mater 317:125876
14. Rathnayaka M, Karunasinghe D, Gunasekara C, Wijesundara K, Lokuge W, Law DW (2024) Machine learning approaches to predict compressive strength of fly ash-based geopolymer concrete: a comprehensive review. Constr Build Mater 419:135519
15. Ma C, Wang W, Wang S, Guo Z, Feng X (2023) Prediction of shear strength of RC slender beams based on interpretable machine learning. In: Structures, Elsevier, p 105171
16. Naser MZ, Alavi AH (2023) Error metrics and performance fitness indicators for artificial intelligence and machine learning in engineering and sciences. Arch, Struct Constr 3:499–517
17. Botchkarev A (2019) A new typology design of performance metrics to measure errors in machine learning regression algorithms. Interdiscip J Inf Knowl Manag 14:45–76
18. Khan AQ, Awan HA, Rasul M, Siddiqi ZA, Pimanmas A (2023) Optimized artificial neural network model for accurate prediction of compressive strength of normal and high strength concrete. Clean Mater 10:100211

Chapter 13
Challenges and Future Directions in 3D Concrete Printing

13.1 Introduction

The construction sector is undergoing a paradigm shift with the integration of automation, digital fabrication, and advanced material technologies. Traditional methods, though widely practiced, are constrained by dependence on extensive formwork, intensive labor requirements, and long construction timelines. These constraints have become increasingly significant in the context of global challenges such as rapid urbanization, demand for affordable housing, skilled labor shortages, and the need for sustainable and resilient infrastructure systems.

Within this landscape, 3D concrete printing (3DCP) has emerged as a potentially transformative technology. The process involves the layer-by-layer deposition of cementitious material, extruded through a controlled nozzle system guided by robotic- or gantry-based mechanisms. This technique minimizes the reliance on formwork and manual labor while enabling the fabrication of non-standard, complex, and geometrically optimized structural forms. The potential advantages of 3DCP extend beyond efficiency, encompassing reductions in construction waste, material optimization, and enhanced architectural flexibility.

Several demonstration projects worldwide have showcased the application of 3DCP in diverse contexts, including residential housing, emergency shelters, bridges, and public infrastructure prototypes. These case studies underscore the promise of 3DCP in addressing pressing global construction challenges. In particular, the technology offers opportunities for rapid deployment in post-disaster reconstruction, remote or high-altitude regions, and areas with severe labor constraints. The adoption of industrial and agro-waste materials within printable mixes also highlights the sustainability dimension of this technology.

Despite these promising developments, the large-scale implementation of 3DCP in mainstream construction practice remains limited [1]. The barriers are multifaceted and span across several domains. On the material side [2], issues such as rheological control, buildability, dimensional stability, interlayer bonding, and long-term

A. Chourasia et al., *Advances and Challenges in 3D Printed Concrete (3DPC)*, Materials Horizons: From Nature to Nanomaterials, https://doi.org/10.1007/978-981-95-8557-1_13

durability remain unresolved. Structurally, printed elements exhibit anisotropy [3] and raise concerns regarding load-bearing capacity and reinforcement integration [4]. Process-related challenges include printer system limitations, operational reliability, and quality control. Moreover, environmental dependencies such as temperature, humidity, and curing conditions significantly affect printing outcomes. Beyond technical considerations, the absence of codified standards, regulatory approval mechanisms, insurance frameworks, and demonstrated cost-effectiveness hinders widespread acceptance.

This chapter systematically examines these challenges, categorizing them into material, structural, and process and equipment, environmental, regulatory, and economic dimensions. It further incorporates case studies, schematic diagrams, and a tabular summary of challenges and potential solutions to provide a comprehensive understanding of the barriers currently limiting 3DCP. The aim is to contribute to the discourse on advancing this technology from research and demonstration stages toward reliable, codified, and industry-ready applications.

13.2 Material-Related Challenges

13.2.1 Rheology and Fresh-State Properties

The fresh-state behavior of 3D printable concrete is one of the most critical aspects determining the success of the printing process. Unlike conventional concrete, which can be vibrated or compacted to achieve workability, printable mixes must inherently satisfy two contradictory requirements [5–8]: they must be sufficiently fluid to allow continuous pumping and extrusion and, at the same time, stiff enough to maintain their shape immediately after deposition. This balance ensures that successive layers can be placed without collapse or distortion, while also enabling the structure to gain stability as it is built up.

Rheological Parameters

The key rheological properties that govern printability are yield stress, viscosity, and thixotropy:

- **Yield Stress**: The static yield stress influences the ability of a deposited layer to resist deformation, while dynamic yield stress controls flowability during extrusion.
- **Viscosity**: Determines ease of pumping and resistance to segregation or bleeding. Excessive viscosity increases pumping effort and risk of nozzle blockage, while low viscosity leads to slumping.
- **Thixotropy**: Allows the material to act as a fluid during extrusion but quickly regain stiffness at rest, providing shape retention and buildability.

If the rheology is not properly tuned, major problems arise. A stiff mix with high yield stress may clog the nozzle and interrupt the print, while a highly fluid mix may collapse under self-weight, leading to layer misalignment or complete failure of the wall. Thus, rheology directly links to extrudability, buildability, and dimensional stability, the three pillars of printability.

Effect of External Factors

Rheological control is further complicated by environmental conditions, since 3DCP is generally carried out in open sites. Temperature, humidity, and wind speed significantly affect hydration kinetics, setting behavior, and surface stability:

- **High temperatures** accelerate hydration, increasing yield stress and reducing open time, which can cause cold joints or blockages.
- **Low temperatures** delay stiffening, leading to excessive sagging of layers under load.
- **Low humidity** enhances evaporation, causing surface drying, plastic shrinkage cracks, and weak interlayer bonding.
- **High humidity** prolongs workability but reduces buildability due to slow stiffening.
- **High wind speeds** promote uneven drying, differential shrinkage, and even physical disturbance of fresh filaments.

In many cases, these conditions act together. For instance, hot, dry, and windy environments accelerate moisture loss, often resulting in cracking and poor adhesion. Conversely, cold and humid conditions may keep mixes workable for too long, delaying strength gain and causing wall deformation. Table 13.1 summarizes the effect of various external factors (e.g., temperature, relative humidity) on rheology along with possible mitigation strategies.

The rheology of 3D printable concrete is highly dynamic, shaped by both material design and environmental exposure. To ensure consistency and reliability, real-time monitoring systems, adaptive mix designs, and site-specific mitigation strategies are essential. Rheology, therefore, forms the foundation upon which the success of 3D concrete printing rests, linking material science with process control and structural performance.

13.2.2 Interlayer Bonding

One of the defining characteristics of 3D concrete printing is its layer-by-layer construction process, which introduces the unique challenge of achieving reliable bonding between successive layers. Unlike conventionally cast concrete, where material is placed monolithically, 3DCP inherently produces interfaces at each deposition step [3, 9]. These interfaces can act as planes of weakness if adhesion is inadequate, thereby compromising the mechanical integrity of the printed element.

Table 13.1 Effect of various external factors on rheology and possible mitigation strategies

External factor	Effect on rheology	Mitigation strategies
High temperature (>35 °C)	Accelerates hydration, increases yield stress and viscosity, reduces open time, risk of nozzle blockage and cold joints	Use retarders, apply surface misting, schedule printing in cooler hours, shading structures
Low temperature (<10 °C)	Slows hydration, delays stiffening, reduces buildability, risk of wall deformation under self-weight	Add accelerators, localized heating, insulated enclosures for print area
Low humidity (<40%)	Rapid evaporation, surface drying, plastic shrinkage cracks, weak interlayer bonding	Surface curing (fogging, curing compounds), reduce layer interval time, use shrinkage-reducing admixtures
High humidity (>80%)	Prolongs workability, delays setting, excessive deformation of base layers under additional weight	Adjust admixture dosage, increase layer curing intervals, controlled ventilation
High wind speed (>20 km/h)	Enhances evaporation, uneven drying, differential shrinkage, risk of layer delamination, physical disturbance of filaments	Install wind barriers, mist curing, adjust mix for faster setting
Low wind speed	Uniform hydration, favorable for bonding, but may prolong open time when combined with high humidity	Controlled ventilation, adjust accelerator dosage

The bond strength between layers is primarily governed by the time gap between successive depositions, surface moisture content, and the rheological properties of the mix. When the interval between two consecutive layers is too long, the lower layer begins to stiffen and lose its surface moisture, reducing its capacity to chemically and mechanically bond with the next layer. This results in the formation of what are termed "cold joints." Such joints are characterized by reduced cohesion at the interface, which in turn leads to a loss of continuity in load transfer.

Inadequate interlayer bonding gives rise to anisotropic mechanical behavior, where the structural performance varies depending on the loading direction. Typically, the compressive strength parallel to the layers remains relatively unaffected, but the tensile and shear strength perpendicular to the layers can be significantly reduced. This anisotropy poses a serious concern for load-bearing elements and structures exposed to dynamic or seismic forces. Experimental studies have consistently shown that printed specimens fail along the weak interlayer zones under flexural and tensile loading, highlighting the need for improved bonding mechanisms.

Beyond time-dependent factors, the fresh-state rheology also plays a crucial role. If the lower layer has excessively high yield stress at the time of receiving the next layer, mechanical interlocking is reduced. Conversely, if it is too fluid, the layer may deform under the load of deposition, causing misalignment. The use of chemical admixtures such as set retarders, viscosity-modifying agents, or surface bonding agents has been explored to regulate this delicate balance.

Environmental conditions further exacerbate the challenge. High temperatures, low humidity, or strong winds accelerate drying of the exposed layer surface, leading to rapid stiffening and poor adhesion. On the other hand, excessively humid conditions may prolong setting, delaying the attainment of interlayer strength.

To address these issues, several strategies have been proposed:

- **Minimizing time gaps** between layer depositions through optimized printing paths and continuous operation.
- **Mechanical surface treatments**, such as roughening or re-extrusion of the surface, to improve physical interlocking.
- **Application of bonding agents or chemical sprays** to reactivate or extend the surface's open time.
- **Robotic control systems** that maintain consistent intervals and ensure precise material deposition.

In summary, interlayer bonding remains a critical determinant of the structural reliability of 3D-printed elements. Without robust adhesion, printed structures are prone to anisotropy and premature failure under service loads. Developing reliable, repeatable strategies to ensure strong interlayer cohesion is therefore central to advancing 3DCP from laboratory-scale demonstrations to full-scale, codified construction practice.

13.2.3 Reinforcement Challenges

A fundamental limitation of 3D concrete printing lies in its ability to incorporate effective reinforcement systems. Concrete, whether conventionally cast or 3D printed, is inherently strong in compression but weak in tension and flexure. In traditional construction, this deficiency is addressed by embedding steel reinforcement bars (rebars) within the formwork prior to casting [10–12]. However, the very advantage of 3DCP—elimination of formwork—makes the integration of conventional reinforcement methods extremely difficult.

During the printing process, material is deposited continuously in thin layers under automated control. Introducing rebars or cages within this sequence requires either frequent interruptions of the printing path or specially designed robotic mechanisms capable of simultaneously placing reinforcement. Both options disrupt the efficiency and continuity of the process. Moreover, the precision required to align reinforcement with narrow extruded filaments adds another layer of complexity.

Various alternative strategies have been investigated to overcome this challenge:

- **Steel Meshes and Cages**: Pre-fabricated meshes or lightweight cages can be inserted at specific heights during printing. However, this often requires halting the printer and risks misalignment between reinforcement and the deposited layers.
- **Murfor Reinforcement Bands**: These prefabricated steel strips can be embedded between layers during printing. While practical for low-rise walls and partitions, they provide limited structural capacity in comparison with traditional rebars.
- **Fiber Reinforcement**: Short fibers such as polypropylene (PP), steel, basalt, or glass can be dispersed within the mix to enhance tensile and flexural strength. Although this approach improves crack resistance and energy absorption, it does not provide the same level of reinforcement as continuous rebars for large-scale load-bearing elements.
- **Post-Tensioning Systems**: Some researchers have proposed threading ducts during printing, which can later accommodate tendons for post-tensioning. This provides substantial load-bearing capacity but complicates the printing process and requires precise coordination.
- **Hybrid Approaches**: Combining external reinforcement frames or modular printed components with embedded reinforcement has shown promise in pilot projects. However, such methods often reduce the efficiency advantages of 3DCP.

The absence of a universally accepted reinforcement technique has significant structural implications. Printed elements without reinforcement are generally limited to non-load-bearing walls, architectural features, or small-scale structures. For structural components subjected to bending, shear, and seismic forces, reliable reinforcement is indispensable. Without it, 3DCP cannot yet compete with conventional reinforced concrete construction in terms of safety and long-term performance.

The challenge is further compounded by the anisotropy of printed elements, which places additional demands on reinforcement to bridge weak interlayer zones. Innovative approaches such as automated bar placement robots, continuous wire embedding, or functionally graded fiber reinforcement are being actively explored, but these remain largely experimental.

In summary, reinforcement continues to be one of the most significant barriers preventing 3DCP from scaling beyond small-scale prototypes and demonstration projects. A robust, standardized reinforcement strategy—capable of providing both structural reliability and compatibility with automated printing—is essential for the transition of 3DCP into mainstream structural applications.

13.2.4 Durability Issues

While much of the research on 3D concrete printing (3DCP) has focused on mix design, printability, and structural performance in the short term, long-term durability remains a relatively underexplored domain. For conventional reinforced concrete structures, durability considerations are well documented in building codes and standards. In contrast, the layered and anisotropic nature of 3D-printed elements introduces new challenges that may alter how these structures perform when exposed to aggressive environmental conditions.

One of the most pressing concerns is the susceptibility of printed elements to shrinkage and cracking. The high surface-area-to-volume ratio of extruded filaments, coupled with rapid evaporation during printing, makes printed layers prone to plastic shrinkage cracks. Over time, drying shrinkage can further contribute to crack propagation, reducing service life and compromising durability. Unlike cast concrete, which benefits from continuous curing in bulk form, the thin layers of 3DCP are more vulnerable to early-age cracking.

Creep and long-term deformation represents another area of concern. Since 3DCP often uses highly thixotropic mixes with modified rheology, their long-term viscoelastic behavior may differ significantly from conventional concrete. Sustained loading on printed walls could lead to increased creep strains, particularly at interlayer interfaces where bonding is weaker.

Environmental exposure adds further complexity:

- **Freeze–Thaw Cycles**: In cold climates, the repeated freezing and thawing of moisture within pores can accelerate deterioration. The interlayer interfaces, which may have higher porosity, are particularly vulnerable to spalling under such cycles.
- **Chloride Ingress**: Structures exposed to marine environments or de-icing salts are at risk of chloride penetration, which accelerates the corrosion of embedded reinforcement. The directional pore structure of printed elements may allow faster chloride migration along layer boundaries.
- **Carbonation**: The layered geometry and potential microcracks at interfaces provide additional pathways for carbon dioxide to penetrate, leading to reduced alkalinity and corrosion risk.
- **Sulfate Attack and Chemical Exposure**: Industrial environments or soils rich in sulfates may also degrade printed mixes differently, depending on the supplementary cementitious materials used.

Another dimension of durability relates to the heterogeneity introduced by printing parameters. Variations in layer height, extrusion rate, and time gaps can create non-uniform microstructures across the printed element. These weak zones, invisible during early use, may accelerate deterioration under sustained exposure.

To mitigate these risks, several strategies are being investigated:

- Optimizing mix design with supplementary cementitious materials (e.g., fly ash, slag, silica fume) to improve chemical resistance.
- Incorporating fibers to limit shrinkage-induced cracking.
- Developing protective surface coatings or sealants to reduce permeability.
- Employing controlled curing regimes, such as fogging, insulated enclosures, or curing compounds, to minimize early-age shrinkage.
- Accelerated durability testing to establish benchmarks specific to 3DCP.

While 3DCP has demonstrated potential in terms of mechanical strength and construction efficiency, the long-term durability of printed structures is not yet fully established. The unique morphology of printed layers, interfacial weaknesses, and directional pore networks introduce new variables absent in conventionally cast concrete. Systematic research, coupled with the development of standardized testing protocols, is essential to evaluate and ensure the resilience of 3D-printed structures under real-world service conditions.

13.3 Structural Challenges

The structural performance of 3D-printed concrete elements remains one of the most critical concerns limiting the widespread adoption of this technology. Unlike conventionally cast reinforced concrete, which is generally homogeneous and isotropic in nature, 3DCP introduces a set of structural complexities arising from its layered deposition process, limited reinforcement options, and geometric constraints. These challenges affect not only the strength and stability of printed elements but also their long-term reliability under service conditions.

13.3.1 Anisotropy of Mechanical Properties

A key structural limitation of 3DCP is its anisotropic behavior, which results from the directional deposition of material. The interfaces between layers often exhibit weaker bonding compared to the bulk of the filament, creating planes of reduced tensile and shear strength. As a result, printed elements tend to perform well under compressive loads parallel to the layers, but show significant reductions in tensile, flexural, and shear capacity perpendicular to the layering direction. This anisotropy not only influences ultimate load-bearing capacity but also dictates failure modes, often leading to premature delamination or cracking along interlayer zones. For structures located in seismic or wind-prone regions, this characteristic poses serious reliability concerns.

13.3.2 Printing of Overhangs and Complex Geometries

While 3DCP is celebrated for its architectural flexibility, its capability to produce overhanging elements, inclined walls, or intricate geometries remains limited. In the absence of formwork or temporary supports, freshly extruded layers must sustain their own weight as well as that of subsequent layers. For large cantilevers or steeply inclined elements, this often leads to sagging, collapse, or loss of dimensional accuracy. Although rheology control, specialized nozzle designs, or temporary support systems can mitigate this problem to some extent, fully realizing the potential of 3DCP for complex forms continues to be a challenge.

13.3.3 Scaling up to Large Load-Bearing Structures

Most successful 3DCP applications to date are restricted to small-scale prototypes or single-storey demonstration units. Scaling up to multi-storey, load-bearing buildings or large-span infrastructure introduces new challenges. These include:

- **Stability During Printing**: Tall structures are prone to buckling or misalignment if lower layers lack sufficient stiffness at the time of deposition.
- **Load Transfer Mechanisms**: Anisotropy and limited reinforcement complicate the ability of printed walls to carry axial, bending, and shear loads safely.
- **Curing Control**: Large elements often exhibit non-uniform moisture distribution and temperature gradients, leading to differential shrinkage and cracking.
- **Construction Logistics**: Transporting and assembling large-scale printers or cable-driven systems on actual construction sites adds further complexity.

In addition, the absence of codified structural design methods for 3DCP means engineers cannot yet rely on established calculation models for safety and serviceability. Existing design equations for reinforced concrete do not fully capture the influence of anisotropy, interlayer bonding, or unconventional reinforcement techniques used in 3DCP.

Structural challenges in 3DCP are therefore multifaceted, involving material anisotropy, geometric limitations, and scale-up issues. Unless these barriers are systematically addressed through advances in reinforcement methods, improvements in material bonding and the development of design guidelines 3DCP will remain confined to non-structural or small-scale applications. Unlocking the full potential of this technology requires robust solutions that ensure both mechanical safety and structural reliability on par with conventional reinforced concrete.

13.4 Process and Equipment Challenges

The efficiency and reliability of 3D concrete printing (3DCP) depend not only on material design but also on the performance of the printing systems and associated equipment. The machinery used in 3DCP must ensure continuous extrusion, precise deposition, and geometric accuracy over extended periods of operation. Unlike traditional construction methods, which rely heavily on manual adjustments and formwork for control, 3DCP places extraordinary demands on automated systems. Current printer technologies, however, face a range of limitations that constrain their large-scale application.

13.4.1 Gantry-Based Systems

Gantry printers are among the most widely used setups in current research and demonstration projects. These systems consist of a rigid frame with linear tracks that guide the print head along three axes. Their primary advantage lies in their stability and precision, making them suitable for producing consistent wall geometries within a fixed workspace. However, they suffer from major drawbacks:

- **Lack of Mobility**: Gantry systems must be assembled on-site and can only operate within their defined frame. Relocating them to another site requires complete disassembly and reinstallation, which is time- and resource-intensive.
- **Space Constraints**: The footprint of the gantry often exceeds the size of the structure being printed, making them impractical for congested or irregular construction sites.
- **Limited Scalability**: Printing very large structures requires proportionally larger gantries, which quickly become uneconomical and logistically challenging.

13.4.2 Robotic Arm Systems

Robotic arms adapted from industrial automation are also widely explored in 3DCP. These systems are highly flexible in movement and capable of printing intricate curves and free-form geometries with ease. They excel in architectural applications where design freedom is critical. However, robotic arm systems face their own challenges:

- **Restricted Reach**: Even large robotic arms can only cover a limited printing area. For larger structures, multiple arms or repositioning systems are required, which complicates the process.
- **Accuracy Limitations**: While precise for small-scale printing, errors accumulate when arms are extended near their operational limits, leading to deviations in geometry.
- **Payload Constraints**: The extruder and pumping systems add weight to the robotic arm, often restricting the size of material hoppers or the flow rate of extrusion.

13.4.3 Cable-Driven Parallel Robots (CDPRs)

Cable-driven parallel robots represent a newer category of large-scale printing systems. In these setups, the print head is suspended and maneuvered by cables anchored to towers or supports positioned around the construction site. The main advantage of CDPRs is their exceptional scalability, as the workspace can be adjusted by changing the cable anchor positions. This makes them particularly attractive for large infrastructure projects. Despite their promise, CDPRs are still largely experimental and face significant technical hurdles:

- **Complex Calibration**: Precise control of multiple cables under varying tension is difficult, and even small errors can lead to print head instability.
- **Environmental Sensitivity**: External factors such as wind can disturb the suspended print head, causing inaccuracies during deposition.
- **Operational Safety**: The risk of cable failure or entanglement adds an additional layer of concern for field deployment.

13.4.4 Ancillary Equipment and Process Reliability

Beyond the printers themselves, 3DCP depends on a network of pumping, mixing, and material delivery systems. These must operate seamlessly over long durations to maintain continuous printing. Issues such as nozzle blockages, pump failures, and material segregation are common, often resulting in stoppages that compromise both the quality and efficiency of the print. Unlike traditional concrete placement, where interruptions can sometimes be managed with compaction or jointing, stoppages in 3DCP frequently lead to cold joints and dimensional inaccuracies. Table 13.2 illustrates various challenges associated with three different types of printer in terms of multiple printing-related factors.

Current 3DCP equipment demonstrates a promising but incomplete picture. Gantry systems offer stability but lack flexibility, robotic arms provide design freedom but limited reach, and CDPRs offer scalability but remain experimental.

Table 13.2 Comparison of 3D concrete printer systems

Criteria	Gantry systems	Robotic arms	Cable-driven parallel robots (CDPRs)
Mobility	Low—fixed frame, difficult to relocate	Moderate—can be repositioned but limited workspace	High—flexible cable setup allows repositioning
Reach	Moderate—limited to gantry footprint	Low—restricted operational envelope	High—can cover very large areas
Scalability	Low—requires very large frames for large structures	Low—not ideal for large-scale structures	High—easily scalable for infrastructure projects
Precision	High—stable and accurate within defined workspace	Moderate—precise for small-scale, accuracy drops at extended reach	Moderate—sensitive to calibration and environmental disturbances
Maturity level	High—widely used in labs and pilot projects	Moderate—used mainly for free-form prototypes	Low—mostly experimental and under research

In addition, reliable pumping and delivery systems are essential yet still prone to failures in field conditions. To move toward large-scale adoption, the development of robust, mobile, and standardized printer systems capable of operating under diverse site conditions with minimal interruptions is crucial.

13.5 Environmental and Site Challenges

The performance of 3D concrete printing (3DCP) is closely linked to its operating environment. Unlike conventional construction, where adjustments can be made during casting or curing, the success of 3DCP relies heavily on maintaining continuous and controlled conditions during the entire printing process. This makes the technology particularly vulnerable to weather fluctuations, site infrastructure limitations, and variability in raw material quality.

13.5.1 Sensitivity to Weather Conditions

One of the most significant challenges in outdoor 3DCP operations is the strong dependence of fresh concrete rheology and early strength development on ambient conditions.

- **Temperature** directly influences hydration rates. High temperatures (> 35 °C) accelerate setting and stiffening, reducing open time and risking nozzle blockages, while low temperatures (< 10 °C) delay strength gain and compromise buildability.
- **Humidity** governs evaporation and interlayer bonding. Low humidity accelerates surface drying, leading to shrinkage cracks and weak adhesion, whereas high humidity prolongs workability but reduces early stiffness.
- **Wind speed** affects both drying rates and mechanical stability. High winds promote uneven evaporation and can physically disturb freshly deposited layers, causing dimensional inaccuracies.

Because 3DCP is often promoted for rapid housing solutions in disaster-prone or remote regions, where environmental control is limited, these weather sensitivities present a substantial barrier to practical deployment.

13.5.2 Infrastructure Requirements

Although 3DCP minimizes the need for formwork and manual labor, it requires a robust and uninterrupted site infrastructure. Continuous power supply, automated batching and mixing facilities, and precise pumping systems are essential for reliable printing. In remote or high-altitude areas where 3DCP could provide the greatest benefits such infrastructure is often lacking. Any disruption in pumping, mixing, or power can result in print stoppages, leading to cold joints, dimensional inconsistencies, and material wastage.

13.5.3 Variability in Industrial and Agro-Waste Materials

A key advantage of 3DCP lies in its potential for sustainability, particularly through the incorporation of supplementary cementitious materials (SCMs) such as fly ash, ground granulated blast furnace slag (GGBS), silica fume, and agro-industrial by-products like bagasse ash or rice husk ash. These materials reduce reliance on Portland cement and lower the overall carbon footprint of construction. However, their use introduces variability in chemical composition, fineness, and reactivity, which can significantly alter rheological behavior, setting times, and long-term durability.

For example, bagasse ash sourced from different sugar mills may vary in unburnt carbon content, while fly ash quality depends heavily on the combustion process and collection system of power plants. Such variability makes it difficult to achieve consistent printability and performance without extensive quality control.

13.5.4 Combined Site Challenges

In many real-world projects, these environmental and infrastructural issues act simultaneously. For instance, printing in high-altitude Himalayan regions faces a combination of low temperatures, variable humidity, strong winds, and limited power supply. These conditions amplify the risk of cold joints, shrinkage cracking, and stoppages, undermining the reliability of the technology.

Environmental and site challenges represent one of the most critical bottlenecks for scaling 3DCP from controlled laboratory demonstrations to real construction sites. While laboratory research has achieved promising results, field conditions expose the technology to external factors that remain difficult to standardize or control. Addressing these issues will require adaptive strategies, such as weather-protected enclosures, site-specific mix designs, mobile batching units, and robust quality control of supplementary materials. Without these measures, the deployment of 3DCP in its intended contexts—remote, resource-limited, or disaster-prone regions—will remain highly constrained.

13.6 Regulatory and Acceptance Challenges

Despite notable advancements in research and pilot-scale demonstrations, the large-scale deployment of 3D concrete printing (3DCP) is still constrained by regulatory and societal challenges. In conventional construction, every stage—from material formulation and testing to structural design and execution—is governed by codified standards. These standards ensure safety, durability, and compliance with established practices, giving confidence to engineers, regulators, and the public. However, 3DCP presently lacks such comprehensive frameworks, making it difficult to obtain approvals, insurance coverage, and user trust.

13.6.1 Absence of Building Codes and Standards

Currently, no universally accepted design codes or construction guidelines exist for 3D-printed structures. Established frameworks such as those of BIS, ASTM, ACI, or Eurocode were developed for cast-in-place or precast concrete and cannot be directly applied to 3DCP due to its layered deposition process, interlayer bonding issues, anisotropy, and reinforcement limitations. This gap makes it challenging for engineers to design, test, and certify printed elements within existing regulatory systems.

In India, significant progress is being made through the combined efforts of CSIR—Central Building Research Institute (CBRI), Roorkee and the Bureau of Indian Standards (BIS). Both institutions are actively engaged in drafting preliminary guidelines and standards for 3D concrete printing. These efforts aim to establish testing protocols for fresh-state properties, interlayer bonding, durability performance, and structural design considerations. The development of such codes is expected to provide a strong foundation for regulatory approval and to facilitate wider adoption of 3DCP in the Indian construction sector.

13.6.2 Insurance and Liability Issues

The absence of codified benchmarks also complicates the role of insurers and financiers. In conventional practice, insurance companies rely on established standards to evaluate risks and issue coverage. Without recognized guidelines, 3D-printed buildings are often considered experimental, leading to hesitation in providing insurance, warranty, or financing. Moreover, liability becomes ambiguous in the event of structural failure: responsibility could be debated among the printer manufacturer, material supplier, designer, or contractor. Until codified standards are in place, such uncertainties will continue to hinder private investment and commercial viability.

13.6.3 Public and Professional Acceptance

Acceptance of 3DCP by engineers, architects, contractors, and the general public remains another challenge. The lack of long-term performance data, absence of codified design methods, and unfamiliarity with the technology generate skepticism within the professional community. End-users may also be hesitant to occupy or purchase 3D-printed buildings without assurances of safety, durability, and regulatory approval. Demonstration projects carried out under the supervision of institutions such as CSIR-CBRI have helped to build confidence, but broader acceptance will only follow once standards are institutionalized and supported by regulatory authorities like BIS.

13.6.4 Pathways Toward Acceptance

To address these barriers, the following steps are crucial:

- **Codification Efforts**: Ongoing initiatives by CSIR-CBRI and BIS must be accelerated to produce official standards covering materials, printing processes, structural design, and durability.
- **Demonstration Projects**: Large-scale, regulator-approved pilot projects should serve as case studies for validation and public confidence.
- **Collaborative Frameworks**: Close cooperation between research institutes, industry partners, and government agencies is essential for standardization.
- **Awareness and Training**: Workshops, training modules, and professional capacity-building programs will be necessary to familiarize stakeholders with the principles and potential of 3DCP.

The absence of comprehensive codes, insurance frameworks, and professional as well as public trust forms one of the most pressing barriers to mainstream adoption of 3DCP. The pioneering efforts of CSIR-CBRI and BIS in India to draft preliminary codes and testing protocols mark an important step forward. However, until these efforts mature into enforceable standards and gain international alignment, 3DCP will continue to face regulatory and acceptance hurdles on its path to becoming a mainstream construction technology.

13.7 Economic and Practical Challenges

While 3D concrete printing (3DCP) is often promoted as a technology capable of reducing construction costs through automation and labor savings, the economic reality at present is far more complex. The capital cost of machinery, the need for specialized operators, and the absence of established supply chains currently make the technology less competitive than conventional construction methods in most contexts.

13.7.1 High Initial Investment

Industrial-scale 3D concrete printers—whether gantry-based, robotic arm systems, or cable-driven setups—require substantial upfront investment. Beyond the printer itself, costs include pumping systems, mixing units, material handling equipment, and site preparation. For small- to medium-scale contractors, these costs are prohibitive unless supported by government programs, research funding, or industry collaborations.

13.7.2 Skilled Workforce and Training Needs

Although 3DCP reduces dependency on manual labor, it creates demand for a new category of skilled professionals capable of operating, maintaining, and troubleshooting advanced printing equipment. Expertise is also required in digital modeling, process control, and material science to ensure printability. At present, such expertise is concentrated in research institutions and a limited number of specialized companies, creating a bottleneck for scaling up.

13.7.3 Material and Supply Chain Limitations

Printable mixes often require precisely graded aggregates, specialized admixtures, and strict quality control, which are not always available in conventional supply chains. The incorporation of industrial and agro-waste materials (e.g., fly ash, slag, bagasse ash) holds promise for sustainability but further complicates consistency and logistics. Until local supply chains evolve to provide ready-to-print material formulations, costs remain elevated due to reliance on specialized production and testing facilities.

13.7.4 Economies of Scale

In traditional construction, costs decrease as projects scale up due to standardized practices and bulk procurement. For 3DCP, economies of scale have not yet been realized. Most projects are demonstration units or prototypes, which means the full cost of machine mobilization, site setup, and customized mix development is borne by a single project. As a result, cost per square meter remains higher than conventional methods. Only with wider adoption, standardization, and mass production of printers and materials can 3DCP achieve the cost efficiencies it promises.

13.7.5 Maintenance and Operational Costs

Beyond capital investment, ongoing maintenance, calibration, and repair costs of advanced equipment are considerable. Pumps and nozzles are prone to wear due to the abrasive nature of cementitious materials. Interruptions during printing also cause material wastage and rework, adding to costs.

Although 3DCP offers the potential for reducing labor dependency and accelerating construction timelines, its current economic feasibility is uncertain. High capital investment, reliance on skilled operators, material supply chain gaps, and the absence of economies of scale restrict its competitiveness against conventional methods. Until standardized processes, mass production of equipment, and large-scale adoption reduce costs, 3DCP will remain primarily confined to pilot projects, research initiatives, and high-profile demonstration structures rather than mainstream construction. Table 13.3 summarizes the various challenges with respect to 3DPC and their potential solutions.

Table 13.3 Summary of challenges and potential solutions

Challenge	Description	Potential solutions
Rheology	Balancing pumpability and buildability	Advanced admixtures, real-time monitoring
Interlayer bonding	Weak vertical strength	Robotic timing control, bonding agents
Reinforcement	Difficult to embed steel	Fibers, meshes, hybrid methods
Durability	Unknown long-term performance	Accelerated testing, protective coatings
Anisotropy	Different properties in each direction	Optimized printing strategies
Overhangs	Limited unsupported prints	Temporary supports, thixotropy tuning
Equipment limits	Gantry size, robotic reach	Modular/mobile printers
QC and standards	No codes or tests	Standardized test methods, codification
Weather sensitivity	Printing disrupted outdoors	Controlled environments, accelerators
Economics	High machine/material costs	Scale-up, local material sourcing

13.8 Future Directions

The future of 3D concrete printing (3DCP) lies not only in overcoming its current limitations but also in establishing a robust ecosystem that enables its integration into mainstream construction. Moving forward, several strategic focus areas must be addressed to transition the technology from laboratory-scale innovations and pilot projects to codified, scalable, and commercially viable solutions [2, 4].

13.8.1 Development of Codes, Standards, and Testing Protocols

The foremost requirement is the formulation of comprehensive codes and design standards that address the unique aspects of 3DCP, such as interlayer bonding, anisotropy, and reinforcement integration. Without codified benchmarks, it is impossible to certify safety, obtain approvals, or gain public trust. Organizations such as CSIR-CBRI, BIS (India), ASTM, and RILEM are already working toward drafting guidelines on fresh-state testing, structural performance, and durability assessment. Accelerating these initiatives and harmonizing them at the international level will be vital.

13.8.2 Reinforcement Strategies

A breakthrough in reinforcement integration is essential for 3DCP to evolve from producing non-structural prototypes to load-bearing and seismic-resistant structures. Potential approaches include continuous wire embedding, hybrid reinforcement systems, functionally graded fiber reinforcement, and robotic bar placement technologies. The development of standardized reinforcement solutions compatible with automated printing will significantly expand the structural applications of 3DCP.

13.8.3 Advancement of Printer Systems

Future printer systems must achieve greater mobility, scalability, and reliability. Portable gantry systems, robotic arms with extended reach, and field-deployable cable-driven parallel robots (CDPRs) will need to be refined for practical construction scenarios. Robust calibration systems, automated error correction, and environmental adaptability (resistance to wind, temperature, and humidity fluctuations) will also be critical to ensure consistency in real-world sites.

13.8.4 Sustainable and Locally Adapted Materials

Integrating sustainable materials into printable mixes will play a central role in reducing the environmental footprint of 3DCP. This includes greater use of industrial by-products (fly ash, GGBS, silica fume) and agro-waste materials (bagasse ash, rice husk ash). Research must focus on stabilizing the variability of such materials while ensuring consistent rheology and durability. Locally adapted mix designs will enable 3DCP to be deployed in diverse geographical and climatic conditions.

13.8.5 Large-Scale Demonstration Projects

Demonstration projects will remain essential to build confidence among policy-makers, industry stakeholders, and the general public. High-profile applications in housing, public infrastructure, and disaster-resilient shelters will serve as proof-of-concept, validating both technical performance and cost competitiveness. Successful case studies supported by regulatory authorities will pave the way for wider acceptance and integration into national housing and infrastructure programs.

13.8.6 Integration with Digital and Smart Technologies

The future of 3DCP will also be shaped by its integration with Building Information Modeling (BIM), digital twins, and real-time monitoring systems. Embedding sensors within printed elements could allow for structural health monitoring, while AI-driven process control may optimize mix designs and printing paths dynamically. Such convergence of construction automation and digital technologies will transform 3DCP into a smart, adaptive system.

13.9 Conclusions

3D concrete printing (3DCP) has moved beyond the realm of futuristic speculation to become a tangible construction technology with demonstrated potential. Successful pilot projects across the globe—ranging from residential units and pedestrian bridges to public infrastructure prototypes—have proven its capacity to deliver design freedom, reduced labor dependency, faster construction timelines, and opportunities for sustainability. These achievements mark an important milestone in the evolution of construction practices.

At the same time, it is important to recognize that the path to mainstream adoption remains complex. 3DCP introduces new materials, processes, and structural systems that cannot be fully addressed by conventional standards. The technology faces substantial challenges across multiple dimensions:

- **Material-related issues**, including rheology, interlayer bonding, reinforcement integration, and durability.
- **Structural concerns**, such as anisotropy, overhang stability, and scalability to multi-storey or large-span structures.
- **Process and equipment limitations**, encompassing reliability of printers, pump and nozzle performance, and environmental adaptability.
- **Environmental and site constraints**, where temperature, humidity, wind, and site infrastructure heavily influence print quality and continuity.
- **Regulatory and acceptance barriers**, with the absence of codified standards, insurance mechanisms, and widespread professional trust.
- **Economic challenges**, including high initial investment, supply chain immaturity, and the lack of economies of scale.

Addressing these barriers requires systematic and collaborative efforts. Research institutions such as CSIR-CBRI in collaboration with BIS in India, as well as international bodies like ASTM and RILEM, are making important strides toward developing codes, guidelines, and testing protocols. Industry partnerships are essential for advancing printer systems, supply chains, and reinforcement technologies. Demonstration projects, particularly in housing, infrastructure, and disaster resilience, will play a crucial role in building public confidence and regulatory acceptance.

Looking ahead, the full potential of 3DCP lies in its ability to combine technical innovation with sustainability goals. The integration of industrial and agro-waste materials can reduce environmental impact, while digital tools such as BIM, AI-driven control systems, and embedded monitoring sensors can create smarter and more resilient built environments.

In conclusion, 3DCP is no longer an experimental novelty but a promising construction method at the cusp of transformation. Its widespread adoption will depend on the industry's ability to systematically resolve its challenges through research, standardization, large-scale demonstrations, and collaborative innovation. If these pathways are pursued, 3DCP can evolve into a mainstream technology capable of reshaping the construction landscape and addressing some of the most pressing global challenges in housing, sustainability, and infrastructure development.

References

1. Xiao J, Ji G, Zhang Y, Ma G, Mechtcherine V, Pan J, Wang L, Ding T, Duan Z, Du S (2021) Large-scale 3D printing concrete technology: current status and future opportunities. Cem Concr Compos 122:104115
2. Zhang C, Nerella VN, Krishna A, Wang S, Zhang Y, Mechtcherine V, Banthia N (2021) Mix design concepts for 3D printable concrete: a review. Cem Concr Compos 122:104155. https://doi.org/10.1016/J.CEMCONCOMP.2021.104155

3. Chourasia A, Pal B, Kapoor A (2025) Influence of printing direction and interlayer printing time on the bond characteristics and hardened mechanical properties of agro-industrial waste-based 3D printed concrete. J Mater Civ Eng ASCE 37:04025065. https://doi.org/10.1061/JMCEE7/MTENG-18196
4. Zhou Y, Jiang D, Sharma R, Xie YM, Singh A (2023) Enhancement of 3D printed cementitious composite by short fibers: a review. Constr Build Mater 362:129763. https://doi.org/10.1016/J.CONBUILDMAT.2022.129763
5. Paul SC, Tay YWD, Panda B, Tan MJ (2018) Fresh and hardened properties of 3D printable cementitious materials for building and construction. Arch Civ Mech Eng 18:311–319
6. Kazemian A, Yuan X, Cochran E, Khoshnevis B (2017) Cementitious materials for construction-scale 3D printing: Laboratory testing of fresh printing mixture. Constr Build Mater 145:639–647
7. Özalp F, Yilmaz HD (2020) Fresh and hardened properties of 3D high-strength printing concrete and its recent applications. Iran J Sci Technol, Trans Civ Eng 44:319–330
8. Le D-H, Sheen Y-N, Lam MN-T (2018) Fresh and hardened properties of self-compacting concrete with sugarcane bagasse ash–slag blended cement. Constr Build Mater 185:138–147
9. Pal B, Chourasia A, Kapoor A (2024) Intricacies of various printing parameters on mechanical behaviour of additively constructed concrete. Arch Civ Mech Eng 24:41
10. Alonso-Canon S, Blanco-Fernandez E, Castro-Fresno D, Yoris-Nobile AI, Castañon-Jano L (2022) Reinforcements in 3D printing concrete structures. Arch Civ Mech Eng 23:25
11. Somireddy M, Singh CV, Czekanski A (2020) Mechanical behaviour of 3D printed composite parts with short carbon fiber reinforcements. Eng Fail Anal 107:104232
12. Geneidy O, Kumarji S, Dubor A, Sollazzo A (2020) Simultaneous reinforcement of concrete while 3D printing. In: Second RILEM international conference on concrete and digital fabrication: digital concrete, pp 895–905

MIX
Papier aus verantwortungsvollen Quellen
Paper from responsible sources
FSC® C105338

If you have any concerns about our products,
you can contact us on
ProductSafety@springernature.com

In case Publisher is established outside the EU,
the EU authorized representative is:
Springer Nature Customer Service Center GmbH
Europaplatz 3, 69115 Heidelberg, Germany

Printed by Libri Plureos GmbH
in Hamburg, Germany